ENVIRONMENTAL CONTAMINATION
FOLLOWING A MAJOR NUCLEAR ACCIDENT

PROCEEDINGS SERIES

ENVIRONMENTAL CONTAMINATION FOLLOWING A MAJOR NUCLEAR ACCIDENT

PROCEEDINGS OF AN INTERNATIONAL SYMPOSIUM
ON ENVIRONMENTAL CONTAMINATION
FOLLOWING A MAJOR NUCLEAR ACCIDENT
JOINTLY ORGANIZED BY THE
FOOD AND AGRICULTURE ORGANIZATION
OF THE UNITED NATIONS,
THE INTERNATIONAL ATOMIC ENERGY AGENCY,
THE UNITED NATIONS ENVIRONMENT PROGRAMME
AND THE WORLD HEALTH ORGANIZATION
AND HELD IN VIENNA, 16–20 OCTOBER 1989

In two volumes

VOLUME 1

INTERNATIONAL ATOMIC ENERGY AGENCY
VIENNA, 1990

ENVIRONMENTAL CONTAMINATION
FOLLOWING A MAJOR NUCLEAR ACCIDENT
IAEA, VIENNA, 1990
STI/PUB/825
ISBN 92-0-020090-7
ISSN 0074-1884

Printed by the IAEA in Austria
June 1990

FOREWORD

Since the beginning of nuclear power production on a large scale, a small number of accidents have been reported in which nuclear facilities have been damaged. During one such accident, at Chernobyl in 1986, significant amounts of radioactive materials were released into the atmosphere and caused contamination of the environment both locally and in other countries. The extent and effects of the potential contamination due to a major accident at a nuclear facility are still matters of public concern. Scientific research on the after-effects of the Chernobyl accident on the environment and on human health has provided new data pertaining to large scale contamination, and much more information is expected to come from Chernobyl related studies in the near future.

The objective of the symposium was to review present knowledge of the extent and magnitude of environmental contamination occurring after a massive release of radioactive materials. Papers and posters covered a wide range of subjects, including: monitoring of radioactive contaminants in the environment, levels of radioactive contamination of farmland, agricultural crops and dairy products in subsequent years, and methods for minimizing contamination of feed and food. A special session on 'hot particles' drew attention to the potential risk from inhaling particles containing high levels of alpha and beta emitting radionuclides, and the importance of setting up valid descriptive radioecological models.

The symposium demonstrated that on technical matters there is a clear and urgent need for international communication and co-operation concerning the harmonization of guidelines and terminology and the adoption of acceptable reference levels for radionuclides in food and feed moving in international trade.

The presentations and discussions showed clearly that national authorities in affected countries had prepared for nuclear accidents and acted accordingly with protective measures in most cases based on sound technical reasoning, ranging from evacuation of people to guidelines on safer preparation of food. However, these measures in turn caused psychological stress and financial losses without proper compensation among the affected and dependent communities. Some of these consequences had been neither foreseen nor prepared for, either nationally or internationally. A new challenge for the international community is thus to determine how to deal with the technically less well defined consequences of large (nuclear) accidents having long term adverse effects on life in affected areas.

The symposium was organized by the International Atomic Energy Agency together with the Food and Agriculture Organization of the United Nations, the United Nations Environment Programme and the World Health Organization, and was attended by approximately 250 participants from some fifty countries.

EDITORIAL NOTE

The Proceedings have been edited by the editorial staff of the IAEA to the extent considered necessary for the reader's assistance. The views expressed remain, however, the responsibility of the named authors or participants. In addition, the views are not necessarily those of the governments of the nominating Member States or of the nominating organizations.

Although great care has been taken to maintain the accuracy of information contained in this publication, neither the IAEA nor its Member States assume any responsibility for consequences which may arise from its use.

The use of particular designations of countries or territories does not imply any judgement by the publisher, the IAEA, as to the legal status of such countries or territories, of their authorities and institutions or of the delimitation of their boundaries.

The mention of names of specific companies or products (whether or not indicated as registered) does not imply any intention to infringe proprietary rights, nor should it be construed as an endorsement or recommendation on the part of the IAEA.

The authors are responsible for having obtained the necessary permission for the IAEA to reproduce, translate or use material from sources already protected by copyrights.

Material prepared by authors who are in contractual relation with governments is copyrighted by the IAEA, as publisher, only to the extent permitted by the appropriate national regulations.

CONTENTS OF VOLUME 1

Part I(b). Air

PART II: MONITORING OF RADIOACTIVITY

Part II(a). General

Part II(b). Strategies and Policy

Part II(c). Methods and Techniques

Poster presentations

Part I

RADIOACTIVE CONTAMINATION
OF THE ENVIRONMENT

(a) General

Invited Paper

STUDY OF RADIOACTIVE CONTAMINATION OF THE ENVIRONMENT CAUSED BY THE CHERNOBYL NUCLEAR POWER PLANT ACCIDENT: MAIN RESULTS

Yu.A. IZRAEHL'
USSR State Committee for Hydrometeorology,
Moscow,
Union of Soviet Socialist Republics

Presented by V.N. Petrov

Abstract

STUDY OF RADIOACTIVE CONTAMINATION OF THE ENVIRONMENT CAUSED BY THE CHERNOBYL NUCLEAR POWER PLANT ACCIDENT: MAIN RESULTS.

The paper presents the main results of research on environmental contamination during the three years following the Chernobyl accident. The author describes the work and names the organizations which took part in it. The information obtained made it possible to solve many urgent scientific problems and provided the basis for extremely important decisions concerning the evacuation of the population, the elaboration of a regime under which contaminated land could be inhabited and used for agriculture, appropriate protective measures and decontamination requirements. The paper includes maps of radioisotopic contamination produced by airborne gamma surveys and isotopic analysis of soil samples. Total gamma activity as well as ^{137}Cs and ^{134}Cs concentrations in areas bordering on the accident zone are given. Calculated doses for internal and external exposure in so-called caesium 'spots' are presented, and the measures taken to prevent human exposure are described. The author describes mathematical modelling of the transport of radioactive materials released, contamination of water bodies and migration of radionuclides in the soil, and indicates the steps which have been taken to prevent any further spread of radionuclide contamination in the environment. The paper ends with a statement of basic conclusions and proposals.

1. INTRODUCTION

As is already known, immediately after the accident at the Chernobyl nuclear power plant, large scale emergency measurements were made of radioactive contamination of the area and of the atmosphere, followed by comprehensive studies of the radioactivity of all natural media, including biota.

3

The earliest measurements of the radiation situation at the Chernobyl site and in the surrounding area were made by troops equipped as for chemical warfare conditions using ground based equipment. The USSR State Committee for Hydrometeorology developed and carried out an aerial gamma survey of the contaminated area and atmosphere (as from 26 April 1986 with a helicopter and from 27 to 29 April with a specially equipped aircraft; subsequently, the number of aeroplanes and helicopters with special equipment, including gamma spectrometers, was increased to ten for the entire contaminated area of the Union of Soviet Socialist Republics). The gamma survey was carried out daily during April and May 1986 and then at regular intervals according to a special timetable.

The first complete map of the short range plume track (up to about 80 km from the site of the accident) was produced on 2 May 1986 and was submitted to the Government Commission on 3 May, immediately after it became possible to differentiate in detail the effects of the airborne and ground level contamination. Up to this point (as from 26 April 1986), data on the radiation situation at helicopter and aircraft flight altitudes and measurement data from meteorological stations were available.

At the same time, with the resources of the USSR State Committee for Hydrometeorology, an airborne gamma survey was carried out over a huge area of the European part of the country (approximately half of it). This made it possible to distinguish additional zones of contamination and individual 'spots'.

These surveys provided basic information for taking fundamental decisions. Subsequently, the V.I. Vernadskij Institute of Geochemistry and Analytical Chemistry (USSR Academy of Sciences) and the Aerial Geology Department of the USSR Ministry of Geology became involved in this work.

Over the same period, large scale sampling of soil from the contaminated region and analysis of the samples in various (gamma spectrometry and radiation chemistry) laboratories were carried out. Participating in this work were the USSR State Committee for Hydrometeorology, the USSR Ministry of Public Health, the USSR Academy of Sciences, the Academy of Sciences of the Ukrainian Soviet Socialist Republic, the Academy of Sciences of the Byelorussian Soviet Socialist Republic, the USSR Ministry of Defence, the USSR Ministry of Medium Machinery Manufacture, the USSR State Committee on the Utilization of Atomic Energy, the USSR Ministry of Atomic Energy and the USSR State Agroindustry.

The present paper makes use of these primary findings although, essentially, it is based on data of the divisions of the USSR State Committee for Hydrometeorology.

The data obtained were used in the solution of a great number of operational and scientific problems and formed the basis for highly important decisions concerning evacuation of the population and for determination of the conditions for staying and operating farms in the contaminated territory, and also of the conditions for taking protective and decontamination measures. Most of the data were

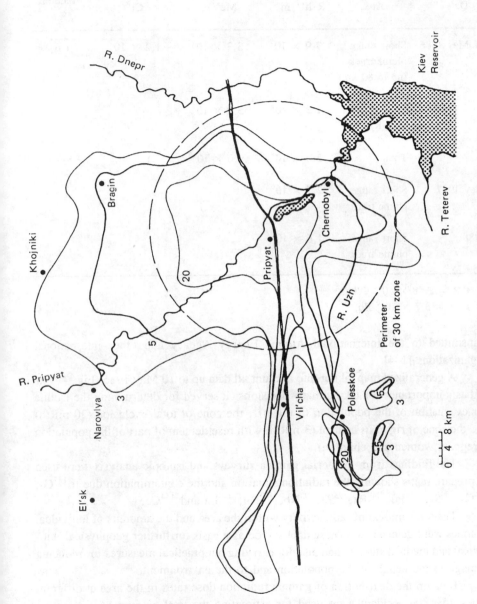

FIG. 1. Generalized map of gamma dose rates (mR/h) for 10 May 1986.

TABLE I. TOTAL GAMMA RADIOACTIVITY IN PLUME TRACK

Date	Area	$R \cdot h^{-1} \cdot m^{-2}$ [a]	$MeV \cdot s^{-1}$	Ci [b]	Fraction (%)
11 May 1986	Short range plume track (up to 80 km)	7.9×10^7	3.3×10^{17}	1.2×10^7	1.6
	Long range (USSR)	1.2×10^8	5.0×10^{17}		
	Total	2.0×10^8	9.0×10^{17}	3.1×10^7	3.5–4.0
May 1987	Short range plume track	2.7×10^6			
1988	Short range plume track	1.6×10^6			

[a] $1 R = 2.58 \times 10^{-4}$ C/kg.
[b] $1 Ci = 3.7 \times 10^{10}$ Bq.

transmitted to the International Atomic Energy Agency and other international organizations [1–4].

A generalized map taking into account all data up to 10 May 1986 (Fig. 1) was of basic importance for taking many decisions: it served for determining the isoline for evacuation of the population (5 mR/h)[1], the zone of total exclusion (20 mR/h) and the zone of rigorous control (3 mR/h) with resettlement of part of the population (pregnant women and children).

The findings from the aerial gamma surveys and isotopic analysis were used to prepare maps showing the radiation situation and the contamination due to ^{137}Cs, ^{134}Cs, ^{90}Sr, ^{239}Pu, ^{240}Pu, ^{95}Zr, ^{95}Nb, ^{103}Ru, ^{140}La and ^{144}Ce.

The total amount of radioactivity within the area and the amounts of individual isotopes were determined. These data served as a basis for further geophysical, biological and medical studies, and also for carrying out practical measures for reducing damage to the health of the population and to the environment.

Data on the distribution of gamma radiation dose rates in the area at different times after the accident were used for estimating the total amount of radioactive products falling in the short range plume track (up to about 80 km), and the variation in this amount over time, due to radioactive decay and other factors.

---

[1] $1 R = 2.58 \times 10^{-4}$ C/kg.

TABLE II. VARIATION IN AREAS (km^2) BOUNDED BY DIFFERENT DOSE RATE ISOLINES IN PLUME TRACK

Dose rate (mR/h)[a]	10 May 1986	29 Nov. 1986	1 May 1987	1 Oct. 1987	15 Apr. 1988
2	8000	560	280	190	130
5	3000	180	70	32	16
20	1100	20	8	6	—

[a] 1 R = 2.58 × 10^{-4} C/kg.

The total amount of gamma radioactive products in the short range plume track in May 1986 is shown in Table I; it amounted to 2.0×10^8 R·h^{-1}·m^{-2} (approximately 3.1×10^7 Ci)^2. A year after the accident, this amount had decreased by a factor of 70; two years after the accident, by a factor of 120.

Presently, radiation levels of more than 10.0 mR/h do not extend beyond the industrial zone of the power plant. Areas bounded by different isolines are given in Table II.

Careful isotopic analysis showed the presence of a large amount of radioactivity from fragments and induced radioactivity.

A characteristic feature is very considerable isotope fractionation (variation in isotope ratios as compared with the initial ratio in the reactor). The fractionation of the volatile products of ^{131}I and ^{137}Cs (relative to ^{95}Zr) in the radioactive air current amounted to a factor of 5 and above (this figure characterizes the enrichment of the volatile isotopes in the current). Significant fractionation was also observed in the area, especially in the 'northern' plume track; there was considerably less in the 'western' one and fractionation was completely insignificant in the 'southern' one.

A particularly large number of anomalies are associated with caesium contamination since the release of volatile caesium from the reactor was due mainly to the high temperature; in the atmosphere, this contamination was linked to the extremely fine (highly disperse) particles and fell on the area mainly together with precipitation — hence the 'spottiness' of the caesium contamination — especially at greater distances from the accident site. The fractionation of ^{137}Cs compared with ^{95}Zr amounts to a factor of 5–8 in the northern direction [2], and in northern 'spots' to even more.

The caesium contamination is spotty in nature owing to both the dynamics of the discharge and the meteorological conditions affecting the distribution of the

^{2} 1 Ci = 3.7 × 10^{10} Bq.

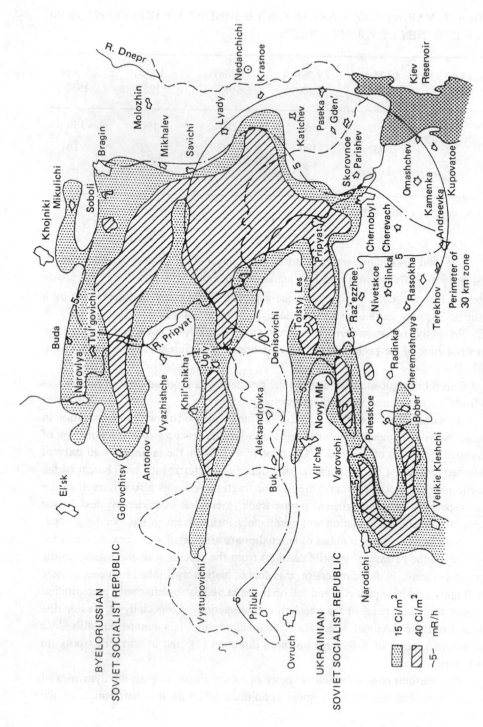

FIG. 2. Map of ^{137}Cs contamination (10 May 1986).

FIG. 3. Map of contamination by ^{90}Sr (thick continuous isoline: >3.0 Ci/km^2) and ^{239}Pu (dotted isoline: >0.1 Ci/km^2) (December 1987).

TABLE III. DISTRIBUTION OF ^{137}Cs IN THE SHORT RANGE AND LONGER RANGE PLUME TRACKS

	MCi[a]	Percentage of total yield
Short range plume track (up to 40 km)	0.28	1.9
Longer range plume track (European part of USSR)	0.7–0.8	4.8–5.4
Beyond USSR	~1.2	8.0
Total	~2.3	~15

[a] 1 Ci = 3.7 × 10^{10} Bq.

radioactive products. Figures 2 and 3 are maps showing the distributions of ^{137}Cs, ^{90}Sr and ^{239}Pu according to sampling data and the aerial gamma survey. The amount of ^{137}Cs falling in the short range plume track, according to the data from the aerial survey and the analysis of soil samples, was approximately 0.28 MCi.

The total amount of gamma radioactive substances which fell on the territory of the USSR beyond the limits of the short range plume track, according to an aerial gamma survey at the end of March 1987, is estimated at $(6–9) \times 10^6$ R·h^{-1}·m^{-2}, as compared with 1.2×10^8 R·h^{-1}·m^{-2} in the first ten days of June 1986.

The total figure for fallout of all isotopes in the short range and longer range areas (in the territory of the USSR) is 3×10^7 R·h^{-1}·m^{-2}, or an average of approximately 3.5–4.0% of the theoretical total energy release of radioactive products in the reactor during this time.

According to the data of the aerial gamma spectral survey, the amount of ^{137}Cs and ^{134}Cs in the European part of the USSR (beyond the short range plume track) is about 0.7–0.8 MCi. The amount of ^{137}Cs in the short range and longer range areas of the European part of the USSR is estimated at about 1.0–1.1 MCi (Table III).

Fallout of radioactive products has been detected in the entire western part of the European region of the USSR. At the stations of the USSR State Committee for Hydrometeorology, fallout or contamination of air by ^{131}I was detected up to 2 May 1986 in the Ukraine (at Kiev, Vinnitsa, Ivano-Frankovsk, Poltava, Krivoj Rog, Zaporozh'e, Donetsk, Kerch, L'vov and Rovno); in Byelorussia (at Bragin, Minsk, Brest, Mogilev and Grodno); and in the Baltic area (at Kaliningrad, Klajpeda, Riga, Vilna and, further away, at Narva and Leningrad). Iodine-131 was detected also not far away from Moscow — at Obninsk: here the concentration of ^{131}I in a filter sample dated 28–29 April amounted to 0.5 Bq/m^3. In most cases, this contamination was short lived and was associated with contamination of atmospheric air.

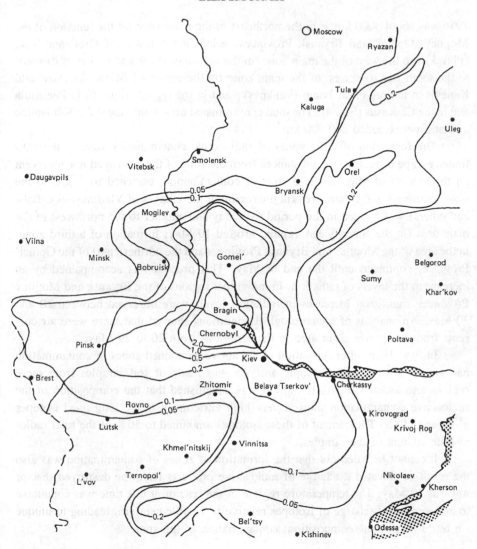

FIG. 4. Map of gamma dose rates (mR/h) in the European part of the USSR (late May to early June 1986).

The amount of ^{137}Cs falling beyond the borders of the USSR was estimated from published data on contamination of the territory of European countries, and also from calculated values obtained by numerical modelling of radionuclide transport in the atmosphere. The total fallout of ^{137}Cs on European territory outside the USSR is estimated at 1.2 MCi (Table III).

A survey of the radiation situation showed zones of contamination (apart from the main one, where the area with radiation levels higher than 5 mR/h on 10 May

1986 was about 3000 km^2): to the northeast of the main zone (at the junction of the Mogilev, Gomel' and Bryansk Provinces, south of the towns of Orel' and Tula (Plavsk)); to the west of the main zone (in the region of Pinsk and in that of Rovno); to the south and southwest of the main zone (in the region of Belaya Tserkov' and Kanev); in the region of Ivano-Frankovsk; and in the region of the Kola Peninsula and in the Caucasus (Fig. 4). The total contaminated area within the 0.2 mR/h isoline at that time exceeded 200 000 km^2.

The formation of large zones of radioactive contamination (due to the continued escape of radioactive substances from the zone of the destroyed reactor) went on through all of May. Thus, a zone of contamination, enriched in ^{137}Cs, to the west-southwest of the main area (in the region of the villages of Vladimirovka, Bobr and others) was formed in the period 6–9 May and another to the northwest of the main area (in the Narovlyansk region) around 10 May; formation of a third zone, in the area of the Mogilev and Bryansk Provinces and the northern part of the Gomel' Province, continued until the end of May. This process was accompanied by an increase in the levels of radiation. In individual regions of the Bryansk and Mogilev Provinces significant increases in radiation levels were observed between 20 and 30 May. An analysis of meteorological conditions showed that there were air currents from the power plant area in these regions from 20 to 29 May.

In June 1986, after formation of the above mentioned zones of contamination had ended, a large scale isotopic analysis was made of soil samples from these regions and as early as 10–15 June it was established that the composition of the radioactive contamination showed very high enrichment in the long lived isotopes ^{137}Cs and ^{134}Cs. The content of these isotopes amounted to 50% of the total radionuclide content of the samples.

It cannot be ruled out that the formation of zones of contamination was also the result of elevated discharge of radioactive products from the destroyed reactor towards 20 May. The temperature regime in the reactor at this time was conducive to an increased discharge of isotopes relative to volatile caesium, leading to unique (in terms of isotopic composition) contamination of the area.

2.   CALCULATING INTERNAL AND EXTERNAL EXPOSURE DOSES AND SETTING UP CONTAMINATION NORMS

Integration of the values for concentrations and fallout of individual radionuclides during the ten day period following the accident afforded a means of estimating the effective dose equivalents of internal and external exposure of human subjects in hundreds of inhabited localities during the first year.

As a result of the particular features of the isotopic composition of the caesium spots, internal exposure doses in the case of the Bryansk and Mogilev Provinces exceed the external dose by approximately a factor of 3, as shown in Table IV. The

TABLE IV. POSSIBLE EXTERNAL AND INTERNAL EXPOSURE DOSES (rem)[a] IN CAESIUM 'SPOTS'

Cs-137 (Ci/km^2)[b]	Calculated dose for 1 year			Calculated dose for 30 years		
	External	Internal	Total	External	Internal	Total
Inhabited locality						
El'nya	26	2.0	4.4	6.4	11.4	31.4	42.6
Tur'ya	6.7	0.5	1.1	1.6	2.9	8.0	10.8
Isoline	40	3.0	6.5	10	18	48	66
Isoline	15	1.0	2.5	3.5	6.0	18	24
Isoline	5.0	0.4	0.9	1.3	2.5	7.0	9.5

[a] 1 rem = 1 × 10^{-2} Sv.
[b] 1 Ci = 3.7 × 10^{10} Bq.

table gives the calculated doses without any kind of prophylactic measures being taken to reduce possible doses.

The following conclusions can be drawn from the table. For the case of living in an area with a ^{137}Cs contamination of 40 Ci/km^2 and over, the total dose will exceed 50 rem for 50–70 years of life.[3] Accordingly, even for ^{137}Cs contamination of more than 15 Ci/km^2, it is necessary to take measures for significantly reducing the internal dose, which is received mainly through food products; in agriculture, measures are required for binding caesium in the soil or replacing it with other elements from food chains (e.g. application of potassium fertilizers) (this reduces the dose by a factor of 2–4), or for eliminating it from the ration of contaminated products (this eliminates the internal exposure dose almost entirely). It is this approach that was chosen in the areas of caesium spots in the Mogilev, Gomel' and Bryansk Provinces and the result was positive: persons who complied with the necessary requirements (and they were the majority) did not receive doses exceeding the permissible ones. Decontamination reduced the external exposure dose appreciably (by a factor of 1.5).

However, in the use of foodstuffs produced in the contaminated areas, in some cases even with contamination levels of 5–15 Ci/km^2 individual norms can be exceeded (especially in milk); this depends on the type of soil, the biomass of the grassy vegetation (fodder) and other factors.

---

[3] 1 rem = 1 × 10^{-2} Sv.

3.   MODELLING ATMOSPHERIC TRANSPORT AND RADIOACTIVE
     FALLOUT

In the interest of gaining a deeper understanding of the overall physical pattern
of radioactive plume track formation and more detailed knowledge of the dynamics
of the variation in the parameters of a source of radionuclide discharge, of estimating
the amount of radionuclides transported beyond the boundaries of the USSR and also
of solving the problems associated with the Convention on Early Notification of a
Nuclear Accident, the Institute of Applied Geophysics and the 'Typhoon' Scientific
and Industrial Association of the USSR State Committee for Hydrometeorology
developed mathematical models of the transport and precipitation of radionuclides
as a result of discharge into the atmosphere. These are intermediate scale and
regional models as well as models for transboundary transport; they cover spatial
scales ranging from tens of kilometres to a few thousand. The models constitute a
description of the transport and precipitation of radioactive products entering the
atmosphere from the accident source over a long period of time.

Let us consider the model describing the precipitation of radioactive products
entering the atmosphere from the damaged unit of the Chernobyl nuclear power plant
over a long time and leading to radioactive contamination in the region of the plant
and up to 100 km away [2, 5].

The full spectrum of particles discharged into the atmosphere from the
damaged unit represents the superposition of a number of spectra, corresponding to
different particle sizes and radioisotopic compositions.

In the model, a simpler representation of the distribution of particles in a
limited range of sizes was used in order that short range radioactive fallout could be
described. It was assumed that an integral distribution of the total gamma activity
of particles in a definite size interval forming the short range plume track is approximated by a log-normal law with the distribution parameters: median diameter ξ and
standard deviation σ.

In addition, the radioactive substances lifted up from the damaged unit by hot
currents to height H are represented in the form of a raised point source of a polydisperse contaminant which is constantly emitting radioactive particles into the
atmosphere, the distribution of activity among which, depending on their size, is
subject to a log-normal law.

The radioactive fallout on the Earth's surface from a given source located at
height H is determined by the kinematics of the gravitational precipitation of particles
in a wind field variable over time, with allowance for horizontal diffusion.

Factual information about this radioactive fallout was provided by data on the
distribution of radiation levels on the Earth's surface, obtained from an airborne
gamma survey.

Integral analysis of the findings obtained from all aerial gamma surveys was
applied. The fallout area in the short range plume track was plotted as a function of

dose rate isolevel bounding the area, with allowance for normalization of the data from the surveys to a single point in time, namely 29 May 1986. Within the area bounded by the isoline, the dose rate was greater than or equal to the given value. The results obtained from different surveys are fairly close to each other. This made it possible to determine values of the log-normal distribution when the model calculated and real relationships between areas limited by dose rate isolines were closest. The proposed model can be used for estimating the possible scale of the nuclear hazard in regions near a nuclear power plant and for creating conditions to ensure the safety of the population in the event of a possible accident.

4.   TRANSFER TO WATER SOURCES

In the case of the Chernobyl accident, conditions developed in which the radioactive products were able to penetrate water bodies because of: direct deposition on the water surface, drainage of water from the contaminated area and migration with underground water.

During the first weeks and months after the accident, the most urgent matter was determining the degree of contamination of the River Pripyat and the Kiev Reservoir. The maximum concentrations during the first period of observations are given in Table V.

Continuous observations showed that after the end of the period of aerosol contamination of the atmosphere, there was a sharp decrease — by two to three orders of magnitude — in the radionuclide concentration of river water. Rainfall of short duration, which occurred on 15 June 1986, led to an increase in the contamination of the Pripyat by approximately a factor of 3. This effect was observed a few times again during June 1986.

TABLE V. MAXIMUM CONTAMINATION LEVELS OF WATER BODIES $(10^{-9}$ Ci/L$)^a$

River	Date, 1986	I-131	Ba-140	Zr-95
Pripyat	2 May	120	60	42
Teterev	3 May	54	34	39
Irpen'	6 May	50	30	22

^{a} 1 Ci = 3.7 $\times$ 10^{10} Bq.

Such marked fluctuations in water contamination were not observed in the Kiev Reservoir. During the first two months after the accident, the total beta activity of the water was within the range of $(1-6) \times 10^{-9}$ Ci/L.

In the entire cascade of Dnepr reservoirs, the content of radioactive substances decreased continuously down the course of the Dnepr water body; in the Kremenchug Reservoir, the concentration of ^{90}Sr was around 5×10^{-12} Ci/L.

An estimate of the contamination of bottom deposits of the Dnepr reservoirs was made in May 1986 and showed that the most contaminated bottom areas of the Kiev Reservoir are in the sector bordering on the mouth of the Pripyat. In the southern part of the Kiev Reservoir, and also in the Kanev Reservoir, the contamination is lower by factors of tens and hundreds. The contamination of the bottom deposits is of the spotty type.

Radioactive contamination of rivers in the territory of the Russian Soviet Federated Socialist Republic and Byelorussia is continuously monitored, especially of those rivers in whose catchment areas there are spots of contamination with ^{137}Cs. In the period of the spring flood (7–12 April 1987), maximum levels of water contamination in terms of the total beta activity were observed; these levels amounted to $(1-1.5) \times 10^{-10}$ Ci/L (Rivers Iput' and Besed').

To prevent further radioactive contamination of the River Pripyat, the following construction work was carried out:

— Dykes and walls were built in the ground to cut off the escape of radioactivity from the short range area of the Chernobyl plant.
— Filtering dykes (131 structures) were built in small rivers to contain the radionuclides.

Verification of the operation of the protective structures showed them to be highly effective in isolating the short range area of the Chernobyl plant, in that they impeded the escape of radionuclides from this area almost completely. During the period from May 1986 to September 1987, 1700 Ci of ^{137}Cs and around 700 Ci of ^{90}Sr were removed by rivers into the Kiev Reservoir.

The maximum recorded level of contamination of the surface layer of water with ^{137}Cs in the Black Sea was 1.4×10^{-11} Ci/L (1986). This contamination was due almost exclusively to atmospheric fallout of accident products on the surface of the water, since outflow from the Dnepr from May 1986 to September 1987 amounted in all to 0.1 of the ^{137}Cs content of the total fallout.

In the Baltic Sea (Gulf of Finland), the content of ^{137}Cs in the summer of 1986 was 1×10^{-11} Ci/L.

In view of the possible increase in radioactivity in surface water during the spring flood, models and forecasts were made of secondary radioactive contamination of rivers. In the calculations, account was taken of the fact that the entire melted outflow from the industrial zone of the Chernobyl plant and the town of Pripyat would be held up by the corresponding reservoir structures, and transport by small rivers was calculated taking into account the variation in effectiveness of filtering dams.

The most important result of the forecasting calculations with respect to the 1987 spring flood was that for not one of the rivers of the area were concentrations expected to exceed the permitted levels. The results of measurements not only confirmed the basic conclusion of the forecast but also showed that the actual removal of radionuclides into the Kiev Reservoir was less than expected. Analysis of the data shows that, on the whole, the hydrological forecast proved to be true.

Calculations show that the increased transfer of radionuclides into the Kiev Reservoir during the flood period was due not only to the washing out of radionuclides from catchment areas but also to an intensification of processes in which radionuclides are washed out from the bottom in a diluted state and to an increase in the radionuclide fraction present in suspensions.

Figures 5(a) and (b) present a forecast and the results of observations of the transfer of ^{90}Sr and ^{137}Cs across the alignment of two points located on the River Pripyat above (village of Benevka) and below (town of Chernobyl) the Chernobyl nuclear power plant [6].

The effectiveness of the washing out of particles of various origins (and at different sites) was determined originally in the '1003' peaceful underground nuclear explosion [7]. In all, only 1–4% of the total activity from the particles was washed out into the water (1970). Applying these coefficients and relationships to the case of contamination of a locality in connection with the Chernobyl accident, as early as May 1986 we made a favourable preliminary forecast that contamination of river water and reservoirs during the 1987 spring flood would not exceed the established norms (subject to isolation of the area of the power plant itself). We also used coefficients obtained earlier for migration with underground water [7].

Ion exchange processes reduce the potential hazard that contamination of underground water will occur. The flow rate of ions of a given type with water is hundreds or thousands of times less than the flow rate of the water. Even in the case of relatively high flow rates for the water current, F_B (about 10 m/d), the migration of ^{90}Sr over a distance of 1 km takes years.

After the accident at Chernobyl the USSR Ministry for Water Management had dozens of holes drilled around the station at distances of several hundred metres (down to the water bearing horizon); these were for monitoring purposes and, where necessary, for pumping out contaminated water. Observations of the ^{90}Sr concentrations in the water of these holes over a period of one and a half years after the accident showed that these concentrations did not exceed the background values (Table VI), which is confirmation of the estimates referred to above.

Thus, the secondary radioactive contamination of rivers in the area of the Chernobyl accident, due to the washing out of ^{90}Sr and ^{137}Cs from the contaminated zone and from the bottom deposits of rivers, has not led during the period which has elapsed since the accident, and will not lead in future, to concentrations

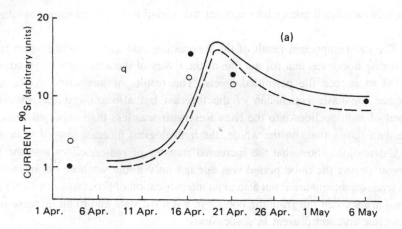

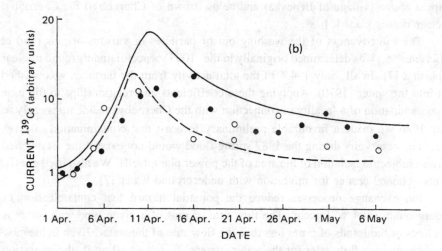

FIG. 5. *Forecast and results of observations of (a)* ^{90}Sr *and (b)* ^{137}Cs *transfer along the River Pripyat for the town of Chernobyl (full line: predicted; full circle: measured) and for the village of Benevka (broken line: predicted; open circle: measured).*

exceeding the maximum permissible levels in the rivers and reservoirs of the Dnepr cascade.

Even lower concentrations of radionuclides were observed in the waters of the Black Sea (in the area of discharge of the Dnepr), and these will decrease in the future.

TABLE VI. ^{90}Sr ACTIVITY OF UNDER-
GROUND WATER IN THE SHORT RANGE
AREA OF THE ACCIDENT (FOR 1987)

Distance from industrial zone of plant (km)	Sr-90 concentration (Ci/L)[a]
0.15	7.2×10^{-12}
1.5	2.5×10^{-11}
2.5	1.0×10^{-12}
6.0	7.2×10^{-12}
PC$_S$ for Sr-90:[b]	4×10^{-10} Ci/L

[a] 1 Ci = 3.7×10^{10} Bq.
[b] Permitted concentration for limited category of population.

5.   MIGRATION IN SOIL

In the plume tracks, a study was made of the migration of radionuclides due to wind transport (horizontal migration) and penetration of isotopes deep into the soil (vertical migration).

The vertical migration of radionuclides in soils of contaminated regions proceeds by two mechanisms, a 'slow' one and a 'fast' one. Between 75% and 95% of the entire radionuclide content is concentrated in the upper 3–4 cm of the soil. The activity decreases exponentially with depth.

Horizontal migration proved to be insignificant as regards the scales of the areas of contamination: there was practically no shift in the isolines as a result of migration, but small areas or individual structures subjected to decontamination in contaminated regions rapidly became contaminated again (owing to transport of radioactivity by wind).

The coefficients for the lifting of radioactivity by wind were insignificant. For this reason, the concentrations of the various isotopes (including plutonium) in the air at a wind velocity of up to 10 m/s, even in the zone of total exclusion, were found practically everywhere to be lower than the maximum permissible values. (An exception was the lifting due to draughts in the operation of vehicles.)

6.   SUMMARY

(a)   In the reactor accident in Unit 4 of the Chernobyl nuclear power plant there was a considerable release of radioactivity into the environment, leading to long range contamination. In terms of the scale of discharge and environmental contamination, this accident was the most extensive of any reactor accidents.

(b)   The area with a contamination level higher than 5 mR/h (on 10 May 1986), which is dangerous for habitation by the population and must remain evacuated for a long time, amounts to somewhat more than 3000 km^2 and extends up to 75 km to the west of the plant and up to 45 km to the north. The zone of total exclusion amounted to around 1000 km^2.

On the basis of surveys of the radiation situation as early as 2 May 1986, a 30 km resettlement evacuation zone was decided on (and not a 10 km one, as originally assumed).

According to calculations, the total activity of substances finding their way into water upon passage of the cloud and the air current was 2×10^5 Ci, of which ^{90}Sr accounted for about 10^3 Ci.

(c)   It was predicted and confirmed that, with reliable isolation of the industrial site (an area of approximately 10 km^2) and the cooling pond, the escape of radioactivity from the entire plume in the course of a year would not exceed, in terms either of total activity or of individual isotopes, the radioactivity which had already entered water during passage of the cloud and the air current in the first days after the accident, and the concentration in water would not exceed the permissible concentrations in the following period either.

(d)   The time norms for density of surface contamination worked out during the first months (mainly in May 1986) on the basis of a study of isotopic composition and gamma fields, which were postulated as the permitted limits for habitation by the population (namely 7–15 Ci/km^2 for ^{137}Cs, 3 Ci/km^2 for ^{90}Sr and 0.1 Ci/km^2 for ^{239}Pu and ^{240}Pu), were confirmed by further studies and are applicable at present. (However, for the regularization of farming practices, further details or clarifications are required.)

Recommendations have been formulated with regard to safe habitation by the population, the use of food products and the practice of farming in areas of increased contamination, including those where contamination with ^{137}Cs is higher than 15 Ci/km^2.

(e)   Continuous monitoring of radioactive contamination of the environment was organized and is still functioning both in the 30 km zone and for the entire extensive area of contamination.

7.   CONSIDERATIONS FOR THE FUTURE

(1)   In 'spots' distant from the centre and in areas where ^{137}Cs contamination is higher than 15 Ci/km^2 (these sites will not change appreciably in the years immediately ahead), continuous implementation of the following measures is required: the fixation and deep burial of caesium in soils and on the surface, and the partial or complete replacement of food products of local origin, along with improvements in farming practice. The partial removal and burial of the contaminated part of the soil is advisable in the most highly contaminated areas. Here a gradual return to customary farm management will be observed.

(2)   Very long term and costly measures are the prerequisites for a return to normal conditions in the zone from which the population has been completely excluded. A more realistic step would be to establish a (permanent) reservation here, which would include a testing ground for scientific research.

(3)   Long term monitoring of contamination of the environment and of food products is required, as well as of underground water in the region of the industrial site, with the possibility of pumping the water into a cooling pond.

(4)   Implementation of a detailed scientific programme covering geophysical, biological and medical aspects is required, including a study of the penetration of radionuclides (with allowance for their disperse form and their accessibility) via the chain:

soil → plant → animal
(water)   ↓
            ↘   man

A study of transfer coefficients in media and the regulation thereof, and the development of detailed standards for permissible concentrations and densities of environmental contamination are also called for.

(5)   Further requirements are the improvement of models of radioactive product distribution and the formulation of international measures for the swift transmission of essential information in the event of an accident (in accordance with the Convention on Early Notification of a Nuclear Accident).

REFERENCES
(in Russian)

[1]   ABAGYAN, A.A., et al., Information on the accident at the Chernobyl nuclear power plant and its consequences, prepared for the IAEA, At. Ehnerg. **61** 5 (1986) 301–320.

[2]   IZRAEHL', Yu.A., et al., Radioactive contamination of the environment in the area of
        the accident at the Chernobyl nuclear power plant, Meteorol. Gidrol. No. 2 (1987)
        5–18.

[3]   ASMOLOV, V.G., et al., The accident at the Chernobyl power plant: A year later, At.
        Ehnerg. 64 1 (1988) 3–23.

[4]   IZRAEHL', Yu.A., et al., Ecological consequences of radioactive contamination of the
        environment in the region of the accident at the Chernobyl nuclear power plant, At.
        Ehnerg. 64 1 (1988) 28–40.

[5]   IZRAEHL', Yu.A., PETROV, V.N., SEVEROV, D.A., Modelling of radioactive fall-
        out in the near zone of the Chernobyl nuclear power plant accident, Meteorol. Gidrol.
        No. 7 (1987) 5–12.

[6]   SEDUNOV, Yu.S., et al., "Modelling and forecasting of secondary radioactive con-
        tamination of rivers in the area of the Chernobyl nuclear power plant accident with long
        lived radionuclides", Radioactive Contamination of Natural Media in the Area of the
        Accident at the Chernobyl Nuclear Power Plant, 3rd edn, Gidrometeorizdat, Moscow
        (1988).

[7]   IZRAEHL', Yu.A., et al., "The washing out of radionuclides with flood water from
        natural catchment areas and migration with underground water, ibid., pp. 30–34.

ENVIRONMENTAL RADIOACTIVITY IN FINLAND AFTER THE CHERNOBYL ACCIDENT

R. SAXÉN, A. RANTAVAARA,
H. ARVELA, H. AALTONEN
Finnish Centre for Radiation
   and Nuclear Safety,
Helsinki, Finland

Abstract

ENVIRONMENTAL RADIOACTIVITY IN FINLAND AFTER THE CHERNOBYL ACCIDENT.

Results of the environmental monitoring programmes carried out in 1986–1988 by the Finnish Centre for Radiation and Nuclear Safety (STUK) are reviewed. Airborne radioactivity, radionuclides in deposits, in situ measurements of external gamma radiation, and radionuclides in foodstuffs and water are discussed. Of the over thirty nuclides initially identified from air samples, only ^{137}Cs and ^{134}Cs can presently be detected. The concentration of ^{137}Cs has decreased slowly, currently being about ten times higher than it was before the Chernobyl accident. In deposition the ratio of ^{137}Cs to ^{90}Sr varied at different sampling stations from 10 to 110, from 3 to 70 and from 5 to 60 in 1986, 1987 and 1988, respectively. On 1 October 1987 the mean value of ^{137}Cs surface activity for the 461 municipalities was 10.9 kBq/m^2, the population weighted mean environmental dose rate being 0.037 μSv/h. The distribution pattern of ^{137}Cs fallout also represents the distribution of other volatile nuclides in Finland. The pattern of non-volatile nuclides is related to the distribution of hot particles. In almost all terrestrial foodstuffs the maximum radiocaesium contents were found in 1986. Since then the annual decrease in agricultural products has been substantial. Efficient uptake from the soil mostly explains the radiocaesium contents of wild berries and mushrooms. In predatory fish maximum values of ^{137}Cs were reached in 1988 at the latest. Milk and beef made the dominant contribution to dietary intakes in the first fallout year. In the second year, the contribution of natural products to intakes almost equalled and in the third year exceeded that of agricultural produce.

1. INTRODUCTION

The regular environmental monitoring programme of the Finnish Centre for Radiation and Nuclear Safety (STUK) was used to study the distribution, behaviour and effects in Finland of fallout from the accident in Chernobyl on 26 April 1986. The ordinary programmes were expanded to meet the requirements of the new radiation situation. This paper is a review of some of the results obtained by the environmental studies during the three years since the Chernobyl accident. The results for 1986 have been published in detail separately, as will be those for 1987 and 1988,

23

in the STUK report series [1–12]. The analysis and counting methods used for the different types of environmental samples are also in Refs [11–13]. Data processing of the results for foodstuffs in 1987 and 1988 is still in a preliminary stage.

Monitoring of airborne radioactivity as well as amounts of deposited ^{137}Cs and ^{90}Sr at different stations is discussed in this paper. Deposition patterns of a non-volatile nuclide, ^{95}Zr, and those of volatile fission products, such as ^{131}I, ^{132}Te, ^{134}Cs and ^{137}Cs, as well as the deposition of ^{103}Ru in Finland are also described. Representative, country-wide monitoring of basic agricultural foodstuffs, milk, grain and meat, as well as of less heavily consumed foodstuffs originating from nature, such as fish, berries, mushrooms and game, and also of drinking water, makes it possible to estimate the impact of the released radionuclides on man. Together with dietary data the country-wide production weighted average activity concentrations of radiocaesium in foodstuffs were used to estimate intakes of radio-caesium by people in Finland. Estimated contributions of different foodstuffs to the total intake during the three years since the Chernobyl accident are given.

2.   AIR

The air surveillance programme was expanded with four new samplers in 1989. Nowadays airborne dust is collected on glass fibre filters with six high volume air samplers. The locations of the samplers are shown in Fig. 1. The capacity of the Nurmijärvi sampler is 750 m^3/h, and of the others, 150 m^3/h. All the samplers are equipped with charcoal cartridges to catch the gaseous fractions of radionuclides as well. The detection limit depends among other things on the nuclide, the air volume and the measurement time. Usually, the minimum detectable activity of ^{131}I and ^{137}Cs is 0.2 μBq/m^3 for the Nurmijärvi sampler and 1 μBq/m^3 for the others.

Over thirty nuclides have been identified from air samples [1]. Most of them were short lived, and now only ^{137}Cs and ^{134}Cs can be detected. Figure 2 shows the monthly mean concentration of ^{137}Cs in air in Nurmijärvi in 1968–1989. Since a pronounced peak in spring 1986 (maximum concentration 10^5 μBq/m^3) the concentration of caesium has decreased slowly, and is now about ten times higher than it was before the Chernobyl accident. The proportion of global fallout from the 1960s cannot be distinguished from the results.

The concentration of radionuclides in air is dependent on the amounts deposited in the area where the sampler is located. The radionuclide concentration in air is higher in areas which had a higher deposition in spring 1986 than in areas where the deposition was lower because of resuspension [1, 2]. On average, in Lapland the concentrations of ^{137}Cs are less than 5 μBq/m^3, in the Helsinki region about 10 μBq/m^3 and in Viitasaari about 20 μBq/m^3. There can be quite large variations in these figures owing to weather conditions, wind, rain, etc.

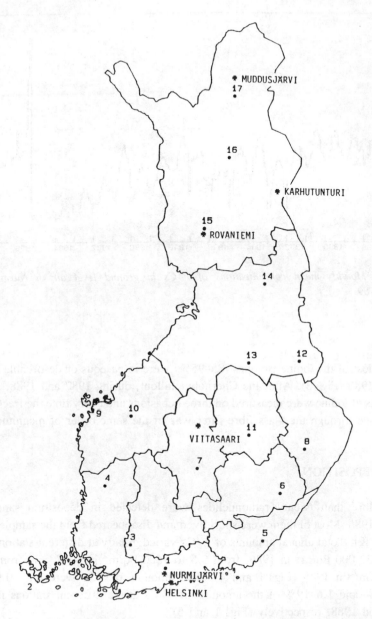

FIG. 1. Sampling stations for air (*) and deposition (•) samples in Finland in 1989. The names of the air sampling stations are shown on the map; the names of the deposition sampling stations are: 1, Nurmijärvi; 2, Maarianhamina; 3, Jokioinen; 4, Niinisalo; 5, Lappeenranta; 6, Savonlinna; 7, Jyväskylä; 8, Joensuu; 9, Vaasa; 10, Kauhava; 11, Kuopio; 12, Kuhmo; 13, Kajaani; 14, Taivalkoski; 15, Rovaniemi; 16, Sodankylä; 17, Ivalo.

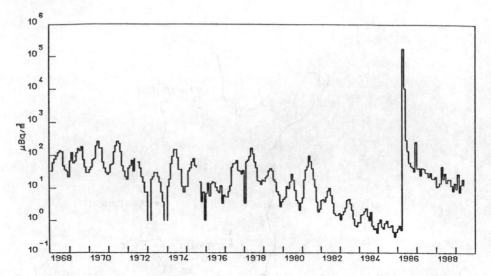

FIG. 2. Monthly mean concentrations of ^{137}Cs in ground level air in Nurmijärvi in 1968–1989.

Most of the iodine isotopes (85–95%) were in gaseous or desorbable form in spring 1986 (Fig. 3). After the Chernobyl fallout, during 1987 and 1988, gaseous fractions of iodine were measured on three occasions and every time the fraction that penetrated through the glass fibre filter was of the same order of magnitude.

3. DEPOSITION

More than twenty radionuclides were detected in deposition samples in spring 1986. Most of them were short lived and disappeared from the samples rather quickly [2]. Total annual amounts of ^{137}Cs varied greatly at different stations, from 140 to 32 000 Bq/m^2 in 1986, from 6.5 to 1500 Bq/m^2 in 1987, and from 3.5 to 480 Bq/m^2 in 1988 (Figs 1 and 4). The amounts of ^{90}Sr were only 0.9–12%, 1.5–38% and 1.6–19% of the amounts of ^{137}Cs at the different stations in 1986, 1987 and 1988, respectively (Figs 1 and 5).

Because the Chernobyl discharge period was short, the radionuclides found in deposition samples in 1987 and 1988 are resuspended, either locally or from farther away, in addition to small amounts of ^{137}Cs and ^{90}Sr originating from the period of nuclear weapons tests. The degree of recirculation was dependent on the total deposition at the sampling station in 1986. At the stations of northern Finland, with the lowest deposition from Chernobyl in 1986, the proportion of ^{137}Cs deposition from

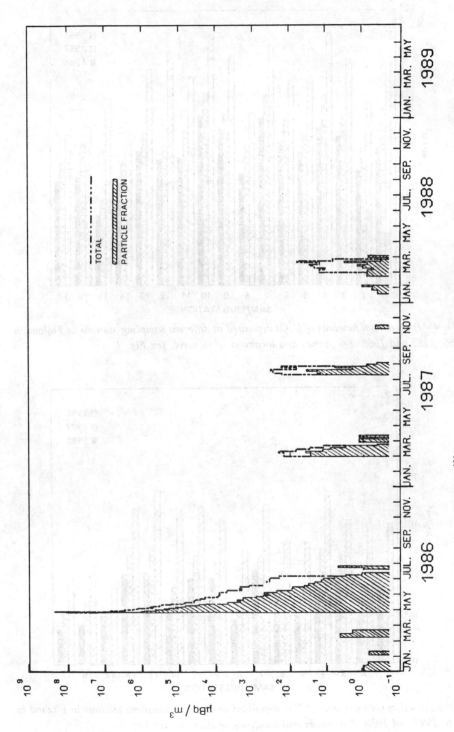

FIG. 3. Activity concentrations of ^{131}I (total and particulate) in ground level air in Nurmijärvi in 1986–1989.

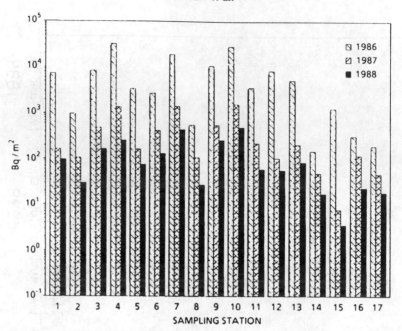

FIG. 4. Total annual amounts of ^{137}Cs deposited at different sampling stations in Finland in 1986, 1987 and 1988. For names and locations of stations, see Fig. 1.

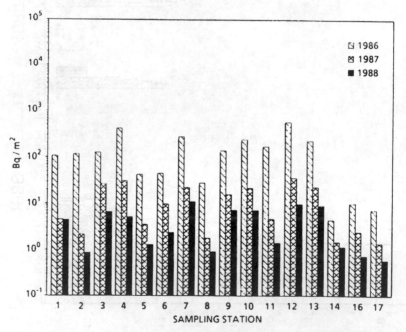

FIG. 5. Total annual amounts of ^{90}Sr deposited at different sampling stations in Finland in 1986, 1987 and 1988. For names and locations of stations, see Fig. 1.

1987 to that from 1986 was 30% ± 10%, while at stations with higher ^{137}Cs deposition in 1986 the corresponding proportion was only 4% ± 1%. The amounts of ^{137}Cs deposited in 1988 were only 1–20% of those deposited in 1986. The total annual amounts of ^{90}Sr detected in deposition samples in 1987 and 1988 were on the average 7% and 4% of the corresponding amounts determined in 1986, the variation being from 2 to 34% and 1 to 26% at different sampling stations.

The dominant alpha emitting transuranic element in the Chernobyl fallout was ^{242}Cm, but the proportion of transuranic elements in the total fallout was small. The highest amounts of ^{239,240}Pu and ^{242}Cm were 0.01% and 0.02%, respectively, of the amounts of ^{137}Cs [2].

4. IN SITU SPECTROMETRIC SURVEY OF GAMMA RADIATION AND FALLOUT LEVELS

The total exposure rate was measured using an efficient Geiger–Müller tube and a high pressure ionization chamber (HPIC). The GM tube was positioned over a car, 2.5 m above the road surface. A Ge spectrometer with a multichannel analyser and HPIC was located inside the car. The instruments measured continuously along each section of the route. The results thus represent the average radiation levels of each section. The final dose rates, for the sections of each route measured using the spectrometer, were based on measurements of ^{134}Cs and on linear correlation of the ^{134}Cs deposition estimates and the dose rate increases calculated from the GM counter results [10]. The final deposition estimates were based on measurements made at calibration sites, where measurements were made in a parked vehicle. The direct spectroscopic estimate of the surface activity was compared with the surface activity calculated from soil sample measurement. The minimum detectable surface activity of ^{134}Cs was better than 100 Bq/m^2. The corresponding dose rate due to caesium nuclides was only 0.0005 μSv/h.

A total of 19 000 km were measured by vehicle. The results comprised 550 gamma measurements with spectroscopy and 450 with only GM measurements. The results present the level of external radiation caused by the Chernobyl fallout on 1 October 1987 (Fig. 6). The dose rate ranges from 0.002 to 0.3 μSv/h, with wide areas having exposure levels exceeding 0.03 μSv/h. Areas exceeding 0.15 μSv/h were very rare. The surface area weighted mean value for the 461 municipalities in Finland was 0.026 μSv/h (range 0–0.16 μSv/h). The population weighted mean value was 0.037 μSv/h.

The deposition of volatile fission products such as ^{131}I, ^{132}Te, ^{134}Cs and ^{137}Cs in Finland was caused by releases from the burning reactor on 26 April 1986, after the initial explosion. The main deposition of volatile nuclides was caused by rain on 28–30 April 1986. Figure 6 shows the estimated ^{137}Cs surface activity. On

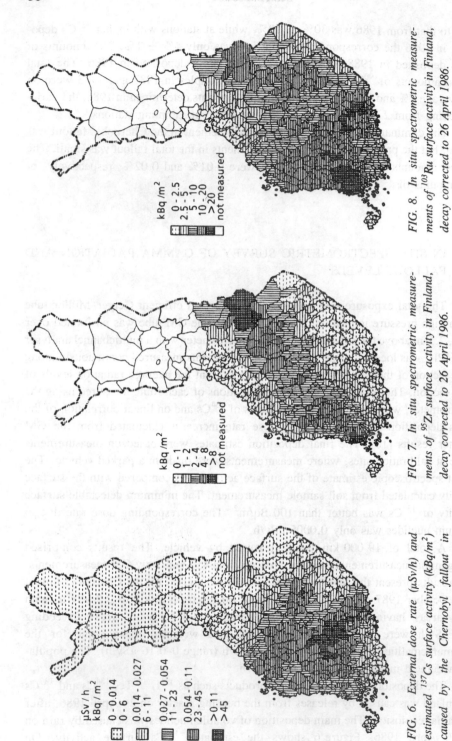

FIG. 6. External dose rate (μSv/h) and estimated ¹³⁷Cs surface activity (kBq/m²) caused by the Chernobyl fallout in Finland, on 1 October 1987.

FIG. 7. In situ spectrometric measurements of ⁹⁵Zr surface activity in Finland, decay corrected to 26 April 1986.

FIG. 8. In situ spectrometric measurements of ¹⁰³Ru surface activity in Finland, decay corrected to 26 April 1986.

1 October 1987 the mean value for the 461 municipalities was 10.9 kBq/m^2; the ^{134}Cs activity was 32% of the ^{137}Cs activity. The correlation between the volatile nuclides ^{131}I, ^{132}Te, ^{134}Cs, ^{140}Ba and ^{137}Cs is strong. Therefore the pattern of ^{137}Cs fallout also represents the distribution of these other nuclides in Finland.

The fallout pattern of ^{95}Zr is shown in Fig. 7. The map is based on 400 spectrometric measurements made south of 65° N. Precipitation on 27 April 1986 strongly affected depositions in the eastern and northern central parts of the fallout area, with the precipitation in the afternoon and evening being most important. Comparisons with dose rate monitoring results and precipitation times show that the cloud from the initial explosion had passed over Finland between the morning and night of 27 April, moving away to the east. Comparisons of the times and intensities of the precipitation and the deposition levels lend weight to the opinion that the deposition pattern is closely related to the trajectory of the cloud. Because of the weak correlation with precipitation intensities the role of dry deposition was most important in the southwesternmost parts of the fallout area.

The deposition pattern of ^{103}Ru shown in Fig. 8 is a combination of the pattern of fallout due to the initial explosion and the reactor fire, as well as to the deposition on 11–12 May 1986, caused by late releases from the reactor. The dust from the initial explosion contained core fragments of the reactor fuel. In Finland hot particles could be readily found in the environment. Comparison between the nuclide compositions shows that the pattern of non-volatile nuclides is related to the distribution of hot particles from the Chernobyl fallout.

5. SURFACE WATER

Ten gamma emitting radionuclides were detected in surface water in 1986 after the Chernobyl accident [5]. The most dominant of these were ^{134}Cs and ^{137}Cs. In May 1986 the activity concentrations of ^{137}Cs were highest (5300 Bq/m^3) at the drainage basin in the area of highest deposition (Fig. 6), where showers washed down radioactive substances into watersheds. The amounts of radiocaesium in surface water decreased quickly, the values at the end of 1987 (after vernal and autumnal turnover of water layers) being only 3–15% of the maximum values in May 1986 in different drainage basins. The decrease was also greatest in the areas which received most deposition.

Activity concentrations of ^{90}Sr were much lower than those of ^{137}Cs. Before the accident the ratio of ^{137}Cs to ^{90}Sr in surface waters was below 0.5. This ratio increased after the accident by a factor of about 100 in the main deposition area and about 10 in the area of lowest deposition. This ratio varied from 3 to 60 and from 3 to14 in March and October 1987, respectively. The change in this ratio reflects the different behaviour of these two radionuclides in the aquatic environment.

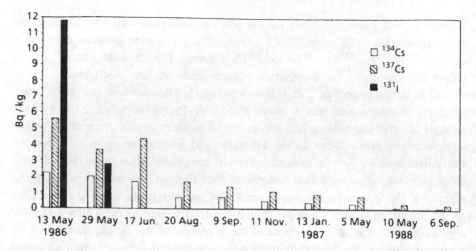

FIG. 9. Activity concentrations of ^{131}I, ^{134}Cs and ^{137}Cs in untreated water of Lake Päijänne. The lake is important as a source of drinking water.

Strontium remains more easily in the water phase, while caesium is removed quickly, sedimenting to the bottom.

As regards drinking water, a significant proportion of the radiocaesium was removed in the surface water treatment process. The amount removed varied, depending on the water quality, but was about 50% [5]. In Lake Päijänne, which supplies drinking water for people in southern Finland, ^{131}I could be detected for a month during 1986. The activity concentration of ^{137}Cs in the water of Lake Päijänne is now about 4% of the maximum concentration in May 1986 (Fig. 9).

6.   FOODSTUFFS

In 1986–1988, about 5000 samples were analysed for gamma emitting nuclides in the foodstuff programme. In the first growing season, the sampling focused on milk and meat and on vegetables, fruit and berries grown in the open [3, 4, 7]. In the following year, the numbers of milk and plant samples decreased, but areally the sampling of foodstuffs was as representative as before. The third year's programme was almost the same as in 1987, but the numbers of samples of game meat and other wild products were increased. The total number of fish samples analysed during 1986–1988 was about 2800. Apart from the biggest lakes, which are the most important for fishing, small lakes of different limnological types were also included in the study. About 190 lakes were sampled in 1987, but fewer in the other years [6].

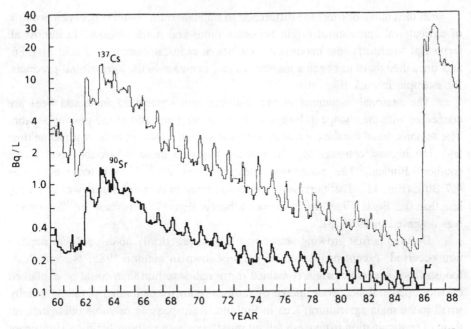

FIG. 10. Nationwide production weighted monthly mean activity concentrations of ^{137}Cs and ^{90}Sr in milk in 1960–1988.

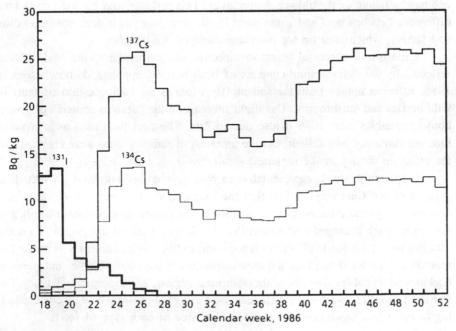

FIG. 11. Weekly nationwide production weighted mean activity concentrations of ^{131}I, ^{134}Cs and ^{137}Cs in milk in May–December 1986.

In the course of time, the difference in radiocaesium contents between produce of agricultural and natural origin becomes more and more obvious. In almost all terrestrial foodstuffs, the maximum contents of radiocaesium were found in 1986, and since then there has been a marked annual decrease in the agricultural products, for example in milk (Fig. 10).

The seasonal variations in the radiocaesium content of milk and beef are connected with the change in forage at the beginning and end of the grazing season. The seasonal trend for the content in beef was similar to that in milk, with some time lag. The highest contents of ^{131}I in milk followed those in ground level air in southern Finland. The nationwide mean content of ^{131}I over ten weeks was 4.7 Bq/L (Fig. 11). The increase in the radiostrontium content of milk was generally less than 0.2 Bq/L (Fig. 10). In cereals a barely detectable increase in ^{90}Sr content was observed in 1986 [3].

During the first growing season in Finland, practically no direct foliar deposition occurred, except on sprouts of cereals sown in autumn 1985. Nevertheless, foodstuffs of vegetable origin contained more radiocaesium than could be explained by root uptake alone. Root uptake of radiocaesium by cultivated plants is usually small in the main agricultural area in Finland. Resuspension or some other indirect form of contamination from fresh fallout must have been responsible for a significant proportion of the first year's caesium content. An exceptional pathway to animal products in the first grazing season might have been the contamination of new grass and hay by fallout on the old vegetation layer. This pathway may have decreased the difference between beef and game meat in the first year, as it does not depend on soil factors, which later on are the main cause of the transfer.

Exposure of perennial plants to direct deposition in the spring of 1986 was reflected by the radiocaesium contents of bush berries, apples and strawberries in 1986. Efficient uptake from the soil mostly explains the radiocaesium contents of wild berries and mushrooms. The slight increase in the caesium contents of greenhouse vegetables after 1986 is also due to this. The peat then used as a growing medium contained new fallout, but the amounts of radiocaesium were checked and the effect on dietary intake remained small.

In spite of the fairly representative sampling, it has been obvious since the first surveys of the Chernobyl fallout that the assessment of dietary intake must take account of regional differences in depositions. The radioiodine contents of milk and vegetables were averaged over subregions for different fallout categories. To make efficient use of the foodstuff surveillance data, radiocaesium content was studied in relation to both local and regional deposition and, if possible, to other independent background variables describing the sampling region, for example soil type. The transfer factor, TF, was used to calculate the mean radiocaesium content corresponding to the actual deposition in the production area of each type of food:

$$TF = \frac{\text{radiocaesium in foodstuff (Bq/kg)}}{\text{deposited radiocaesium (Bq/m}^2)} \tag{1}$$

Areal production figures were used for weighting when nationwide mean contents of radionuclides in different types of foodstuffs were calculated. For game meat, the weightings were based on the annual game bag, and for fish, on the catches of different fish species in different drainage basins.

The transfer of fallout radiocaesium to milk in the different sampling areas at different times of the year varied between 0.0005 and 0.0022 m^2/L in 1986. For beef, the nationwide production weighted transfer factor in May–December 1986 averaged 0.0043 m^2/kg. The mean TF values, weighted for areal production, for wheat, rye, oats and barley were 0.0002, 0.0012, 0.0005 and 0.0002 m^2/kg, respectively.

In the produce from forest environments, the mobility of ^{137}Cs from nuclear weapons test fallout does not differ significantly from that of the new caesium. This is shown by transfer factors calculated via ^{134}Cs for both old and new ^{137}Cs. In most cases no clear increase in radiocaesium content since 1986 has been found. A possible exception is the *Lactarius* mushrooms of 1987, in which the caesium contents were roughly 1.5–1.8 times as high as in 1986. When assessing annual changes in mean contents, the origins of the samples should be carefully compared. The three successive growing seasons all had different weather. This has probably increased the ranges of radiocaesium content in individual sample types, and so added to the uncertainty of the annual means. Compared with agricultural products, the transfer factors for wild produce could be one to two orders of magnitude higher [14].

Increased activity concentrations of radiocaesium were found in fish in June 1986. At first the contents were elevated in non-predatory fish, which fed on directly contaminated plankton. Concentrations of ^{137}Cs in predatory fish reached maximum values in 1988 at the latest (Fig. 12). The kind of nutrition affects the speed of radionuclide accumulation, which also depends on the size of the fish. The activity concentrations of radiocaesium were higher in the smaller lakes than in larger ones in the same deposition area. The proportion of runoff is higher in smaller lakes than in larger ones. An example of the changes in activity concentrations of ^{137}Cs in different fish species in one lake is given in Fig. 13. There are, however, large differences between different lakes, both in activity concentrations and in the speed of accumulation in a particular deposition category.

The transfer of radiocaesium from deposition to fish was estimated for various drainage basins using the average deposition of ^{137}Cs and the weighted average activity concentrations of ^{137}Cs in fish catches from the same areas. The transfer factors varied from 0.02 to 0.05 m^2/kg and from 0.05 to 0.2 m^2/kg in the different areas in 1986 and 1987, respectively. The transfer factor for the whole country in 1987 was on average double that for 1986.

The dietary intake of radiocaesium was calculated for an average Finnish diet, and the diet was kept the same each year. The reduction of radiocaesium during food processing was only taken into account for cheese making, wheat milling and parboiling of mushrooms. There are other factors which also reduce the dietary

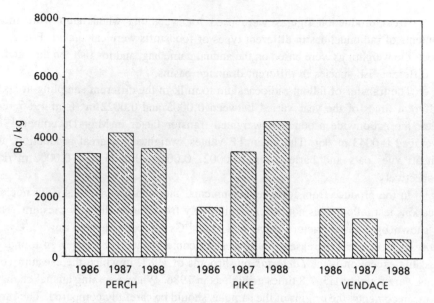

FIG. 12. Average activity concentrations of ^{137}Cs (Bq/kg fresh weight) in perch (Perca fluviatilis), pike (Esox lucius) and vendace (Coregonus albula) in the area of highest deposition in Finland during 1986, 1987 and 1988.

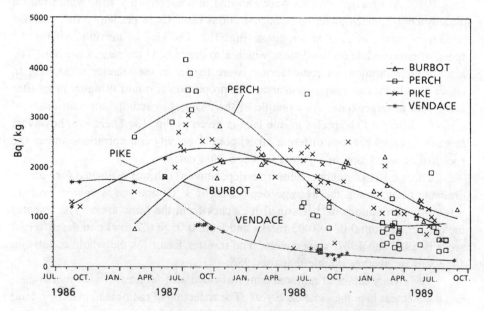

FIG. 13. Activity concentrations of ^{137}Cs (Bq/kg fresh weight) in different fish species in a lake (area about 100 km^2) located in the area of highest deposition in Finland.

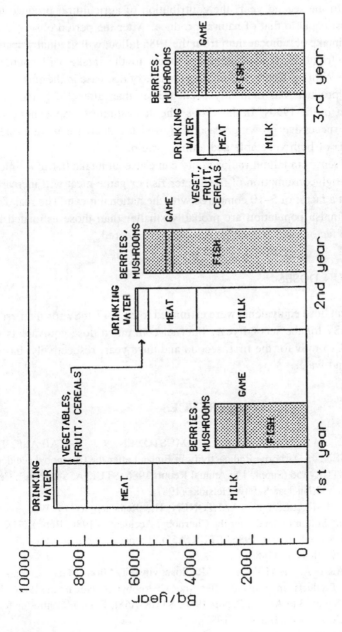

FIG. 14. Dietary intake of ^{137}Cs via foodstuffs of agricultural and non-agricultural origin in Finland in the three years after April 1986.

intake but they have been omitted from the computations. Comparisons of the radio-caesium contents of food and the discussion on pathways point to the conclusion that milk and beef made the dominant contribution to dietary intake in the first fallout year (Fig. 14). In the second year, the contribution of agricultural produce to the intake was almost equal to that of natural produce. After the period observed here, the total dietary intake of radiocaesium from the 1986 fallout will gradually decrease. The relative contribution of agricultural products to the intake will continue to decrease as well. With the Chernobyl fallout, the rate of decrease in the annual intake via agricultural products has evidently been greater than after 1964, the year of maximum fallout in the 1960s. In the 1960s, the deposition of stratospheric radio-nuclides during successive growing seasons slowed the decrease in radiocaesium contents of foods of both vegetable and animal origin.

Areal differences in fallout radioactivity can cause an intake 0.1 to 4 times the national mean. High consumption of freshwater fish or game meat will increase the intake by at most a factor of 5–10 compared with the national mean. The real dietary intakes of the Finnish population are probably smaller than those estimated here, since not all the factors that reduce intake were considered.

7.   RADIATION DOSES

The external dose equivalents were estimated to be 0.17 mSv for the first year [15] and 0.09 mSv for the second year. Internal committed dose equivalents were 0.31, 0.25 and 0.15 mSv for the first, second and third year, respectively, based on intake of food and water.

REFERENCES

[1] SINKKO, K.,   AALTONEN, H.,   MUSTONEN, R.,   TAIPALE, T.K., JUUTILAINEN, J., Airborne Radioactivity in Finland after the Chernobyl accident in 1986, Rep. STUK-A56 (Suppl. 1 to Annual Report 1986, STUK-A55), Finnish Centre for Radiation and Nuclear Safety, Helsinki (1987).

[2] SAXÉN, R., TAIPALE, T.K., AALTONEN, H., Radioactivity of Wet and Dry Deposition and Soil in Finland after the Chernobyl Accident in 1986, Rep. STUK-A57 (Suppl. 2 to Annual Report 1986, STUK-A55), Finnish Centre for Radiation and Nuclear Safety, Helsinki (1987).

[3] RANTAVAARA, A., HAUKKA, S., Radioactivity of Milk, Meat, Cereals and Agricultural Products in Finland after the Chernobyl Accident in 1986, Rep. STUK-A58 (Suppl. 3 to Annual Report 1986, STUK-A55), Finnish Centre for Radiation and Nuclear Safety, Helsinki (1987).

[4] RANTAVAARA, A., Radioactivity of Vegetables and Mushrooms in Finland after the Chernobyl Accident in 1986, Rep. STUK-A59 (Suppl. 4 to Annual Report 1986, STUK-A55), Finnish Centre for Radiation and Nuclear Safety, Helsinki (1987).

[5]   SAXÉN, R., AALTONEN, H., Radioactivity of Surface Water in Finland after the Chernobyl Accident in 1986, Rep. STUK-A60 (Suppl. 5 to Annual Report 1986, STUK-A55), Finnish Centre for Radiation and Nuclear Safety, Helsinki (1987).

[6]   SAXÉN, R., RANTAVAARA, A., Radioactivity of Freshwater Fish in Finland after the Chernobyl Accident in 1986, Rep. STUK-A61 (Suppl. 6 to Annual Report 1986, STUK-A55), Finnish Centre for Radiation and Nuclear Safety, Helsinki (1987).

[7]   RANTAVAARA, A., NYGRÉN, T., NYGRÉN, K., HYVÖNEN, T., Radioactivity of Game Meat in Finland after the Chernobyl Accident in 1986, Rep. STUK-A62 (Suppl. 7 to Annual Report 1986, STUK-A55), Finnish Centre for Radiation and Nuclear Safety, Helsinki (1987).

[8]   ARVELA, H., BLOMQVIST, L., LEMMELÄ, H., SAVOLAINEN, A.L., SARKKULA, S., Environmental Gamma Radiation Measurements in Finland and the Influence of Meteorological Conditions after the Chernobyl Accident in 1986, Rep. STUK-A65 (Suppl. 10 to Annual Report 1986, STUK-A55), Finnish Centre for Radiation and Nuclear Safety, Helsinki (1987).

[9]   FINNISH CENTRE FOR RADIATION AND NUCLEAR SAFETY, Studies on Environmental Radioactivity in Finland 1986, Annual Report 1986, STUK-A55, Finnish Centre for Radiation and Nuclear Safety, Helsinki (1987).

[10] ARVELA, H., MARKKANEN, M., LEMMELÄ, H., BLOMQVIST, L., Environmental Gamma Radiation and Fallout Measurements in Finland, 1986–87, Rep. STUK-A76 (Suppl. 2 to Annual Report 1986, STUK-A74), Finnish Centre for Radiation and Nuclear Safety, Helsinki (1989).

[11] SINKKO, K., Computer Analysis for Gamma-Ray Spectra in Sample Measurements, Licentiate Thesis, Univ. of Helsinki (1981) (in Finnish).

[12] SINKKO, K., AALTONEN, H., Calculation of the True Coincidence Summing Correction for Different Sample Geometries in Gamma-Ray Spectroscopy, Rep. STUK-B-VALO 40, Finnish Centre for Radiation and Nuclear Safety, Helsinki (1985).

[13] TAIPALE, T.K., TUOMAINEN, K., Radiochemical Determination of Plutonium and Americium from Seawater, Sediment and Biota Samples, Rep. STUK-B-VALO 26, Finnish Centre for Radiation and Nuclear Safety, Helsinki (1985).

[14] RANTAVAARA, A., "Transfer of radiocesium through natural ecosystems to foodstuffs of terrestrial origin in Finland", paper presented at Workshop on Transfer of Radionuclides in Natural and Seminatural Environments, Udine, Italy, 1989.

[15] BLOMQVIST, L., PAAKKOLA, O., SUOMELA, M., "Summary and dose assessment in environmental radioactivity in Finland 1986", Annual Report 1986, STUK-A55, Finnish Centre for Radiation and Nuclear Safety, Helsinki (1987) 75–89.

STUDIES OF ^{131}I, ^{137}Cs AND ^{134}Cs IN AIR, MILK AND WATER IN ANKARA FOLLOWING THE CHERNOBYL ACCIDENT

G.A. AYÇIK, T. GÖLGE
Ankara Nuclear Research and Training Center,
Beşevler, Ankara,
Turkey

Abstract

STUDIES OF ^{131}I, ^{137}Cs AND ^{134}Cs IN AIR, MILK AND WATER IN ANKARA FOLLOWING THE CHERNOBYL ACCIDENT.

In Ankara, after the Chernobyl nuclear reactor accident, a large number of Chernobyl radionuclides were detected in samples of air, milk, tap water and rain water. But the most radioecologically important radionuclides, i.e. ^{131}I, ^{137}Cs and ^{134}Cs, are discussed in the paper. Identification and evaluation of radionuclides were carried out by photopeak area without any chemical separation by using a γ spectrometer. Air samples were collected by an air pump which was placed on the roof of the laboratory building. Air was sampled for a period of five hours twice a day (in the morning and in the afternoon); 250–300 m^3 of air per day passed through the Whatman 42 filter papers. Milk samples were investigated in two groups: (a) pasteurized milk collected daily from two dairy companies of the State farms, AOÇ and SEK; this group contains milk from the eastern and western regions of Ankara, respectively; (b) unpasteurized milk collected daily from five local dairies located in different directions from Ankara. Milk samples were counted in plastic containers of 100 mL and were placed directly against the detector. Rainwater samples were collected by a polythene covered funnel (area 1 m^2) mounted above a polythene vessel placed on the laboratory building. All water samples were counted directly as were the milk samples. The radioactive debris was first detected in an air sample taken on 30 April 1986. Iodine-131 and ^{137}Cs reached highest values respectively on 5 and 6 May. The counting of the milk samples began the first week of May and reached the highest activities from 12 to 15 May 1986. After the accident the first rain in Ankara fell on 20 May and the rain water had the highest value of ^{131}I, but the highest value of ^{137}Cs on 28 May. The ratio of ^{134}Cs/^{137}Cs remained nearly constant for all samples. Its value was around 0.54 ± 0.04 (30 April 1986).

1. INTRODUCTION

Following the nuclear accident in the Chernobyl nuclear power station on 26 April 1986, a large number of radionuclides were released into the atmosphere as fission products.

In Ankara ^{131}I, ^{134}Cs, ^{137}Cs, ^{103}Ru, ^{106}Ru, ^{140}La, ^{140}Ba, ^{132}Te, ^{95}Nb, ^{95}Zr, ^{141}Ce and ^{144}Ce were the radionuclides detected. But radioecologically the most important radionuclides in Ankara generated by the Chernobyl accident were ^{131}I, ^{137}Cs and [134]Cs. There were two major discharges into the atmosphere from the Chernobyl accident; one was on 26 April, the other on 5 May. The released radionuclides which rose because of their high temperatures created a radioactive cloud above 1000–1500 m in the atmosphere. This cloud first drifted through the northern areas and then through southern and eastern Europe depending on the meteorological conditions. In Turkey, except for Thrace, especially within Edirne and the coast of the eastern Black Sea, the levels of contamination were very low.

In Ankara, the first Chernobyl contamination was detected in air samples collected on 30 April. On that date and thereafter for three months, the radionuclides in samples of air, milk, tap water and rain water were analysed by using γ spectrometry.

This paper covers the behaviour of the radionuclides ^{131}I, ^{137}Cs and [134]Cs in these samples in Ankara during the period of May to June 1986.

2.   EXPERIMENTAL METHODS

2.1. Sampling

2.1.1. Air sampling

Air samples were collected from only one station. Sampling was carried out by an air pump which was placed on the roof of our laboratory building. On 30 April and thereafter, air was sampled for a five hour period twice a day (in the morning and in the afternoon). It was calculated that 250–300 m^3 of air per day passed through the Whatman 42 filter papers during 15 d from 30 April. After 14 May, a sample was taken for a five hour period once a day.

2.1.2. Milk sampling

Chernobyl radionuclides entered the human body mainly through the human food chain, either directly from the consumption of green vegetables and plants or indirectly from the consumption of milk. Milk is one of the few foods produced over large areas and collected on a daily basis. For this reason after the first week in May the concentrations of ^{131}I, ^{137}Cs and [134]Cs began to be measured in samples of cow milk. Samples collected daily were investigated in two groups: (a) pasteurized milk from two dairy companies of the State farms, AOÇ and SEK, which includes milk from the eastern and western regions of Ankara, respectively; (b) unpasteurized milk collected from five local dairies located in different directions from Ankara.

2.1.3.   Water sampling

Samples of rain water were collected by a polythene covered funnel (area $1 m^2$) mounted above a polythene vessel. To avoid contamination of these samples by airborne soil and surface dust, the sampling apparatus was placed on the roof of our laboratory building.

Tapwater samples were collected at the same time daily from the public water supplies in our laboratory. To obtain a reliable sample the pipe was flushed 5 min prior to sample collection. Since there was the possibility that the radioiodine could be lost, no evaporation process was used during the liquid sampling.

2.2.   Radioactivity measurements

After air sampling, the filter papers were compressed to provide a standard counting geometry, and were counted directly by placing them against the detector. Liquid samples were counted in 100 mL plastic screw capped containers which were placed directly against the detector. The dimensions of the containers were suited to the size of the detector and this type of container was used for all milk and water samples.

Samples were first counted for about 2 to 12 h in the early days, and then were recounted for 2 to 3 d, 2 to 3 months after the accident. Those counts were then used to quantify radionuclides present in samples. The counts were obtained with a Canberra 4096 channel analyser coupled with a $50 cm^3$ coaxial type Ge(Li) detector (FWHM 1.86 keV for the ^{137}Cs 0.661 MeV line). Because of the ability to measure the γ emitter radionuclides directly in the original sample, identification and evaluation of radionuclides were carried out by photopeak area without the need for chemical separations. Efficient calibrations of the detector for all geometries (both air sample and liquid sample geometries) were performed by using certified reference materials from Amersham which were in the same geometries. Also the geometries were calibrated for the density of the samples of interest as a function of γ ray energy. Efficiency calibrations were checked at intervals with a standardized solution of ^{137}Cs and ^{131}I to ensure that the obtained data were correct.

Background measurements were taken as frequently as was practicable and for as long as the samples in order to obtain good counting statistics and correct results.

3.   RESULTS AND DISCUSSION

3.1. Air

Following the Chernobyl nuclear accident, radionuclides were first detected in air in northern Europe, because air samples give the first indication of radiation.

TABLE I. DETECTION LIMITS FOR DIFFERENT GEOMETRIES AND MEASURING TIMES

Radionuclide	Measuring time (s)	Limit of detection (Bq)	
		Liquid geometry	Compressed filter paper geometry
Cs-137	10 000	1.17	0.26
	200 000	0.30	0.09
I-131	10 000	0.81	0.18
	200 000	0.26	0.07

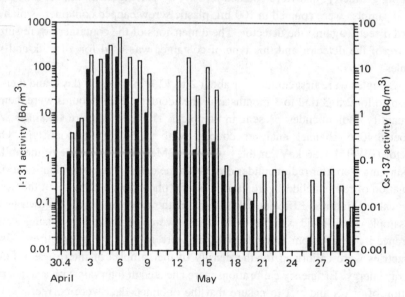

FIG. 1. Histogram showing the activity concentrations of I-131 and Cs-137 in air samples (■ indicates I-131 and □ Cs-137 concentrations in Figs 1–4).

Also in Ankara, the radioactive debris was first detected in the air sample taken on the morning of 30 April 1986. On 30 April and thereafter a rapid increase was measured in air samples collected in our sampling station. Iodine-131 and ^{137}Cs reached the highest values on 5 and 6 May, respectively. Their peak activities were around 203 and 46 Bq/m^3, respectively. The air activity gradually declined from 15 May until 17 June, when an indeterminate peak was observed. On 17 June and thereafter, the Chernobyl radioactivity was below the detection limit of compressed filter paper geometry and showed a stable trend at 200 000 s measuring time (1.0×10^{-2} Bq/m^3 for ^{131}I, 3.0×10^{-2} Bq/m^3 for ^{137}Cs) (Table I). The activity concentrations of ^{131}I and ^{137}Cs in air samples from 30 April to the end of May, when a small peak was observed, are shown in Fig. 1. Two major peaks on 1 to 9 May and on 12 to 16 May can be identified in this figure. As is shown in Fig. 1, the ^{137}Cs activity in air samples was less initially than the ^{131}I activity. Even though ^{131}I declined with a uniformly rapid rate, ^{137}Cs declined slowly.

After 5 May, ^{103}Ru appeared in our samples. Its activity was 79 Bq/m^3. Tellurium-132 and ^{140}La also appeared with less activity. There were, in addition, indications of the presence of ^{140}Ba, ^{144}Ce, ^{141}Ce and ^{95}Nb in air samples collected after 5 May. However, since these nuclides appeared only for a few days, we did not take them into consideration.

3.2. Milk

The maximum level of ^{131}I in cow milk was from a dairy in the vicinity of Haymana, 25 km south of Ankara. The level was measured as 8000–8200 Bq/L on 13 to 15 May. In these milk samples, ^{137}Cs reached its highest value of 570 Bq/L on 13 May. The minimum level of ^{131}I in cow milk was from a dairy in the vicinity of Çubuk, 20 km north of Ankara. The level was measured as 170 Bq/L on 13 May (Fig. 2). In these milk samples the highest value of ^{137}Cs was 65 Bq/L on 29 May. There were important differences in activity concentrations of those two adjacent regions' milk samples. There appeared to be a correlation with the rainfall during the preceding days. The radioactivity in the pastures bore a close relationship to that in the rain. In Haymana, it rained from 3 to 5 May, but in Çubuk it rained on 21 and 22 May, eight days after the passing of the Chernobyl cloud over Ankara. Thus, in Haymana the rain fell with contamination from the air, because during those days there was notable contamination in the air. But in Çubuk, the air contamination decreased until 21 May so pastures were less contaminated by the rain.

Cow milk taken from the State farms was derived from the mixing of milk supplies from many dairies, so that from individual dairies any samples which might have been highly contaminated would have been diluted considerably. Hence, the milk samples from State farms were representative of the concentrations in Ankara and the surrounding area of all the radionuclides originating from Chernobyl. These samples came from the bottled milk consumed especially by children in Ankara (Fig. 3).

Milk samples collected from five different dairies and from two State farms were monitored from 12 May to July. The concentration of ^{137}Cs, not only in air samples, but also in milk samples, was initially much less than that of ^{131}I and declined more slowly than ^{131}I; by 5 June it accounted for the greater part of the total activity. For example in the AOÇ State Farm milk samples, ^{131}I concentration

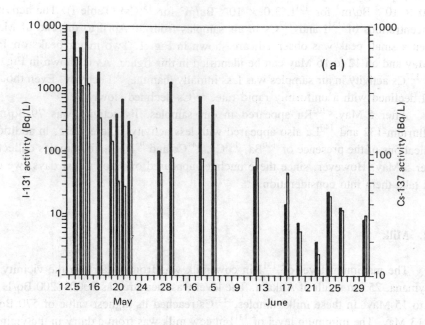

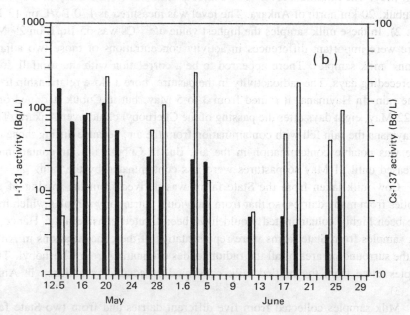

FIG. 2. Histogram showing the activity concentrations of I-131 and Cs-137 in milk samples collected from dairies in (a) southern and (b) northern areas of Ankara.

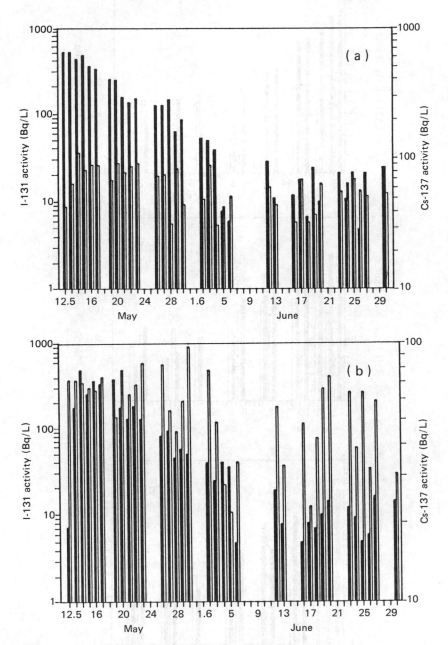

FIG. 3. Histogram showing the activity concentrations of I-131 and Cs-137 in milk samples from (a) AOÇ and (b) SEK State farms.

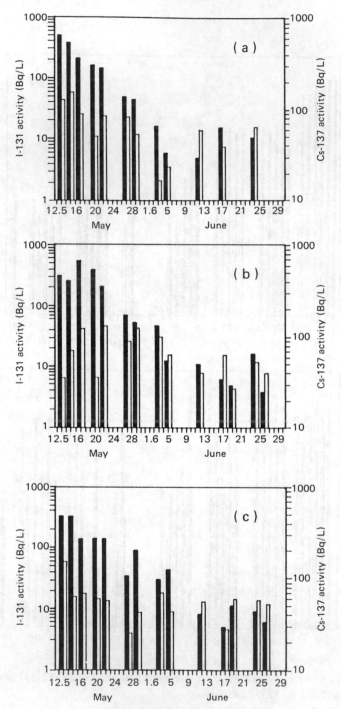

FIG. 4. Histogram showing the activity concentrations of I-131 and Cs-137 in milk samples collected from dairies in (a) southwestern, (b) eastern and (c) western areas of Ankara.

TABLE II. ACTIVITY CONCENTRATIONS (Bq/L) IN RAIN WATER COLLECTED IN MAY AND JUNE 1986

Date	I-131	Ru-103	Cs-137
20 May	260 ± 21	143 ± 9	75 ± 11
21 May	78 ± 13	75 ± 7	41 ± 8
22 May	23 ± 5	—[a]	37 ± 7
26 May	16 ± 4	37 ± 6	34 ± 7
27 May	15 ± 4	16 ± 4	22 ± 6
3 June	93 ± 14	21 ± 4	218 ± 18
5 June	53 ± 7	—	68 ± 11
12 June	15 ± 4	—	40 ± 7
13 June	5 ± 2	—	29 ± 6

[a] 'Not found'.

decreased with an effective half-life of 4.0 ± 0.4 d, corresponding to an environmental half-life of 8.0 ± 1.0 d[1]. Similarly the concentration of ^{137}Cs declined to 10–15 Bq/L in approximately one month (the effective half-life was 23.0 ± 3.0 d, environmental half-life was 23.1 ± 3.0 d). But this decline was less uniform in ^{131}I. It was found that every milk sample collected from a different dairy had a different effective half-life. These half-lives changed from 3.0 ± 0.4 d to 6.9 ± 0.7 d for ^{131}I and from 23.0 ± 3.0 d to 48.3 ± 5.1 d for ^{137}Cs from May to June.

Even though monitoring of samples took place from 12 May to 30 July, only the results of samples collected between 12 May and 30 June were investigated here. After 30 June the changes in measurements or the changes in sample activities were not very substantial. The levels in cow milk from all other dairies and farms are shown as bars in Fig. 4.

In this paper only cow milk samples are mentioned because it was not possible to collect goat milk in a daily sequence. However, it should be noted that the ^{131}I concentration in goat milk was found to be about three times that in the milk from

---

[1] The effective half-life contains components for the radiological (T_R) and the environmental (T_E) half-lives, which are related by the equation $1/T_{eff} = 1/T_R + 1/T_E$. Values for the environmental half-lives give a more direct indication of the clearance of the nuclide from the system being studied [1].

cows on the same pasture. Another point was that ^{131}I was the most important source of activity both in cow and goat milk in Ankara and its surroundings.

3.3. Water

Rain water is also an early indicator of radioactive contamination if it rains soon after an accident. After the Chernobyl accident, at about 11.00 a.m. on 20 May, a heavy rain began to fall on Ankara, amounting to precipitation of 36 mm within a few hours. Because the Chernobyl cloud had passed over Ankara and the contamination in the air in Ankara had decreased, the levels of radioactivity in that rainfall were not very important. Thus, in that rain water the highest activity value of ^{131}I was 260 Bq/L. After that, the concentration of ^{131}I activity decreased gradually until the last sampling date, 23 June; ^{137}Cs had the highest value on 28 May of 218 Bq/L. The changes in activity concentrations in rain water which fell on Ankara in the area surrounding our laboratory during May and June are shown in Table II.

Because there was less of an increase in the rain water's activity, the Kurtboğaz Dam, from which the tap water was pumped, had a total activity of 180 Bq/L on 21 May. Thus, in tapwater samples in Ankara, the Chernobyl radionuclides were found in a small fraction of the maximum permitted levels for drinking water allowed by the Commission of the European Communities and the World Health Organization. The tap water had its highest values on 19 May of 149 Bq/L and 56 Bq/L for ^{131}I and ^{137}Cs, respectively. Tapwater samples collected from 10 May to 13 June showed a uniform decrease in ^{131}I and [137]Cs. On 13 June and thereafter, ^{131}I was no longer detectable and ^{137}Cs reached a stable value of about 7 Bq/L.

At some of these sampling stations, collecting of air, milk and water samples continued, but no Chernobyl radionuclides had been detected through August 1989.

For all samples, the [134]Cs/^{137}Cs ratio remained nearly constant. This ratio is around 0.54 ± 0.04 and is in line with most of the values given in the literature [2].

REFERENCES

[1]   MARTIN, C.J., HEATON, B., ROBB, J.D., Studies of I-131 and Ru-103 in milk, meat and vegetables in northeast Scotland following the Chernobyl accident, J. Environ. Radioact. **6** (1988) 247–259.

[2]   AARKROG, A., The radiological impact of the Chernobyl debris compared with that from nuclear weapons fallout, J. Environ. Radioact. **6** (1988) 151–162.

VALIDATION OF MODELS FOR THE TRANSFER OF RADIONUCLIDES IN TERRESTRIAL, URBAN AND AQUATIC ENVIRONMENTS AND ACQUISITION OF DATA FOR THAT PURPOSE

G.S. LINSLEY, H. KÖHLER, D. CALMET
Division of Nuclear Fuel Cycle
  and Waste Management,
International Atomic Energy Agency,
Vienna

M. CRICK
National Radiological Protection Board,
Chilton, Didcot, Oxfordshire,
United Kingdom

Abstract

VALIDATION OF MODELS FOR THE TRANSFER OF RADIONUCLIDES IN TERRES-
TRIAL, URBAN AND AQUATIC ENVIRONMENTS AND ACQUISITION OF DATA
FOR THAT PURPOSE.

The radioactive fallout from the Chernobyl accident has offered the opportunity to test
the predictions of dose assessment models using measurements made under real conditions.
A joint co-ordinated research programme of the International Atomic Energy Agency and the
Commission of the European Communities seeks to take advantage of the 'natural laboratory'
created as a result of the Chernobyl release. Three working groups have been established concerned with model validation for the terrestrial, urban and freshwater aquatic environments;
in addition a fourth group, the Multiple Pathway Assessment Working Group, will seek to
validate models for estimating overall transfer from the atmosphere and ground deposits to
human intake. The paper outlines the plans for the study, which started in 1988 and will continue until 1992.

1. INTRODUCTION

The radiological impact of radionuclide releases to the environment is normally evaluated using mathematical models. Ideally the models should be developed
and tested using data on the transfer of the specific nuclides and physicochemical
forms being modelled, obtained in the actual environment being considered.
However, this is rarely possible, especially for accident situations. Therefore, use
has been made in the past of the information obtained from the fallout after the
nuclear weapons testing in the 1950s and 1960s or from laboratory experiments.

However, it has always been recognized that there may be differences between the physicochemical forms of the radionuclides from these sources and of those that could be released from nuclear installations. Furthermore, weapons testing fallout occurred over a long period; it did not provide a single pulse, which is more useful for testing models that predict time dependent behaviour of radionuclides in the environment. The Chernobyl accident created such a pulse, which was detected and measured in a variety of environments, mainly in the European parts of the Union of Soviet Socialist Republics and elsewhere in Europe.

An international programme has been established to collate data from Member States of the International Atomic Energy Agency (IAEA) and to use the data in a co-ordinated programme of model testing. The possibilities for data acquisition and model testing in the 'natural laboratory' created by the Chernobyl release were recognized at the Post-Accident Review Meeting, held in Vienna from 25 to 29 August 1986. In the summary report of that meeting, prepared by the International Nuclear Safety Advisory Group [1], it was recommended that:

"In order to improve predictions of the consequences of accidental releases of radioactivity, the IAEA should, in collaboration with WMO, review and inter-calibrate models of atmospheric transport of radionuclides over short and long distances and of radionuclide deposition on terrestrial surfaces (soils, vegetation, buildings, etc.) and establish a database for validation studies on such models. In addition, it should carry out similar activities with regard to models of the transfer of radionuclides through the terrestrial environment and in food chains, their transfer through surface waters (fresh water and sea water) and their transfer in urban environments."

Following these recommendations, the IAEA established a co-ordinated research programme (CRP) in 1988 on Validation of Models for the Transfer of Radionuclides in Terrestrial, Urban and Aquatic Environments and Acquisition of Data for that Purpose (VAMP) [2].

In January 1989, the Commission of the European Communities (CEC) joined the IAEA in sponsoring VAMP as a means of ensuring that international programmes in this subject area are co-ordinated and that overlap is minimized.

VAMP is concerned with models and data relevant to the terrestrial, urban and aquatic environments. It is not concerned with models for atmospheric transport, which is the subject of a related IAEA activity. VAMP will, however, consider the interactions of aerosols in the near surface air with terrestrial, urban and aquatic surfaces.

2.   SCOPE AND OBJECTIVES

The principal objectives of the VAMP CRP are:

(a)   To facilitate the validation of assessment models for radionuclide transfer in the terrestrial, urban and aquatic environments. It is envisaged that this will be achieved by acquiring suitable sets of environmental data from the results of the national research and monitoring programmes established following the Chernobyl release.
(b)   To guide, if necessary, environmental research and monitoring efforts to acquire data for the validation of models used to assess the most significant radiological exposure pathways.
(c)   To produce a report or reports reviewing the current status of environmental assessment modelling, including a review of the improvements achieved as a result of post-Chernobyl validation efforts and identification of the principal remaining areas of uncertainty in models used for radiation dose assessments.
(d)   To run test scenarios for model validations, these scenarios being selected for their importance in relation to radiation dose assessments. In selecting scenarios and processes for model validations it is necessary to bear in mind that there should be a clearly demonstrable need to improve the reliability of predictions of radionuclide transfer in the pathways chosen.

Four working groups on terrestrial, urban and aquatic environments and multiple pathway analysis have been established.

Two main approaches to model validation are being adopted. The first approach, that of formal model validation, consists of formulating scenarios to test models; modellers then perform calculations, which are subsequently analysed and compared with the observed data. The second approach involves expert review of the available data associated with processes simulated in a model, examination of the concepts and assumptions of the model, and consideration of whether such concepts and assumptions are consistent with the data.

2.1. Terrestrial Working Group

Potential issues for investigation by the Terrestrial Working Group were determined at the first full meeting of the CRP in May 1988. They included the following questions:

(1)   Are the accepted transfer factors from vegetation to animal products too high for ^{131}I and ^{137}Cs?
(2)   What are the transfer mechanisms from soil to pasture/sheep on upland pastures?

(3)   What are the typical losses during food preparation for different nuclides and food products?

(4)   How does the nature of a terrestrial watershed affect contamination levels in the freshwater aquatic environment?

(5)   What are the processes leading to elevated radionuclide concentrations in some foods, e.g. mushrooms and wild berries?

(6)   How does crop contamination vary with the season of the year in which a release occurs?

(7)   How effective are agricultural countermeasures in reducing radionuclide transfer to man?

(8)   How does rainfall affect the deposition and retention of radionuclides on crops?

(9)   What has the Chernobyl experience taught us about the resuspension of radionuclides from terrestrial surfaces into the atmosphere?

(10)   How has our understanding of the behaviour of radionuclides in forest ecosystems improved since Chernobyl?

(11)   What was the impact of the 'hot particles', detected in some countries, on food chain contamination?

For the first year of the study items (3), (6), (9) and (10) were chosen for analysis. The analysis is by means of expert review. Firstly, information on each of the four issues was acquired by means of questionnaire; secondly, experts were asked to review the available information, including the recent literature, and to produce preliminary analysis reports on each of the four topics. The reports will be discussed and critically reviewed by the Terrestrial Working Group and by all CRP participants at the next CRP meeting in December 1989. By this means it is hoped to produce a report on each of the topics in which the current best understanding of the process will be expressed, if possible quantitatively.

2.2.   Urban Working Group

Key questions for initial analysis by the Urban Working Group decided upon at the first CRP meeting in May 1988 were as follows:

(1)   What are the processes involved in the dry deposition of radionuclides on to trees, lawns and gardens, roofs and internal surfaces of buildings?

(2)   What are the processes involved in the wet deposition of radionuclides and subsequent runoff from trees, lawns and gardens, roofs and buildings?

(3)   How does the concentration of radionuclides on urban surfaces change as a function of time?

(4)   To what extent is the Chernobyl release unique in terms of the behaviour of the radioactive material in the urban environment?

(5)   Can the Chernobyl data be used for the validation of models for:
       — dose rates above lawns/soil/gardens?
       — dose rates inside buildings, taking into account shielding?
       — dose rates obtained after various means of decontamination?
       — dose rates obtained by handling air-conditioning filters?
(6)   Can ^{7}Be act as an analogue for caesium?

The Urban Working Group is considering these questions initially by obtaining information on available data and modelling capability by means of questionnaires sent to selected participants. The first questionnaire was sent out in the early part of 1989. It requested data of use for model validation, examined the understanding of potential participants of the theoretical concepts involved in urban modelling and requested information on available models for predicting the processes of wet and dry deposition. The replies are currently being assessed by a subgroup of experts. During 1990 the the Urban Working Group will be turning its attention to the question of estimating exposure outdoors, the shielding effects of buildings, the effectiveness of decontamination and dose rates from air-conditioning filters.

As well as attempting to validate models for the urban environment, and to understand the processes involved, the group has as one of its aims to stimulate interest in the urban problem, an area which has, perhaps, been given insufficient attention in the past.

2.3. Aquatic Working Group

The Aquatic Working Group will focus its attention on freshwater ecosystems. The proposed area of work will cover:

(1)   The description of factors regulating the transfer of radionuclides from land to water systems,
(2)   The major processes involved in radionuclide transfer in freshwater systems (rivers and/or lakes),
(3)   The causal relationships regulating uptake and loss of radionuclides in fish,
(4)   Possible remedial measures to reduce the concentration of caesium in lakes and fish.

A questionnaire has been distributed to screen individuals and institutions with an interest in the study of the dynamic behaviour of radionuclides in watersheds. The information collected on the models being used and measurements made by various laboratories after the Chernobyl accident will be analysed by experts. On the basis of this analysis a programme will be developed for model validation for freshwater aquatic ecosystems.

2.4. Multiple Pathway Assessment Working Group

A validation exercise that considers all pathways of exposure to human populations on the basis of data from the Chernobyl accident was proposed as an overall test for assessment models. The objectives of the proposed exercise are as follows:

(1)   To test predictive capability on a regional scale for multiple pathways of exposure,

(2)   To evaluate the importance of individual pathways and processes as contributors of radiation dose to regional populations and selected population subgroups,

(3)   To determine information important for collective dose assessment and for assessing the distribution of doses in selected population subgroups.

The validation exercise will include three components: (a) data collection and scenario formulation, (b) calculations by modellers, and (c) analysis of results. Questionnaires on the feasibility of such a study have been sent to many institutes. Responses received so far indicate that at least ten countries have data of potential use for multiple pathway model validation. The data from six countries have been chosen for more detailed study. It is intended finally that the test scenarios will be based on the data obtained from no more than three or four countries.

Once the test scenarios are designed, modellers will be invited to calculate whole body concentrations of radiocaesium for specified population groups and will be provided with appropriate site specific data. Input data will comprise information on radionuclide concentrations in air, deposited amounts of radioactivity, meteorological characteristics, environmental information on soil, farming, etc., population statistics concerning residential habits, and patterns of food production, distribution and consumption.

The exercise will be a blind test. Modellers will obtain observed values of whole body concentrations of ^{137}Cs only after their calculations are submitted to the scientific secretary. The scientific secretary will document the initial results submitted by each individual or organization that has performed calculations. Upon receipt of these results, measured values of whole body concentrations and concentrations in relevant environmental samples will be distributed so that each participant will be able to compare observations and predictions. This exercise is likely to provide insights into the relative importance of pathways and processes and of the influence of site specific factors, such as food supply patterns and preparation methods. It should result in an improved appreciation of the realism of the assumptions commonly made in dose assessment modelling.

REFERENCES

[1]   INTERNATIONAL NUCLEAR SAFETY ADVISORY GROUP, Summary Report on
        the Post-Accident Review Meeting on the Chernobyl Accident, Safety Series
        No. 75-INSAG-1, IAEA, Vienna (1986).
[2]   INTERNATIONAL ATOMIC ENERGY AGENCY, On the IAEA Co-ordinated
        Research Programme on Validation of Models for the Transfer of Radionuclides in
        Terrestrial, Urban and Aquatic Environments and Acquisition of Data for that Purpose,
        Progress Report No. 1, VAlidation of Model Predictions, IAEA, Vienna, 1988.

FRESHWATER AND ALPINE ECOSYSTEM RESPONSE TO CHERNOBYL FALLOUT IN NORWAY

T.B. GUNNERØD, I. BLAKAN, O. UGEDAL, T. SKOGLAND
Norwegian Institute for Nature Research,
Trondheim, Norway

Abstract

FRESHWATER AND ALPINE ECOSYSTEM RESPONSE TO CHERNOBYL FALLOUT IN NORWAY.

Environmental and agricultural research programmes were initiated in Norway in June 1986 following the Chernobyl accident. The main fallout in Norway occurred on 28 and 29 April and had an extremely patchy distribution between 60 and 66° N. In Lake Høysjøen in the county of North Trøndelag, high radioactivity was found in the sediment and in all levels of the food chains from vegetation to fish, including phytoplankton, zooplankton, bottom animals, brown trout and Arctic char. Food chains for both wild and domestic animals are studied in the Alpine areas of central Norway. High seasonal variation has been found in the radiocaesium load of reindeer. Such a variation is a result of the high radioactivity (up to more than 100 kBq/kg dry weight) in lichens, the only winter food for reindeer. Also there are indications of increased calf mortality and chromosomal aberrations in the blood cells of the reindeer in nutrient poor areas with heavy fallout. In mountain pastures where sheep and goats graze, a sharp increase in meat and goat milk radiocaesium occurred during the late pasture season in 1988. This was caused by the large number of mushrooms with high levels of radioactivity. Present research in Norway will have to be continued in order to develop further national and international co-operation among scientists and institutions with whom biologists had little or no contact previously.

1. INTRODUCTION

From all points of view, Norway was completely unprepared for the nuclear fallout from the Chernobyl accident in April/May 1986. This is surprising since Norway has been leading in the study of long range transport of pollutants, through analysis of the acid rain problems.

No environmental or agricultural research activities within the field of nuclear fallout were in progress in Norway at the time of the Chernobyl accident. During June 1986 two major research programmes were initiated, one under the Norwegian Agriculture Research Council (NLVF) and one under the research unit of the Directorate for Nature Management (DN), now the Norwegian Institute for Nature Research (NINA). Both programmes are in progress and will be completed by 1990–1991.

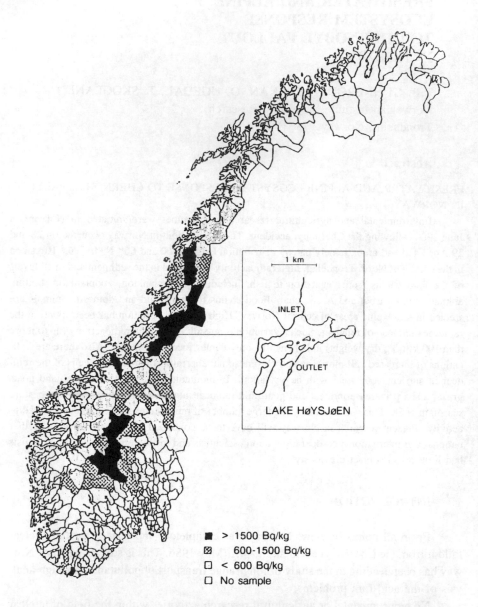

FIG. 1. *Average radiocaesium (^{134}Cs and ^{137}Cs) in Bq/kg wet weight in freshwater fish (mainly brown trout and Arctic char) from 573 lakes sampled from 161 Norwegian municipalities in 1986.*

2.   THE FALLOUT

As in central Sweden, Norway was affected mainly by the first days' releases from Chernobyl. The Norwegian Meteorological Institute has studied the air transport over the nuclear plant on 26 April and the days following. Two–three days after the accident, from 0700 Greenwich mean time (GMT) on 28 April to 0700 on 29 April, these air masses produced locally heavy precipitation in central Norway between 60 and 66° N, as snow in the mountain areas of east Jotunheimen and Rondane, and as rain in the counties of Trøndelag and in the southern part of Nordland [1, 2]. All later measurements of radiation on the ground, in soil samples, and in samples of plants and animals show the same patchy geographical distribution, both horizontally and vertically, as the 28–29 April precipitation. An example from freshwater fish is given in Fig. 1.

3.   FRESHWATER ECOSYSTEMS

At the time of the nuclear fallout in Norway, a scientist at NINA was conducting an acid rain research project in the Atna River, which drains from the Rondane Mountains. This project involved storing his daily water samples from the river as well as his two week precipitation samples.

Analysis of those water samples showed no traces of radionuclides from 25, 26 or 27 April, but more than 10 kBq/kg of ^{131}I and 1 kBq/kg of ^{134}Cs plus ^{137}Cs on 28 April, and still higher values the following day [3]. From 30 April radioactivity in the water samples fell, with a slight increase again by 5 May. Five days later only traces of radioactivity were found in the river water. The 260 mL of precipitation collected from 23 April to 6 May contained 11.6 kBq ^{131}I, 0.3 kBq ^{140}Ba, 1.1 kBq ^{134}Cs and 1.9 kBq ^{137}Cs per kg. In 550 mL of precipitation between 6 May and 23 May, Cs values were down to 1/10, with only traces of ^{131}I and no ^{140}Ba.

More than 600 samples of freshwater fish were analysed for radioactivity in Norway from June to September 1986. The samples, which were split into two parts, showed an actual increase in Cs content before and after 10 July. That was surprising since the radioactivity content of water was extremely low by that time and we assumed that fish took up radioactive Cs through the gills, replacing potassium in the 'oxygen pump' of the fish.

The explanation for the increased radioactivity in freshwater fish was found in our extensive studies in Lake Høysjøen in North Trøndelag [3]. By 16 July we found there an entire ecosystem being polluted by radiocaesium, except for the lake water itself (only 2–3 m above bottom at the deepest part of the lake 23 Bq/kg was found in the water). Sediments contained 7740 Bq/kg, phytoplankton 830 Bq/kg, zooplankton 50–180 Bq/kg, bottom animals (mayflies) 4900–18 500 Bq/kg, brown trout flesh 9027–25 456 Bq/kg with stomach contents (mainly bottom animals) 19 286 Bq/kg, and Arctic char 2994–7731 Bq/kg with stomach contents (mainly zooplankton) 6270 Bq/kg, all measured as wet weight. Submerged plants (*Juncus bulbosus*) and semi-submerged plants (*Carex rostvata*) had 24 900–42 500 Bq/kg and 2500–13 000 Bq/kg dry weight, respectively. In short, the freshwater fish in Lake Høysjøen, and most likely other places, were feeding themselves on increased radioactivity until mid-July 1986.

From July 1986 through 1988, radiocaesium in brown trout and Arctic char dropped, more slowly in char than in trout [4]. By late 1988 both species still had an average radiocaesium value of about 2 kBq/kg wet weight. Their major food organisms (bottom animals and zooplankton) also dropped in radiocaesium contents during late 1986, but stayed fairly constant during 1987 and 1988, especially in bottom animals. The sustained level in bottom animals reflects the high level in the lake sediments [4].

4.   TERRESTRIAL ECOSYSTEMS

The uptake, distribution and removal of radiocaesium in terrestrial ecosystems have been studied extensively both in the agricultural and the environmental research programmes since 1986.

The mountainous region of southeastern Norway and the counties of Trøndelag and south Nordland had the highest deposition of radioactivity, ranging from 2 to 360 kBq Cs/m^2, with a geometric mean of 23 [5]. These areas are natural pastures both for wild and domestic animals, including reindeer, sheep and goats. A total of 126 different places were sampled for vegetation in 1986–1989 as part of the agricultural programme.

In 1989, most of the radioactivity was found in the upper 4 cm of the soil and in the live and dead vegetation. In the mountainous region of southeastern Norway, the average value of total vegetation radioactivity fell by about 70% from 1986 to 1987 (from 2350 to 680 Bq $^{137}Cs/kg$ dry weight), but stayed fairly constant in 1988 and 1989. In the central part of Norway, with lower deposition (mean of 16 kBq Cs/m^2), vegetation activities stayed relatively constant from 1986 to 1989. A transfer coefficient between vegetation and soil of $(28–727) \times 10^{-3}$ has been computed.

A high variation in radioactivity was found between groups of plants and also between species of plants within each group. The following general sequence was found:

leaves < grass, half-grass < herbs ≪ lichens, mosses < mushrooms

Whereas radioactivity decreased from 1987 to 1988 in leaves and grasses by 25–30%, it increased in herbs by 30% [6]. On cultivated land, fertilization by potas-

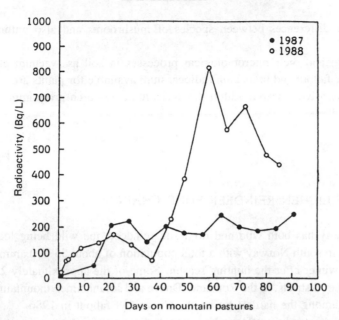

FIG. 2.   Radiocaesium in goat milk from mountain pastures in 1987 and 1988, Griningsdalen, south Norway [7].

sium gave reduced radioactivity in vegetation, whereas excess nitrogen gave increased crops and radioactivity [7].

From an economic point of view, the necessary removal of products from the market owing to high meat radioactivity in sheep, reindeer and some cattle in 1986, and also to the prolonged feeding with caesium binders, has been the most direct and costly effect of the Chernobyl fallout in Norway.

The level of radioactivity in domestic animals on mountain pastures was slightly lower in 1987 compared with 1986, and a further decrease was expected in 1988. However, by the end of the 1988 grazing season, a sharp increase in both meat and other products from the animals occurred. An example is given in Fig. 2 showing the sharp increase in radioactivity in goat milk in 1988 from about 50 to 800 Bq/L after a grazing period of about 40 d on mountain pastures [7].

The year 1988 was an unusually rich year for mushrooms in the Norwegian mountains. Knowing that both domestic and wild animals feed heavily on several of the mushrooms species when available, researchers formulated the hypothesis that it was here that the source of the increased radioactivity in 1988 was located. Extensive sampling of mushrooms, soil and green vegetation from the same locations was conducted. Bakken and Olsen [8] found a 10 to 150 times higher content in the fruit bodies of the fungi (i.e. the mushrooms) than in plants at the same location. There were wide differences between species of mushrooms and also within the same genus.

Fungi and their microbiological processes in soil as 'vacuum cleaners' for radioactive fallout and in making radiocaesium available for plants are now the subject for extensive research within the agricultural research programme [9].

5.   THE LICHEN–REINDEER FOOD CHAIN

Norway has both wild and domestic reindeer, the wild being located in the mountains in south Norway with a total population of about 35 000 animals remaining in the winter after the hunting season. Some of the approximately 25 separated wild reindeer districts, in the Rondane, Dovre and Jotunheimen mountains for example, were among the hardest hit by the Chernobyl fallout in 1986.

To understand the effect and seasonal variation of fallout radioactivity in reindeer, the following facts must be known:

(1)   When snow falls in the mountains, and for a period of about 200 d/a, reindeer feed entirely on different species of lichens. Daily intake of about 3 kg lichen (dry weight) in winter has been estimated. For the rest of the year, in late spring and summer, green vegetation is the main food, but the animals maintain bacterial fauna in their stomachs by eating 20–25% lichens.

(2)   Lichens have no root systems, take up all nutrients for growth from the air, and conserve all of these extremely well. Lichens are therefore extremely susceptible to air pollution, including radionuclides from fallout. Values well above 100 kBq/kg lichen dry weight of ^{134}Cs and ^{137}Cs were measured in Norway after the Chernobyl fallout. Caesium in lichens has previously been reported to have a biological half-life of about 10 a, with an effective half-life of about 7 a due to physical decay.

(3)   The biological half-life of radiocaesium in reindeer has been recorded in the literature to be about 18 d in the winter and as low as 6 d in the summer owing to the high metabolic rate in the animals.

These facts indicate a strong seasonal variation in the body load of radiocaesium in reindeer, with peak values in late winter and low values in late summer. In the nutrient poor mountain area of Rondane in central Norway, with heavy fallout, average values in the muscles of wild reindeer ranged from less than 10 kBq/kg wet weight in the month of August to nearly 50 kBq/kg in March (Fig. 3). In domestic reindeer in central Sweden, Åhman [10] reported still higher variations, from about 0.5 kBq to nearly 40 kBq ^{137}Cs/kg muscle wet weight.

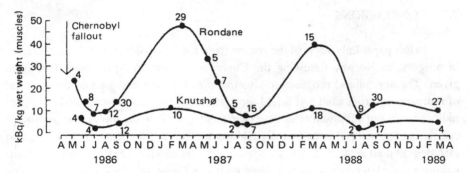

FIG. 3. Seasonal variation in radiocaesium in two wild reindeer populations in Norway from 1986 to 1989. The Rondane Mountain region is an extremely nutrient poor area with high Chernobyl fallout; Knutshø Mountain is more nutrient rich with less fallout. Numbers in figure indicate size of sample.

6. HAS THE CHERNOBYL FALLOUT HAD DETRIMENTAL EFFECTS ON REINDEER IN NORWAY?

In their status report from the environmental research programme for 1986 to March of 1989, Skogland and Espelien [11] make the following statements:

"Following the accident at Chernobyl the radiation dosages received by reindeer during gestation when they forage on a lichen diet are within the range of low dosage values considered to be of detrimental effect to mammals. The mortality and population dynamics of the high exposure (Rondane north and Rondane mid) and low exposure (Knutshø) herds have been monitored before and after the radiation accident. A 25% decline in neonatal survival of reindeer calves has been shown for the two reproductive seasons in the two high exposure herds, but no statistically significant change has been found in the Knutshø herd, or other adjacent low exposure herds. Whether this effect is connected to the Chernobyl accident is still an open question.

"Chromosomal aberrations in blood samples from culled pregnant females, foetuses, calves and male reindeer from Rondane and Knutshø have been studied since the winter of 1987. Aberrations varied with season (highest in winter), sex, age and size (highest in foetuses and calves, and higher in pregnant females than males). From a sample of 12 adult females, a significant correlation between radiocaesium concentration and chromosomal aberrations was found (radiation coefficient $r = 0.57$, $p < 0.05$)."

7. CONCLUSIONS

In this paper only a few of the results from the two major research programmes in progress in Norway following the Chernobyl fallout in April 1986 have been given. The agricultural programme also includes research and tests of the removal of radioactivity from solid and liquid food products, the development and testing of radiocaesium and strontium binders, experimental work on the biological half-life of caesium in fish, as well as basic analytical work. The environmental programme also includes detailed studies of the distribution of radiocaesium in different plant communities and of uptake within individual plants and mountain food chains, including different species of birds and mammals, from mice to wolverines.

All of this research has two main objectives:

(a) To study the effects of the Chernobyl fallout in Norway in natural and domestic food chains, and to reduce the costs of keeping the agricultural products within the accepted level of radioactivity for human consumption;

(b) To increase the general level of scientific knowledge in Norway in case a new nuclear accident occurs so that we will be better prepared to handle the situation than we were in 1986.

To reach these goals our present research will have to be continued after the two current programmes are completed in 1990 and 1991. In this work we will have to develop further national and especially international co-operation among scientists and institutions with whom we as biologists previously had little or no contact. Our own and some of our colleagues' participation in this symposium is an expression of this need.

REFERENCES

[1] SALTBONES, J., The Nuclear Power Accident in Chernobyl: Atmospheric Transport and Distribution of Radioactive Materials, Norwegian Meteorological Institute, Oslo (1986) (in Norwegian).

[2] NORWEGIAN OFFICIAL REPORT, Actions against Nuclear Power Accidents, Part 1: Experiences from the Nuclear Power Accident in Chernobyl, Oslo University Press, Oslo (1986) (in Norwegian).

[3] JENSEN, B.M. (Ed.), Annual Report from the Environmental Research Programme, 1986, Directorate for Nature Management, Trondheim (1987) (in Norwegian).

[4] GAARE, E., UGEDAL, O., Annual Report from the Environmental Research Programme, 1988, Norwegian Institute for Nature Research, Trondheim (1989) (in Norwegian with English abstracts).

[5] HAUGEN, L.E., "Radiocesium in natural pastures in Norway after the Chernobyl fallout in 1986", paper presented at Nordic Agricultural Science Association Sem. on Deposition and Transfer of Radionuclides in Nordic Terrestric Environment, Beitostølen, Norway, 1989, abstract appears in Nordisk Jordbrugsforskning 1990 1.

[6]   GUNNERØD, T.B., GARMO, T.H., Annual Report from the Agricultural Research
      Programme, 1988, Rep. 1-1989, Federal Agricultural Information Service, Oslo
      (1989) (in Norwegian).

[7]   GARMO, T.H., HOVE, K., PEDERSEN, Ø., STAALAND, H., Radioactivity in
      goatmilk 1988, Sau og Geit 3 (1989) 170–173 (in Norwegian).

[8]   BAKKEN, L.R., OLSEN, R.A., "Accumulation of radiocesium in different fungal
      species, compared with concentrations in plant and soil", paper presented at Nordic
      Agricultural Science Association Sem. on Deposition and Transfer of Radionuclides in
      Nordic Terrestric Environment, Beitostølen, Norway, 1989.

[9]   OLSEN, R.A., JONER, E., BAKKEN, L.R., "Soil microorganisms and the fate of
      radiocesium in the soil ecosystems", ibid.

[10] ÅHMAN, B., "Effects of bentonite and zeolite on the accumulation and excretion of
      radiocesium in reindeer", ibid.

[11] SKOGLAND, T., ESPELIEN, T., Status Report from the Environmental Research
      Programme, 1986–1989, Norwegian Institute for Nature Research, Trondheim (1989)
      (in Norwegian).

POSTER PRESENTATIONS

IAEA-SM-306/71P

IMPLICATIONS OF LARGE SCALE
ENVIRONMENTAL CONTAMINATION IN AUSTRIA
AFTER THE CHERNOBYL ACCIDENT

M. TSCHURLOVITS, H. BÖCK, K. BUCHTELA,
F. GRASS, E. TSCHIRF, E. UNFRIED
Atominstitut der Österreichischen Universitäten,
Vienna, Austria

1.   INTRODUCTION

This presentation deals with a discussion of practical issues raised in Austria after the Chernobyl accident as seen by the staff of a university institute not involved in the official hierarchy of those responsible for emergency planning and for establishing countermeasures. The first response to vague information was the creation of an extended air monitoring programme. Complete air monitoring was carried out from the Atominstitut der Österreichischen Universitäten, indicating that such an institution might have a faster response than other bodies. In addition, the first prediction of dose was also issued by the institute soon after the event. The results of these measurements and estimates are shown below.

In general it was found that the non-scientific influences had the greatest impact on the communication with the public. This led to the situation in which the public response was very intensive in the early phase of the accident (which led to a small contribution to the total dose), and faded to little or no attention at a time when the large dose fraction was present owing to ingestion of long lived radionuclides. This situation also resulted from the lack of an established system to provide well founded public information, an exponential increase of self-assigned experts on radiation protection, and the hysterical response of people trying to take advantage of an emergency for their personal promotion.

This discussion is therefore concerned with a realistic consideration of the sequence of exposure pathways and their relevance for total exposure, as well as for the monitoring strategy. Table I shows in brief the relation among different phases, techniques and possible countermeasures after an accident.

TABLE I. COMPILATION OF KEY ISSUES

Pathway	Phase	Measuring technique	Key nuclide	Possible counter-measure	Contribution to total dose
Inhalation	Early	Complete sampling at short intervals	I-131	Sheltering	Small
	a1	a2	a3	a4	a5
External exposure	Early	Continuous gamma counting	Dependent on composition	Sheltering	Small
	b1	b2	and age b3	b4	b5
Ingestion	Early c1	Gamma spectroscopy	I-131	Banning of food	
	Late c2		Cs-137	None c3	Large

Notes:

a1: During passage of the contaminated cloud.

a2: For determination of the activity concentration in different air masses.

a3: Led to approximately 60% of the dose in this pathway.

a4: Boundaries for decision:

        Upper limit: threshold of non-stochastic effects in the thyroid.

        Lower limit: radon exposure in rooms with low ventilation rates.

        Dose reduction factor for sheltering, approximately 0.3.

a5: About 6% of the total dose in 1986.

b1: Owing to short lived radionuclides, long lived radionuclides predictable.

b2: Serves mainly as prewarning system, because contribution to the total dose is small.

b3: Dependent on concentration and age of the fission products.

b4: Dose reduction factor dependent on construction of building, wide range.

b5: About 20% of the total dose in 1986.

c1: Early phase: deposition on leafy vegetables, pasture–cow–milk chain;

c2: Late phase: deposition and uptake both dependent on season.

c3: Banning of foodstuffs reasonable only for a short period and a few pathways.

2.   ESTIMATES AND MEASUREMENTS

Experience has shown that even in the early phase following an accident, a prediction at least of the order of magnitude of dose was possible on the basis of the existing air and external dose rate monitoring results. Therefore, some countermeasures which were implemented were not justified at all, including the banning of foodstuffs, decontamination procedures and other measures.

Furthermore, some confusion was generated in the rather arbitrary setting of limits instead of determining reference levels in terms of activity concentration. This is demonstrated by the variability of data. Further it was found that the limitation of the predicted dose is more in line with radiation protection standards than the use of an auxiliary quantity such as activity concentration in air or in foodstuffs.

Another issue was the measurement by a few experienced but overburdened measuring institutes of a flood of samples requested by the public. Even if the use of numerous, unsuitable and uncalibrated measuring devices is disregarded, the measurements were still a problem. It was found that the reduction of data by measuring a few samples selected at random and the prediction of a distribution and hence a range of figures are sufficient for a reasonable estimate. An important management problem was the measurement of an enormous number of, in some cases, biased samples. This led to useless data, prevented the measurement of reasonable samples and did not improve the quality of the measurement.

3. CONSIDERATIONS FOR HANDLING OF LARGE SCALE CONTAMINATION ACCIDENTS IN THE FUTURE

Regarding information to be provided to the public, the conversion of measuring results in terms of dose instead of publication of the auxiliary quantity 'activity concentration of radionuclide i', which gives no information on the actual dose, seems necessary. This might lead to a simpler, but more understandable impression by use of the external dose rate and the dose due to inhalation and ingestion.

Simple but reasonable cost–benefit analysis has to be implemented in advance to take into account also the costs of countermeasures, including not only monetary cost but also costs of exposure and other detrimental actions associated with the countermeasures.

Samples taken for a preplanned official dose assessment system have to be considered with a higher priority than additional samples satisfying only more or less prominent individuals.

Cost–benefit considerations have also to be carried out in selecting appropriate measuring techniques; also the number of samples to be counted has to be considered. In addition, a preplanned system of reporting data including uniform units has to be established.

Some other issues arising after the event demonstrated that a large scale contamination unmasked the lack of communication among different groups. An improvement of the skills of decision making authorities and the fast response of actually competent bodies might help in the future to minimize the consequences of such events. In addition, some development of measuring techniques appropriate for the given purpose, e.g. to identify the type of accident by analysing the ratio of radionuclides, has also to be undertaken. This analysis might include a check of the completeness of the fission products as well as other issues.

IAEA-SM-306/80P

UPTAKE OF ALPHA AND GAMMA EMITTING RADIONUCLIDES BY MARINE BIOLOGICAL MATERIALS

A. NOUREDDINE
Laboratoire d'environnement,
Haut commissariat à la recherche,
Alger, Algeria

1.   SURVEY METHODS

The work presented involved a small survey of radionuclide concentrations in a number of marine biological materials including seaweeds, mussels and winkles from the Solway coast of southwest Scotland. Samples of plants (*Salicornia dolichostachya*), seaweeds (*Fucus vesiculosis, Ascophyllum nodosum, Fucus spiralis, Pelvetia canaliculata*), mussels (*Mytilus edulis*) and winkles (*Littorina littorea* L.) were collected in April 1986 from four sites on the northern Solway coast, some 60–100 km from the British Nuclear Fuels plc Sellafield plant discharge point as shown in Fig. 1.

The samples of *Salicornia dolichostachya* were collected from an area of intertidal sediment near Carsethorn; the other species (mussels, seaweeds and winkles) obtained from the other sites were all firmly attached to rocks.

The samples were analysed by both direct gamma spectroscopy and radio-chemical separation of Pu and ^{210}Po, followed by alpha spectrometry. For gamma analysis, samples of the ashed, dried and powdered material were put into polythene containers of reproducible geometry for gamma spectroscopy analysis, which was performed using a well shielded Ge(Li) detector with an active volume of 130 cm^3 and a resolution of about 1.9 KeV at 1333 KeV.

In order to calculate the concentrations of the different natural and artificial radioelements found in the samples, the gamma detection efficiency characteristics of the system were determined, using standard solutions covering the main energies (^{109}Cd, ^{57}Co, ^{139}Ce, ^{203}Hg, ^{85}Sr, ^{113}Sn, ^{137}Cs, ^{88}Y, ^{60}Co) and dried on cellulose in the same geometry as the samples. The main gamma emitting radionuclides detected in the samples were radiocaesium isotopes and ^{241}Am. Polonium-210 was extracted from dissolved samples by spontaneous deposition on copper foil, using a ^{208}Po yield tracer [1].

Plutonium isotopes were extracted from dissolved samples by anion exchange separation followed by electrodeposition on stainless steel discs [2]. Both plutonium and polonium were then determined using a surface barrier detector.

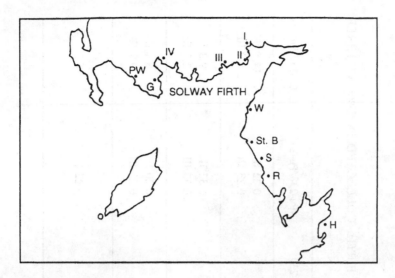

FIG. 1. Map showing the different sampling sites surrounding Sellafield: I: Carsethorn; II: Southerness; III: Kippford; IV: Carlsuith; PW: Port William; G: Garliestown; W: White-haven; St B: Saint Bees; S: Sellafield; R: Ravenglass; H: Heysham.

In addition to these chemical analyses, in some cases only total alpha particle determination or distribution may be required and alpha particle track analysis can be of great use in such cases. Alpha particle track studies were employed in the present work by using a CR-39 plastic track recorder.

2. RESULTS AND DISCUSSION

Of the radionuclides measured in this work it was found that the artificial radionuclide activities were dominated by the Sellafield plant's discharged nuclides.

Initially it is useful to consider the radionuclide activity ratios for the samples as presented in Table I.

It is apparent that for samples of the same species from different areas, the $^{137}Cs/^{134}Cs$ ratios are relatively constant. In contrast, when these ratios are compared for different species, markedly different values are observed. The highest value of 62 was observed for Salicornia dolichostachya with values of 20 to 35 for mussels and winkles and lower values of 7 to 8 for mussels. The latter value is close to the Sellafield discharge value for the preceding years, including little decay of the ^{134}Cs, whereas the other values suggest more substantial decay of ^{134}Cs and may possibly indicate the influence of sediment contribution.

TABLE I. ACTIVITY RATIOS IN BIOLOGICAL SPECIES FROM SOLWAY FIRTH, COLLECTED IN 1986 AT THREE LOCATIONS: I: CARSETHORN; II: SOUTHERNESS; III: KIPPFORD

Sample	Cs-137/Cs-134	Cs-137/Am-241	Am-241/Pu-239	Pu-238/Pu-239	Cs-137/Pu-239
Seaweeds (Fucus vesiculosis)	24 II 20 III	19 II 11 III	0.75 II 0.46 III	0.22 II 0.23 III	14.5 II 5.2 III
Mussels (Mytilus edulis)	8.1 II 6.9 III	4 II 0.9 III	2.2 II 2.9 III	0.32 II 0.33 III	9.3 II 2.5 III
Winkles (Littorina littorea L.)	30 II 35 III	4.5 II 6.1 III	1.8 II 1.6 III	0.20 II 0.23 III	8.1 II 9.7 III
Plants (Salicornia dolichostachya)	62 I	5.6 I	1.1 I	0.22 I	6.3 I

TABLE II: RANGE OF THE ACTIVITY RATIOS PROVIDED BY THE MINISTRY OF AGRICULTURE, FISHERIES AND FOOD; SAMPLES ARE TAKEN FROM SIX LOCATIONS: S: SELLAFIELD; H: HEYSHAM; PW: PORT WILLIAM; G: GARLIESTOWN; R: RAVENGLASS; St B: SAINT BEES

Sample	Cs-137/Cs-134	Cs-137/Am-241	Am-241/Pu-239	Pu-238/Pu-239	Cs-137/Pu-239
Seaweeds (*Fucus vesiculosis*) From S,H,PW,G	17.3–26.9	14.26–60.9	0.30–0.41	0.22–0.26	5.7–22.2
Mussels (*Mytilus edulis*) From R and St B	17.0–21.8	0.75–0.80	0.93–2.6	0.22–0.26	0.7–1.1
Winkles (*Littorea littorea* L.)	47.5–15.9	5.3–3.6	1.3–1.4	0.26–0.23	6.8–4.8

TABLE III. RADIONUCLIDE CONCENTRATIONS (Bq/kg DRY WEIGHT) IN SAMPLES TAKEN IN THREE LOCATIONS: I: CARSETHORN; II: SOUTHERNESS; III: KIPPFORD

Sample	Cs-134	Cs-137	Pu-238	Pu-239	Am-241	Po-210
Mussels (Mytilus edulis)	4.24 III 7.15 II	57.22 III 65.7 II	7.2 III 2.3 II	22.12 III 7.03 II	64.5 III 15.65 II	86.38 III 59.38 II
Winkles (Littorina littorea L.)	5.17 III 7.85 II	266.13 229.9 II	6.2 III 5.6 II	27.5 III 28.36 II	43.3 III 51.38 II	23.90 III 49.51 II
Seaweeds (Fucus vesiculosis)	5.28 III 3.11 II	159 III 149 II	7.03 III 2.23 II	30.4 III 10.26 II	14 III 7.70 II	5.81 III 3.09 II
Plants (Salicornia dolichostachya)	19.2 I	1315.4 I	45.04 I	207.50 I	234.20 I	223.00 I

An analogous variation is seen for the ^{238}Pu/^{239}Pu ratio and once more the mussel samples reflect recent discharge values. The ^{137}Cs/^{241}Am and ^{137}Cs/^{239,240}Pu ratios differ from species to species and also from region to region. However, all of the samples exhibited values of less than 10 for these ratios except the seaweeds where higher values of up to 19 were observed. These values indicate a greater influence of sea water than that of sediment as the source of the radionuclides in this situation.

Except for seaweeds, ^{137}Cs/^{239}Pu ratios are higher than those of ^{137}Cs/^{241}Am, because ^{241}Am is more insoluble in sea water than ^{239}Pu and ^{240}Pu. The variation of ratios from Southerness to Kippford is in agreement with the nature and distribution of sediment in the sampling area [3].

A similar pattern can be observed in the ^{241}Am/^{239}Pu ratios where the seaweed once more shows lower values (indicating a seawater influence) compared with the other samples which have values similar to those of sediments. The different activity ratios observed in Table II are in good general agreement with those provided by the Ministry of Agriculture, Fisheries and Food [4]. From these results the general conclusion can be drawn that the seaweeds show values closest to those of sea water, while the other species show ratios more similar to those of sediments in this area. The radionuclide activities measured from the various species are shown in Table III. The most striking point of the results is the high concentration of each radionuclide by *Salicornia dolichostachya*. It appears that the uptake of radionuclides by the plant would merit more intensive study. The presence of such areas of high alpha activity in the matrix of these plants has been established by the use of an alpha particle track detector (CR-39).

From the concentration tables it can be seen that radiocaesium activities are very similar for individual species from different sampling points consistent with the relatively uniform distribution of radiocaesium in this part of the Irish Sea. It is also notable that natural ^{210}Po is present at an activity level similar to that of the pollutant actinides and is a significant contributor to the background radiation level.

3.   CONCLUSIONS

The results obtained showed that the marine biological species can incorporate various natural and artificial alpha and gamma emitting radionuclides and that the uptake pattern is generally consistent with the established marine biogeochemical behaviour of these nuclides. Furthermore, relatively high concentrations of all the radionuclides investigated have been measured in a type of marine moss, *Salicornia dolichostachya*, which would appear to be worthy of further study.

REFERENCES

[1]   MACKENZIE, A.B., SCOTT, R.D., Separation of bismuth-210 and polonium-210 from aqueous solutions by spontaneous adsorption on copper foils, Analyst **104** (1979) 1151–1158.

[2]   COLEMAN, G.H., The Radiochemistry of Plutonium, Rep. Subcommittee on Radio-chemistry, National Academy of Sciences and National Research Council, Washington, DC (1965).

[3]   HETHERINGTON, A., "The behaviour of plutonium nuclides in the Irish Sea", Environmental Toxicity of Aquatic Radionuclides: Models and Mechanisms (MILLER, M.W., STANNARD, G.N., Eds), Ann Arbor Science Publishers, Ann Arbor, MI (1975).

[4]   MINISTRY OF AGRICULTURE, FISHERIES AND FOOD, Annual Reports for Aquatic Environmental Control, 1977–1985, MAFF, London (1985).

IAEA-SM-306/150P

STUDY OF AIRBORNE AND FALLOUT RADIOACTIVITY IN HUNGARY FOLLOWING THE CHERNOBYL ACCIDENT

P. ZOMBORI, I. FEHÉR, M. LŐRINC
Central Research Institute for Physics
   of the Hungarian Academy of Sciences,
Budapest

E. GERMÁN, L. KEMENES
Paks Nuclear Power Plant,
Paks

Hungary

After the Chernobyl accident the composition and concentration of the airborne radioactivity as well as of the fallout were regularly measured in two different places in Hungary. Air samplers of the same type (pumps sucking air through a plastic fibre filter with an average flow rate of 60 m^3/h) operated simultaneously on the territory of the Central Research Institute for Physics (Budapest) and at eight measuring stations around the Paks Nuclear Power Plant. Daily (later weekly) fallout samples were collected on a 0.2 m^2 area at all nine sites. Both aerosol and fallout samples were measured directly by high resolution gamma spectrometry and the integrated fallout radioactivity was studied also by means of in situ gamma spectrometric measurements [1].

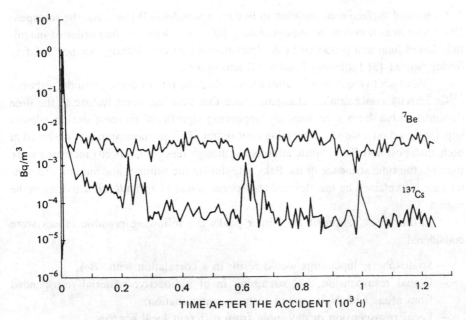

FIG. 1. *Time variations of* ^7Be *and* ^{137}Cs *activity concentrations in ground level air in Budapest.*

The primary objectives of these investigations were to follow the variations of the radioactivity in ground level air, to describe its decrease in an analytical way and to find a possible explanation for the observed time dependence and fluctuations. Our main interest was focused on the behaviour of the longest lived gamma emitter, ^{137}Cs, though other nuclides (especially ^{131}I, ^{106}Ru and ^{134}Cs) were also considered.

The variation of the ^{137}Cs aerosol concentration was followed for three years both in Budapest and at Paks (results of measurements in Budapest are shown in Fig. 1). At both places rapid clearance of the atmosphere was observed. The half-lives were found to be 5.7 and 7.7 d, respectively, and the influence of resuspension could also be followed after the 50th day. In the period of 50–200 days after the accident the time dependence of the ^{137}Cs concentration in Budapest can be described by the following relationship [2]:

$$C_a = 8.0 \times 10^{-3} \exp\left(-\ln(2)\,\frac{t-15}{5.7}\right) + 5.6 \times 10^{-4} \exp\left(-\ln(2)\,\frac{t-15}{95}\right)$$

where C_a is in becquerels per cubic metre and t in days. The first term corresponds to rapid clearance, the second to the fast component of the resuspension factor. The

^{137}Cs ground surface concentration in Budapest was 5400 Bq/m^2, and the resuspension factor was found to be approximately 10^{-7} m^{-1}, which is one order of magnitude lower than that proposed in the International Atomic Energy Agency's Safety Series No. 81 [3] following Linsley [4] and others.

About six to eight months after the accident the rate of decrease of the airborne ^{137}Cs activity concentration changed. Since that time the main feature of the time dependence has been a periodically appearing significant increase during winters superimposed on a slowly decreasing continuum. This phenomenon was observed at both Budapest and Paks, while additional (though less pronounced) increases were found in the time sequence of the Paks data during the summer and autumn. The latter case is explained by the elevated resuspension due to agricultural activities in the region.

In the investigation of the winter peaks the following possible causes were considered:

— Stratospheric input (this would result in a correlation with ^{7}Be);
— Global resuspension, i.e. streaming in of radioactive material resuspended from areas with high ground surface contamination;
— Local resuspension or any input from different local sources.

Though none of these causes can be completely excluded, the analysis of the different data supports the assumption that the variation of the ^{137}Cs activity concentration in ground level air is governed mainly by local resuspension.

REFERENCES

[1] ANDRÁSI, A., et al., Monitoring the Radiation Consequences due to the Disaster at the Chernobyl Nuclear Facility from April 28 to June 12, 1986, Preliminary Report KFKI-1986-49/K, Central Research Inst. for Physics, Budapest, 1986.

[2] FEHÉR, I., Experience in Hungary on the radiological consequences of the Chernobyl accident, Environ. Int. **14** (1988) 113–135.

[3] INTERNATIONAL ATOMIC ENERGY AGENCY, Derived Intervention Levels for Application in Controlling Radiation Doses to the Public in the Event of a Nuclear Accident or Radiological Emergency — Principles, Procedures and Data, Safety Series No. 81, IAEA, Vienna (1986).

[4] LINSLEY, G.S., Resuspension of the Transuranium Elements: A Review of Existing Data, Rep. NRPB-R75, Natl Radiological Protection Board, Chilton, UK (1978).

IAEA-SM-306/149P

CAESIUM RADIONUCLIDE BODY CONTENTS AND RADIATION DOSES TO RESIDENTS IN THE RSFSR TERRITORY AFFECTED BY RADIOACTIVE CONTAMINATION FOLLOWING THE CHERNOBYL ACCIDENT

Yu.O. KONSTANTINOV, G.Ya. BRUK, T.V. ZHESKO,
N.F. KORELINA, O.S. MOSKALEV, V.I. PARKHOMENKO,
P.V. RAMZAEV, M.S. SOLOV'EV
Research Institute of Radiation Hygiene,
RSFSR Ministry of Public Health,
Leningrad,
Union of Soviet Socialist Republics

In planning and implementing the radiation protection measures taken after the Chernobyl nuclear plant accident, the areas with a ^{137}Cs contamination level (w) of ≥ 0.56 MBq/m^2 (15 Ci/km^2) were specified as strict control zones (SCZs). Within the boundaries of the Russian Soviet Federated Socialist Republic (RSFSR) a part of the Bryansk Province of 2000 km^2 populated by 112 000 residents happened to be in an SCZ [1, 2]. The contamination level was w = 0.56–3.0 MBq/m^2 (15–80 Ci/km^2). The most informative index of an actual internal dose from ^{137}Cs and ^{134}Cs is the body content of the radionuclides in residents. This index reflects the influence of natural and social factors (including protective measures) on the internal dose formation in the inhabitants of an area with a fixed level of initial contamination (w).

A radiometric examination of the population to determine the body content of caesium radionuclides (A_{Cs}) was undertaken in SCZs after the Chernobyl accident. More than 90 000 people were examined, most of them twice and some of them three times or more. The value of A_{Cs} averaged over all the population in the SCZs decreased from June 1986 more than twofold to the end of that year, threefold to the spring of 1987 (February through April) and sixfold to the spring of 1988. This decrease is due to natural and social factors: a natural decontamination of plant cover during the first vegetation season after the accident, agrotechnical measures directed to the reduction of caesium intake by agricultural food products, and actions to exclude from the human diet those foodstuffs contaminated above the regulated permissible levels.

Table I gives the mean values of A_{Cs} for adults in the periods of the most extensive mass examination. The distribution of the individual A_{Cs} values has a log-normal pattern. The tail of the distribution (with the highest A_{Cs} figures) is

TABLE I. MEAN VALUES OF A_{Cs} (^{134}Cs AND ^{137}Cs) IN ADULT RESIDENTS OF SCZs (kBq)

Population group	Aug.–Sep. 1986	Feb.–Apr. 1987	Feb.–Apr. 1988
Rural residents	210	89	40
Urban residents	63	24	14

associated with individuals who did not obey the restrictions established regarding the consumption of contaminated foodstuffs, primarily of local milk.

The internal dose equivalents from radiocaesium (D_{Cs}) for adult rural residents of the SCZs were 8 mSv in the first year after the accident, and 4 mSv in the following two years (until May 1989). The values for urban residents were 2.5 and 1.3 mSv. For children of various ages, the mean values of D_{Cs} were 0.6–1.0 of those for adults. The actual values of A_{Cs} and D_{Cs} were lower by an order of magnitude than those projected on the basis of w values. This decrease is due to the reasonable conservatism of the prognostic model of the radiocaesium time course in the human diet [3] and also mainly to the effective implementation of protective actions for exclusion or limitation of the consumption of contaminated foodstuffs.

The total dose to the whole body (including external gamma radiation) over three years was 12–140 mSv in various population centres; nowhere did this total exceed the annual dose limits by the USSR Ministry of Public Health and their sum for this period: $100 + 30 + 25 = 155$ mSv [4]. The total dose averaged over the whole population in the SCZs was 21 mSv in the first year and 36 mSv for three years, i.e. 21–23% cent of the above mentioned limits.

The results of these studies were used in the estimation and prediction of the individual and collective doses, in decision making on the relevant protective measures, in the assessment of countermeasure efficiency and in justification of the subsequent radiation monitoring programme.

REFERENCES

[1]   IZRAEHL', Yu.A., Pravda No. 79/25979 (20 March 1989) 4 (in Russian).

[2]   IL'IN, L.A., Meditsinskaya Gazeta No. 50/4911 (26 April 1989) 3 (in Russian).

[3]   BARKHUDAROV, R.M., et al., in Medical Aspects of the Chernobyl Nuclear Power Plant Accident, Zdorov'ya, Kiev (1988) 111 (in Russian).

[4]   BULDAKOV, L.A., et al., ibid., p. 53 (in Russian).

Part I

RADIOACTIVE CONTAMINATION
OF THE ENVIRONMENT

(b) Air

MODELLING OF THE TRANSPORT AND FALLOUT OF RADIONUCLIDES FROM THE ACCIDENT AT THE CHERNOBYL NUCLEAR POWER PLANT

Yu.A. IZRAEHL', V.N. PETROV, D.A. SEVEROV
USSR State Committee for Hydrometeorology,
Moscow,
Union of Soviet Socialist Republics

Abstract

MODELLING OF THE TRANSPORT AND FALLOUT OF RADIONUCLIDES FROM THE ACCIDENT AT THE CHERNOBYL NUCLEAR POWER PLANT.
The paper models the transport and deposition of radionuclides from the atmosphere on an intermediate scale and on a regional scale beyond the zone closest to the Chernobyl nuclear power plant accident zone. One type of Lagrangian model is examined — the trajectory model for transport and deposition of contaminants from a continuous source dispersed in different directions by a variable wind, taking into account vertical and horizontal diffusion. The model consists of three parts: calculation of the trajectories of airborne particles from the area of the source, calculation of the vertical flux of polydisperse contaminants onto the underlying surface, and calculation of the integral fallout at the nodes of a rectangular grid on the stereographic projection of the Earth's surface. The continuous source of contaminant particles is located at a height H. The distribution of radionuclide activity over the particles as a function of their dimension is approximated by a log-normal law which makes it possible to describe the radionuclide fallout over a wide range of a few dozen to several thousand kilometres. New data are presented on the daily release of total gamma activity of ^{131}I and ^{137}Cs which differ from earlier published data. The meteorological conditions of the transport of airborne particles at various levels are analysed on the basis of data from the Hydrometeorology Centre of the Union of Soviet Socialist Republics. In order to compare the model calculations with the actual values, data are presented on the distribution of the radiation levels and radioactive fallout of radionuclides from the Chernobyl nuclear power plant as well as data on the dynamics of the change in fallout density and radionuclide concentration at various centres in the USSR. The cumulative fallout fields and the maximum concentrations for Europe are also calculated. Comparison of the calculated and the measured total fallout shows that they differ to within a factor of 3.

1. INTRODUCTION

Reference [1] considered the modelling of radioactive fallout in the near zone of the Chernobyl nuclear power plant accident, and introduced some data on the source of the release and the meteorological conditions in the region of the accident.

This paper examines one type of Lagrangian model — the trajectory model for transport and deposition of polydisperse contaminants from a continuous source in a variable wind field on the mesoscale and regional scale. The particle motion trajectories are calculated from the wind field at levels of 1000, 925 and 850 mbar[1] according to data from the Hydrometeorology Centre of the Union of Soviet Socialist Republics.

The model consists of three separate parts: calculation of the trajectories of airborne particles from the area of the source, calculation of the vertical flux of polydisperse contaminants onto the underlying surface, and calculation of the integral fallout at the nodes of a rectangular grid on the stereographic projection of the Earth's surface using data obtained in the first two parts. Transport of contaminants is modelled in the flux direction at levels of 1000, 925 and 850 mbar in relation to the height of the source.

The behaviour of the contaminants released from a continuous source is traced over a grid of 150 km × 150 km squares for a given time interval, which makes it possible to calculate atmospheric concentrations, dry and potentially wet fallout of radionuclides and their dynamics for each portion of the contaminants at a given grid point.

2.   MODEL DESCRIPTION

The radioactive substances raised from the damaged reactor block to a height H by thermal currents are presented in the form of an elevated point (or linear) source of polydisperse contaminants continuously releasing radioactive particles into the atmosphere. The distribution of the radionuclide activity among the particles in relation to their size is approximated by a log-normal distribution with median diameter ξ and standard deviation σ as distribution parameters:

$$a(\delta, \sigma_{10}, \xi) = \frac{dA}{d\delta} = \frac{0.434}{\sqrt{(2\pi)}\sigma_{10}\delta} \exp\left(-\frac{[\lg(\delta/\xi)]^2}{2\sigma_{10}^2}\right) \tag{1}$$

The initial quantity of contaminants in the source is normalized from Eq. (1) by introducing the factor $(\int_{\delta_1}^{\delta_2} a(\delta)\, d\delta)^{-1}$ for the limiting spectrum of particle dimensions $(\delta_1 \leq \delta < \delta_2)$.

---

[1] 1 bar = 10^5 Pa.

Within the semiempirical theory of turbulent diffusion, the non-steady-state process of vertical diffusion of a monodisperse contaminant is described by the following equation:

$$\frac{\partial q}{\partial t} - w \frac{\partial q}{\partial z} = k \frac{\partial^2 q}{\partial z^2} \tag{2}$$

which satisfies the initial conditions:

$$q(z, 0) = \delta(z - H) \tag{3}$$

and the boundary conditions:

$$k \frac{\partial q}{\partial z} + wq = \beta q_{|z = 0} \tag{4}$$

where q is the volumetric concentration of monodisperse contaminants deposited with velocity w, k is the vertical diffusion coefficient and β is the coefficient characteristic of an interaction of contaminant particles with the underlying surface; the total quantity of contaminants released by the source is equal to 1.

The vertical flux of a monodisperse contaminant at the underlying surface is $\Pi = \beta q(0, t)$, where we assume that $\beta = w + \beta_0$, i.e. at high values of w, $\beta \simeq w$, and at low values, $\beta \simeq \beta_0$. For a polydisperse contaminant, the vertical flux is determined by the integral:

$$\Pi(t) = \int_{\delta_1}^{\delta_2} \beta(w) \, q(0, t, w, H) \, a(\delta, \sigma_{10}, \xi) \, d\delta, \qquad w = w(\delta) \tag{5}$$

Over the time interval 2–144 h the contaminant flux (on a logarithmic scale) is approximated by a cubic spline function. The spline function is plotted from 16 interpolation points with relative error checking at three intermediate points between the grid points. The interpolation error in the calculations for $\Pi(t)$ is less than 1%, and less than 2% for the surface air concentration. The programme considers three versions of possible temporal changes of the input parameters in Eq. (5), i.e. three spline functions.

The plume of contaminants from a continuous source formed in a variable wind field is approximated by a line whose break points lie on the trajectories of propagation of instantaneous releases of contaminants arriving from the source with periodicity Δt. Thus, the configuration of the plume in the horizontal plane is determined by the location of each release at a given moment (Fig. 1). The quantity of contaminants in each rectilinear plume segment at the time of formation is equal to the product

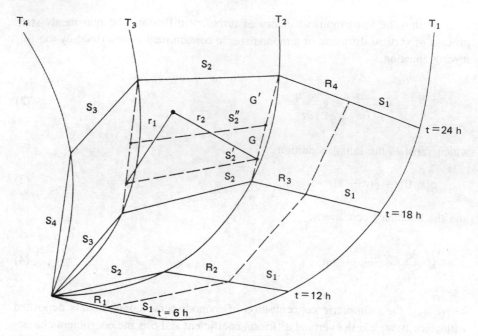

FIG. 1. *Diagram of the plume at various times t. T_1–T_4 are the displacement trajectories of instantaneous releases, 1–4, of contaminants; S_1–S_4 are plume segments; S_2', S_2'' are intermediate positions of segment S_2; G, G' are dispersions for different positions of the segment; and R_1–R_4 is the path travelled by segment S_1.*

of the source intensity and Δt. A uniform distribution along the segment length is assumed.

Horizontal diffusion of contaminants from the segment is taken into account by the law:

$$\frac{1}{2\pi\sigma^2} \exp\left(-\frac{r^2}{2\sigma^2}\right) \frac{dl}{l}$$

where l is the length of the segment, r is the distance from the segment element dl and the dispersion parameter is $\sigma = 0.08R$, where R is the distance travelled by the centre of the segment at the time of calculation. The concentration in the segment from a unit quantity of contaminants is:

$$f(r_1, r_2, l, \sigma) = \frac{1}{2\sigma l\sqrt{(2\pi)}} \exp(-h^2/2\sigma^2) \left(\operatorname{erf} \frac{l^2 + r_2^2 - r_1^2}{2\sigma l\sqrt{2}} \right.$$

$$\left. + \operatorname{erf} \frac{l^2 + r_1^2 - r_2^2}{2\sigma l\sqrt{2}} \right)$$

(6)

where

$$h^2 = r_1^2 - \frac{(l^2 + r_1^2 - r_2^2)^2}{4l^2}$$

and r_1 and r_2 are the distances from the point of observation to the extremities of the segment. It is assumed that within the limits of the time interval Δt, the velocities of displacement of the extremities are constant. The intermediate positions of the segment are given at intervals of σ along the trajectory. The fallout of contaminants from the plume segment at a given point in the standard grid (150 km × 150 km at a latitude of 60° N) for a transit time $(t, t + \delta t)$, with account taken of Eqs (5) and (6), is

$$\Delta\Pi = Q(\overline{t_0}) \, \Delta t \, \Pi(t) \, f(r_1^2, r_2^2, l, \sigma) \, \delta t \tag{7}$$

where $Q(\overline{t_0})$ is the source strength at the initial moment of formation of the given segment, $(t_0, t_0 + \Delta t)$.

3. DAILY RELEASE OF RADIOACTIVITY FROM THE DAMAGED REACTOR BLOCK

Reference [1] considered meteorological features of the transport of radioactive substances from the reactor zone. Further characteristics of the dynamics of the release were obtained from a comparison with the distribution pattern of radioactive fallout. In Fig. 2, the boxed sectors indicate (in degrees) the mean wind direction in layers from ground level up to 500 and 1000 m, at which transport and fallout of plume particles took place at various times after the start of the accident.

Accordingly, the chart of radiation level distribution in the near zone indicates specific sectors where the quantities of fallout have been estimated, both as total gamma activity and as the activity of individual isotopes. Table I shows the results of calculations of the daily discharge into the atmosphere of the total gamma radioactivity and of individual radionuclides in relation to the total radioactivity discharged into the atmosphere or deposited in the near zone over the first five days.

References [3, 4] show that the quantity of ^{137}Cs deposited in the near, intermediate and far zones on the territory of the USSR was 0.22, 0.67 and 0.575 MCi or 15, 46 and 39%.[2] The aerial gamma survey shows that the quantity of gamma activity deposited in the near and intermediate zones during the first ten days of June 1986 was 3×10^4 and 5×10^4 mR·h^{-1}·km^2 [3, 4].[3] By analogy with the percentage distribution of ^{137}Cs fallout, the total gamma activity in the territory of the USSR is estimated at 1.2×10^5 mR·h^{-1}·km^2 for the time indicated.

---

[2] 1 Ci = 3.7×10^{10} Bq.

[3] 1 R = 2.58×10^{-4} C/kg.

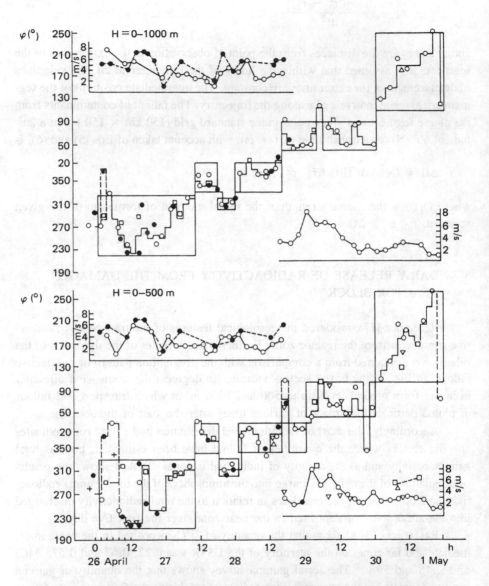

FIG. 2. *Mean values of wind direction and speed from 26 April to 1 May 1986 in the layers 0–500 and 0–1000 m in the region around the Chernobyl nuclear power plant. + Kiev (radiosonde), ○ Kiev (airport), ● Borispol', △ Mozyr', □ Gomel', ▽ Chernigov.*

TABLE I. RELATIVE DISTRIBUTION OF THE DISCHARGE OF RADIOAC-
TIVE PRODUCTS INTO THE ATMOSPHERE IN RELATION TO THE TOTAL
QUANTITY DISCHARGED INTO THE ATMOSPHERE [2] OR DEPOSITED IN
THE NEAR ZONE OVER THE FIRST FIVE DAYS

April 1986	Relative discharge [2]	Relative discharge (this paper)					
		Total radioactivity			Ce-144	Cs-137	I-131
		H = 1000 m	H = 500 m	Mean			
26	0.32	0.17	0.17	0.17	0.19	0.1	0.09
27	0.24	0.29	0.25	0.27	0.28	0.3	0.31
28	0.19	0.29	0.29	0.29	0.3	0.4	0.42
29	0.14	0.14	0.15	0.14	0.12	0.18	0.15
30	0.11	0.11	0.14	0.13	0.11	0.02	0.03

By integration of the gamma activity measured at locations in various areas of
the track (Fig. 3), it is estimated that 25% of the total gamma activity, or
3.0×10^4 mR·h^{-1}·km^2, occurred in the southern sector, where the fallout took
place after 29 April 1986.

Table II shows the calculated daily releases of total gamma activity due to ^{131}I
and ^{137}Cs. The left hand columns indicate the values obtained using data in
Ref. [2], and the right hand columns the values obtained using data from Table I for
the period from 26 to 30 April 1986. For 1 May onwards, the relative values of the
daily release were calculated according to the formula:

$$Q(t) = 0.09 \exp[0.35(t - 5)]$$

where $t_{day} = 5, 6,...9$, and based on 25% of the total ^{131}I and ^{137}Cs activity
released. The total release of ^{137}Cs deposited in the Northern Hemisphere is estimated at approximately 2.0 MCi. The daily releases of ^{131}I are calculated for the
date of release, allowing for decay.

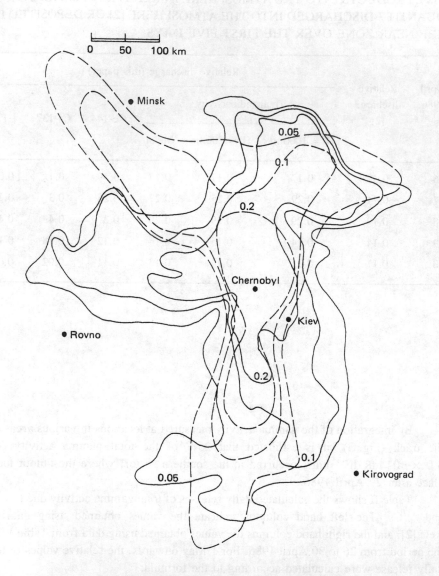

FIG. 3.   Calculated (broken lines) and measured (full lines) distribution of dose rates (mR/h) at the ground surface on 10 June 1986 (1 R = 2.58 × 10⁻⁴ C/kg).

TABLE II. DAILY RELEASES OF GAMMA ACTIVITY DUE TO ^{131}I AND ^{137}Cs

1986	Gamma activity[a] $(10^4 \text{ mR} \cdot \text{h}^{-1} \cdot \text{km}^2)$ Data from:		I-131 (10^6 Ci)[b] Data from:		Cs-137 (10^5 Ci)[b] Data from:	
	[2]	Table I	[2]	Table I	[2]	Table I
26 Apr.	3.0	1.53	5.2	0.9	6.0	1.5
27 Apr.	1.0	2.43	1.5	3.0	1.9	4.5
28 Apr.	0.84	2.61	1.1	4.1	1.5	6.0
29 Apr.	0.66	1.26	0.76	1.46	1.1	2.6
30 Apr.	0.5	1.17	0.51	0.29	0.8	0.3
1 May	0.46	0.27	0.45	0.29	0.8	0.5
2 May	0.92	0.38	0.75	0.41	1.5	0.6
3 May	1.15	0.5	0.8	0.59	1.7	0.9
4 May	1.6	0.77	0.95	0.84	2.2	1.3
5 May	1.9	1.1	0.96	1.18	2.4	1.8

[a] For the first ten days of June (1 R = 2.58×10^{-4} C/kg).
[b] 1 Ci = 3.7×10^{10} Bq.

4. RESULTS OF CALCULATIONS AND COMPARISON WITH EXPERIMENTAL DATA

The input parameters of the model are the meteorological trajectories of particle transport at various levels in the variable wind field, cloud elevation, the values ξ and σ of the log-normal distribution and the daily radioactivity release. The actual radiological information used consisted of data on the distribution of radiation levels and radioactive fallout from the Chernobyl nuclear power plant, data on the dynamics of fallout density variation, and the maximum values of radionuclide concentrations at individual points on the territory of the USSR.

The results of model calculations and their comparison with radiological measurements in the USSR and in Europe were used to estimate ξ and σ and also to correct the plume elevation on various days after the accident.

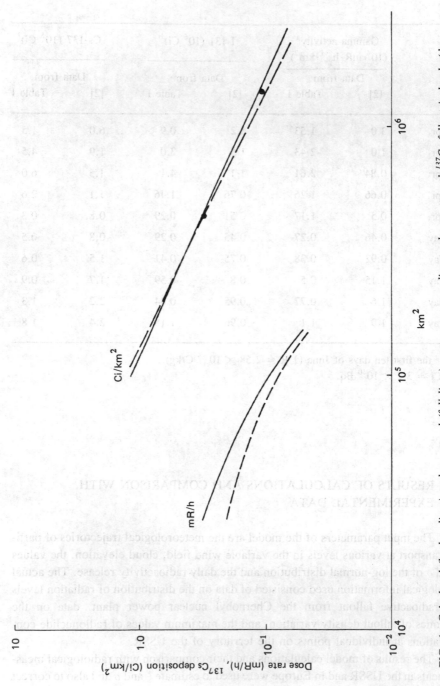

FIG. 4. Calculated (broken lines) and measured (full lines) areas corresponding to dose rate and ^{137}Cs fallout isolevels (dose rate data are for 10 June 1986) (1 R = 2.58 × 10^{-4} C/kg; 1 Ci = 3.7 × 10^{10} Bq).

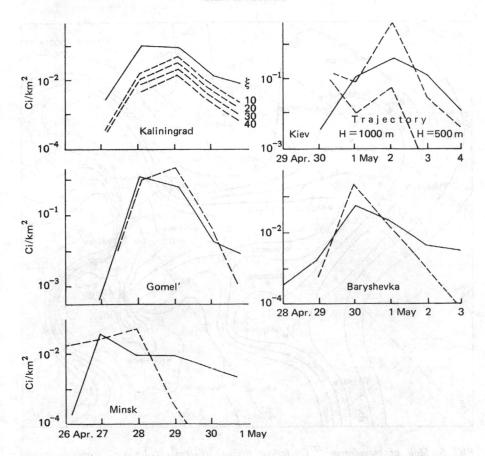

FIG. 5. Dynamics of ^{137}Cs deposition in towns in the USSR (full lines, measured; broken lines, calculated) (1 Ci = 3.7 × 10^{10} Bq).

Figure 3 shows the gamma field distribution over the territory contiguous to the accident zone. The broken lines represent model calculations for the fallout fields using the data in Tables I and II. The parameters ξ and σ were chosen so that the areas of fallout limited by the isolines indicated were as close as possible to those measured in Fig. 4. The quantities thus obtained were $\xi = 20$ μm, $\sigma = 0.4$. It was established from the fallout model in the near zone that $\xi = 50$ μm and $\sigma = 0.25$ [1]. This is to be expected as the particle size spectrum is displaced towards lower values with increasing distance from the source. In general there is a bimodal probability distribution which is confirmed by the experimental data.

Figure 5 shows the data for fallout density measurements on different days after the accident in Minsk, Gomel', Kaliningrad, Kiev and Baryshevka. It also gives

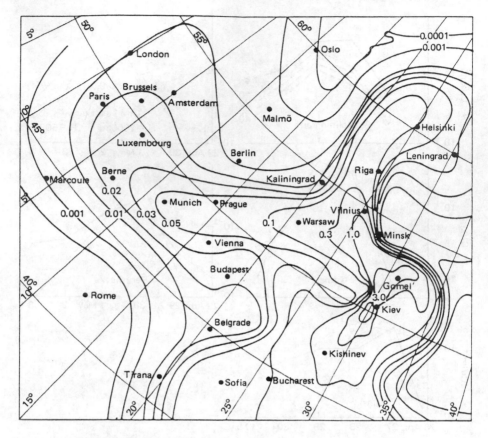

FIG. 6. Map of cumulative ^{137}Cs fallout density (Ci/km^2) calculated from the model
(1 Ci = 3.7 × 10¹⁰ Bq).

the corresponding fallout values calculated from the model and shows examples of
how the dynamics of fallout change in relation to various values of ξ and H. On the
whole there is definite agreement between the measurements and the calculated
values, although in individual cases there are significant discrepancies. On the average, the calculated and measured values of the total fallout at the points indicated
agree to within a factor of 3.

Cumulative fallout fields were calculated, as well as maximum concentrations
in European territory. Figure 6 shows a map of ^{137}Cs fallout density calculated
from the model. For the far zone, $\xi = 10$ μm and $\sigma = 0.25$. It should be noted that
the source height and the dynamics of activity release were not adjusted to the optimum agreement between theoretical and measured values of radioactive fallout and
radionuclide concentration, but were taken from the distributions indicated above,
obtained independently of the model calculations.

TABLE III. MEASURED AND CALCULATED VALUES FOR ^{137}Cs AND ^{131}I GROUND CONCENTRATIONS AND ACCUMULATED ^{137}Cs FALLOUT AT EUROPEAN STATIONS AND IN THE USSR

Place of observation	Concentration (Bq/m^3)				Cs-137 fallout (kBq/m^2)	
	Cs-137 Meas.	Cs-137 Calc.	I-131 Meas.	I-131 Calc.	Meas.	Calc.
Finland (Helsinki)	3.7	1.3 (5.3)[a]	20.7	8.2 (46)[a]	6.3	3.7 (13)[a]
Germany, Fed. Rep. (Munich)	1.7	1.5 (0.7)	4.5	10.0 (5.6)	4.1	1.9 (0.9)
Austria (Vienna)			3.5	7.0 (3.6)	16.2	1.5 (0.6)
Hungary (Budapest)			3.0	3.5 (1.7)	1.8	0.8 (0.4)
France (Paris)	0.41	0.6 (0.3)	7.4	3.7 (2.6)	1.3	0.7 (0.3)
UK	0.9	0.5 (0.2)	5.4	3.2 (2.0)	1.0	0.4 (0.2)
Switzerland					5.5	0.7 (0.4)
Poland					30–1.6	7.8 (3.1)
Czechoslovakia (Prague)	3.3	2.3 (1.3)	70–7.5	15.0 (11)	1.5	1.8 (1.1)
Bulgaria					1.1	2.2 (2.0)
Yugoslavia					1.1	1.1 (0.5)
Belgium					0.8	0.7 (0.4)
Luxembourg					2.8	1.0 (0.5)
Italy (Milan)	0.7	0.3 (0.1)	17.0	2.2 (1.2)		
Greece					3.7	1.8 (1.2)
Netherlands (Amsterdam)			7.0	5.2 (3.8)		
Minsk		3.1 (1.4)	32.0	21.0 (12)	2.6	3.3 (2.6)
Leningrad	1.1	0.7 (3.0)	1.1	4.4 (26)		
Riga	0.85	1.4 (5.5)	3.1	7.4 (47)		
Vilnius	5.9	9.6 (20)	28.0	64.0 (170)		
Baryshevka	81.4	20.0 (7.8)	30.7	59.0 (27)	3.7	8.9 (3.4)
Rakhov	6.7	0.8 (0.3)	9.2	5.2 (2.7)		
Kaliningrad					7.4	1.7 (1.5)
Gomel'					78	100 (57)
Kiev					22	130 (290)

[a] Values in brackets calculated using data in the left hand columns of Table II.

Table III shows the results of various model calculations for cumulative fallout and maximum concentrations of ^{137}Cs and ^{131}I, and also quantities measured at European stations according to published data (e.g. Ref. [5]) as well as on the territory of the USSR.

For the majority of stations, the agreement between calculated and observed values is quite acceptable with respect to both the amount of activity and the dynamics, which indicates the usefulness of the model for describing the general pattern of concentration and fallout fields. The mean deviation of the measured from the theoretical data is within a factor of 3 when the data in the right hand columns of Table II are considered. In the majority of cases the calculation results for ^{137}Cs fallout density are lower than the measured values, which is clearly related to a greater discharge into the atmosphere. The total amount of ^{137}Cs deposited in the Northern Hemisphere is 2.0–2.5 MCi according to our minimal estimates.

There are significant differences in the measurement data given in the various references, sometimes reaching an order of magnitude.

REFERENCES

[1]   IZRAEHL', Yu.A., PETROV, V.N., SEVEROV, D.A., Modelling of radioactive fallout in the near zone of the Chernobyl nuclear power plant accident, Meteorol. Gidrol. No. 7 (1987) 5–12 (in Russian).

[2]   ABAGYAN, A.A., et al., Information on the accident at the Chernobyl nuclear power plant and its consequences, prepared for the IAEA, At. Ehnerg. **61** 5 (1986) 301–320 (in Russian).

[3]   IZRAEHL', Yu.A., et al., Radioactive contamination of the environment in the area of the accident at the Chernobyl nuclear power plant, Meteorol. Gidrol. No. 2 (1987) 5–18 (in Russian).

[4]   IZRAEHL', Yu.A., PETROV, V.N., "Atmospheric transport and dynamics of radioactive product discharge", Radionuclides in Food Chains (Proc. Symp. Laxenburg, 1987), Springer-Verlag, Berlin (1988).

[5]   BAILEY, M.R., et al., Measurements of the Body Content of Radioactive Caesium in Residents of Yorkshire, Rep. NRPB-R-R21, Natl Radiological Protection Board, Chilton, UK (1987).

USE OF MATHEMATICAL MODELLING TO ESTIMATE SCALES AND STUDY PATTERNS OF FORMATION OF CONTAMINATED AREAS RESULTING FROM A NUCLEAR ACCIDENT

Yu.S. SEDUNOV, V.A. BORZILOV,
N.V. KLEPIKOVA, E.V. CHERNOKOZHIN
'Typhoon' Scientific and Industrial Association,
USSR State Committee for Hydrometeorology,
Obninsk,
Union of Soviet Socialist Republics

Presented by V.N. Petrov

Abstract

USE OF MATHEMATICAL MODELLING TO ESTIMATE SCALES AND STUDY PATTERNS OF FORMATION OF CONTAMINATED AREAS RESULTING FROM A NUCLEAR ACCIDENT.

The 'Typhoon' Scientific and Industrial Association has developed a series of physicomathematical models to describe atmospheric transport and fallout of gaseous and aerosol-form radioactive contaminants for the purpose of studying the processes by which contaminated zones form following discharges into the atmosphere during a nuclear power plant accident. These models are for intermediate range and regional transport and transboundary transport, and they work for distances of ten to several thousand kilometres; they were used to analyse the situation which arose after the Chernobyl accident. These modelling techniques were used to study the space–time characteristics of formation of contaminated areas, to reconstruct the dynamics of the change in parameters of the radionuclide discharge source, and to evaluate the quantity of radionuclides carried beyond the boundaries of the Union of Soviet Socialist Republics after the Chernobyl accident. The source parameters were reconstructed by solving the inverse problem using experimental data on the contamination density at different points. The source strength was evaluated for ^{131}I, ^{137}Cs, ^{90}Sr, ^{239}Pu and ^{240}Pu, and the fallout fields were modelled for separate radionuclides.

1. INTRODUCTION

In the event of a nuclear accident with the escape of a large quantity of radionuclides into the atmosphere, one of the most important tasks is to estimate scales and obtain maps of the contaminated areas. Stationary and mobile monitoring systems have the main role in this work. Especially in the case of large scale contamination, optimizing the operation of these systems is essential for the purpose of very fast detection of the most highly contaminated regions.

99

Substantial help can be provided in this respect by methods of mathematical modelling, which permit a fairly rapid study to be made of the patterns of formation of the contamination of the underlying surface. For this, it is essential to have a set of models of different scales of atmospheric transfer and deposition of a contaminant, as well as a technology for the rapid input of initial meteorological information and source characteristics into the models. The study of the latter factors — height, intensity, and radionuclide and disperse composition of the discharge — is a very complex task from the technical and organizational points of view. Consequently, this gives rise to the need to supplement direct measurements of the parameters of a source with the results of the solution of the inverse problem of source reconstruction on the basis of factual data concerning the contamination of an area, obtained by means of monitoring.

Thus the procedure for rapidly calculating the nature of the contamination of an area involves the interrelated use (by the feedback method) of mathematical modelling methods (solution of direct and inverse problems of atmospheric transport of radionuclides) and of the results of measurements by the monitoring network. Below, we give an example of the use of such a procedure in work connected with studying the radioactive contamination of an area caused by the accident at the Chernobyl nuclear power plant.

The following set of mathematical models was used for describing the different scales of radionuclide transfer processes: an intermediate scale model of transport up to 200 km (MESO) and two regional models of transport up to 2000 km (REGION) [1] and up to 4000 km (STOCH). They can be used to calculate the concentration of a contaminant in air and the distribution of the density of contamination in an area due to a source of any form and having any action whatever. The input used in all three models is identical meteorological information, compiled twice a day in the databank of the Hydrometeorology Centre of the Union of Soviet Socialist Republics from objective analyses of current information arriving from radiosonde and meteorology centres over combined telecommunication channels. Inserting meteorological information into a model from a databank by means of magnetic tape or by intermachine exchange channels ensures the essential speed of calculation.

The findings presented below were obtained in the first days (for the regional scale) and months (for the intermediate scale) after the accident, when there was practically no information about the source, and for this reason the source was determined only by solution of the inverse problem. Although we sought in this process to achieve a regime close to the 'real time scale', as a rule this did not work out, owing to the natural lateness of data on the radiation situation from the monitoring network.

In the very first days (April and early May), use was made of data on the strength of the gamma field and the radionuclide composition of a very limited number of soil samples from the 30 km zone surrounding the plant, and subsequently fallout data on topographical maps from the stationary monitoring network. By means of these data, it was possible to reconstruct the characteristics of the source of the different radionuclides which caused the contamination of the zone more distant from the plant. In addition, maps showing calculations of contamination were prepared, indicating the presence of large scale caesium 'spots'. Modelling of the contamination process in the near zone (up to 100 km) was carried out later, when detailed maps had already been obtained showing the contamination of the area by various radionuclides, on the basis of measurement data. Here the only problem was that of determining the parameters of the source responsible for the contamination of the near zone, which is in itself of interest for an understanding of the physics of the processes originating in the destroyed reactor.

All the results of the calculations are presented in the form in which they were obtained in 1986, and for this reason they are open to criticism from the point of view of present knowledge. However, we consider it appropriate to do this, for the purpose of demonstrating the practical feasibility and usefulness of the proposed methodology.

2. MODELS OF RADIONUCLIDE DISTRIBUTION IN THE ATMOSPHERE

The models consist of two basic units: a unit for calculating the characteristics of a boundary layer and a diffusion unit for calculating the transport of a contaminant in the atmosphere. The MESO and REGION models for the atmospheric diffusion of a cloud of a polydisperse contaminant are based on the numerical solution of a non-steady-state polyempirical equation for turbulent diffusion of monodisperse particles. This equation, with corresponding initial and boundary conditions, was solved by the method of splitting, with the use of a hybrid scheme [2, 3] and a special substitution of variables. In the case of the MESO model, the domain of calculation is fixed in space. The horizontal spacing increases as the distance from the source increases from 1 km to 5 km, and the vertical spacing increases with height from 5 m at the Earth's surface to 100 m at the upper boundary of the calculation domain.

In the REGION model, to obtain a good spatial resolution of a contaminant cloud from the moment of its formation to its deposition on the underlying surface, a spatially mobile calculation grid is introduced with a variable spatial resolution, determined by the current dimensions of the cloud. The continuous action of the source is modelled by a succession of instantaneous discharges. The concentration in air of a polydisperse contaminant and its flux at the underlying surface are determined by superposing particle concentrations and flows of different sizes.

The STOCH model is based on the solution by the Monte Carlo method of a system of stochastic equations for the movement of individual particles transported in a field of wind velocity and subject to random deviations.

The methods of determining mean wind velocity fields and turbulent diffusion coefficients are different in the models under consideration. The MESO model is based on a model [4] in which the vertical structure of the boundary layer of the atmosphere is determined by the input data used — temperatures and pressures near the Earth's surface, geopotential altitudes, real wind velocities and temperatures on standard isobaric surfaces (925, 850 and 700 hPa) at the nodes of a 150 km × 150 km grid. At these nodes we solve a system of equations consisting of steady state one dimensional equations of motion, the balance of the kinetic energy of turbulence and the rate of dissipation, the Kolmogorov correlation for the coefficient of turbulence and the interpolation expression for turbulent heat flow. Spatial interpolation (between grid points) and temporal interpolation (between entry times) of input data are a means of obtaining the necessary values for the velocities and coefficients of turbulent diffusion for all nodes of the calculation domain. In the REGION and STOCH models, the vertical structure of the boundary layer of the atmosphere is determined by using the similarity theory correlation together with data obtained by the Wippermann model [5] and empirical laws of resistance and heat exchange.

3.   METHOD OF RECONSTRUCTING SOURCE PARAMETERS

The method of reconstructing the parameters of a source consists essentially in the following: the source is sought in the form of a sum L of elementary sources, each of which is characterized by a fully determined duration of action $(t_l, t_l + \Delta t_l)$, distributed over height, by a sedimentation rate and by an indeterminate activity $\alpha_l \geq 0$. The latter are determined from the condition of a minimum difference in some metric between experimentally determined fallout densities $P(\vec{x})$ and calculated fallout densities $P_l(\vec{x})$, caused by elementary sources. By the use of one of the variants of Tikhonov's regularization method [6] (for overcoming possible instabilities), the problem can be reduced to the minimization of the functional:

$$\phi_\mu(\alpha) = \sum_j \left(\sum_{l=1}^{L} \alpha_l P_l(x_j) - P(x_j) \right)^2 + \mu \sum_{l=1}^{L} \alpha_l^2$$

which is accomplished by the method of co-ordinate descent [7]. Here μ is a regularization parameter dependent on measurement and calculation errors. Regularization permits the selection of the 'smoothest' solution from the solutions resulting from all possible superpositions of elementary sources and leading to the discrepancy $\phi(\alpha) = \epsilon$.

For a unique solution of the inverse problem, it is necessary that the contributions of the different elementary sources be split. In the case of the Chernobyl accident, the problem was made easier by the fact that during the period in which the source was active, there was a substantial change in weather conditions and there was a change through approximately 360° of the direction of transport. An additional factor removing uncertainty consisted in reduction of the possible set of elementary sources through the use of additional a priori information obtained from observational data and physical considerations.

On the basis of observations [8, 9], the range of variation in the height of discharge was chosen as follows: from 26 to 29 April 1986, 500–2000 m; and from 29 April to 6 May, 100–500 m. On the basis of physical considerations, it was assumed that the discharge was distributed, uniformly in intensity and disperse composition, in layer (H − ΔH, H), where ΔH is the width of a flare at height H of its ascent [10]:

$$\Delta H \approx 0.2(H + 5h)$$

where h is a linear dimension of the destroyed reactor. The sedimentation rate varied from 0 to 20 cm/s. Particles having a higher sedimentation rate were responsible in the main for the contamination of the closest, 10 km zone, where local-scale models have to be used.

4.   PARAMETERS OF THE SOURCE OF CONTAMINATION OF THE
     100 km ZONE BY LONG LIVED RADIONUCLIDES

The direct problem for the 100 km zone was solved by using the MESO model. By solving the inverse problem, we obtained the following characteristics of the distribution function of the particles which caused contamination of this zone: a near-monodisperse aerosol with a sedimentation rate of 10 cm/s for ^{90}Sr and ^{239,240}Pu; a bimodal distribution function for ^{137}Cs with a fairly broad, coarsely disperse mode approximating two fractions with sedimentation rates of 5 and 10 cm/s. The temporal behaviour of the source activity for each of these radionuclides is shown in Fig. 1, which also shows the temperature dynamics in the reactor according to data in Ref. [8]. The distinctive features of the action of the source over time are as follows:

— There is a correlation in the time of the rise in activity of the discharge of various radionuclides.
— There is a rise in the intensity of radionuclide discharge in the second half of 27 April 1986, apparently associated with the particular features of processes due to the burning of graphite; and on 2 and 4 May — with a rise in temperature in the reactor after it was covered over without heat removal. The maps for this source which calculate contamination in the 100 km zone are in satisfactory agreement with measurement data. By way of an example, Fig. 2 presents calculated isolines for ^{137}Cs contamination. It should be noted that the MESO model cannot describe small scale non-uniformity of fallout fields,

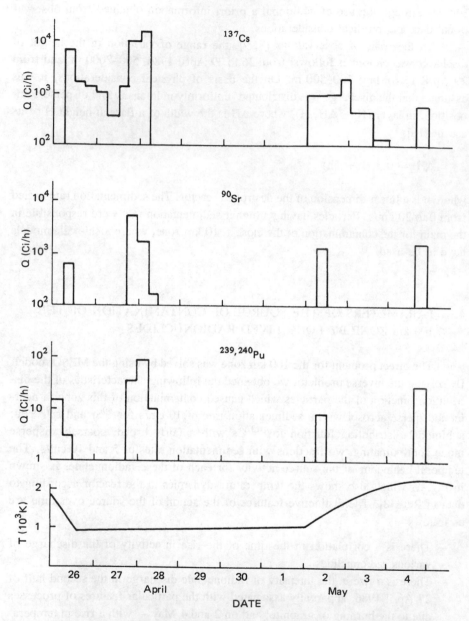

FIG. 1. Temporal behaviour of the temperature and intensity of the source accounting for contamination of the near zone by long lived radionuclides (1 Ci = 3.7 × 10¹⁰ Bq).

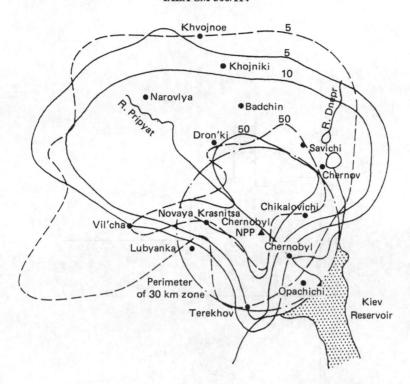

FIG. 2. Calculated (full lines) and measured (broken lines) isolines of ^{137}Cs fallout density (Ci/km^2) in the near zone (1 Ci = 3.7 × 10^{10} Bq).

which is essential for the field of ^{137}Cs contamination in the zone under consideration. The structure of the temporal behaviour of the ^{137}Cs discharge in Fig. 1 differs from that obtained in Ref. [11], owing to the use of different a priori information concerning the length of time for the formation of contamination in the 100 km zone: in Ref. [11] it was assumed that the contamination was completed on 1 May 1986; in the present study, a date of 6 May 1986 was assumed.

5. MODELLING CONTAMINATION IN THE FAR ZONE

Reconstruction of the source and modelling of the fields of contamination in the far zone were carried out with the use of the REGION model for ^{131}I and ^{137}Cs. In the case of ^{131}I, no calculation was made of transport in the aerosol fraction, the activity of iodine in which is estimated to have been 10–20%. It was assumed that

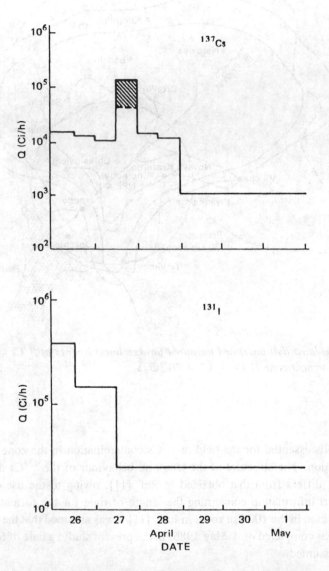

FIG. 3. *Temporal behaviour of the intensity of the source accounting for contamination of the far zone by* ^{137}Cs *and* ^{131}I *(full line: dry fallout; broken line: allowance for moist washout)* $(1 Ci = 3.7 \times 10^{10} Bq)$.

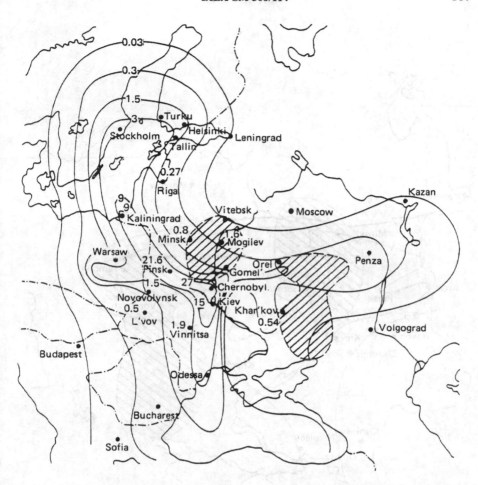

FIG. 4. Calculated density (isolines) and measured cumulative density (individual figures) of ^{131}I fallout (Ci/km^2) on the tenth day after the accident (without allowance for decay) (1 Ci = 3.7 × 10^{10} Bq). Shading indicates regions where there was simultaneous passage of atmospheric precipitation and a radioactive flare.

^{131}I is transported on weightless particles having a rate of surface capture of 1 cm/s. The finely disperse mode of the distribution function for ^{137}Cs is approximated by two fractions with sedimentation rates of 0.5 and 1 cm/s. In May 1986, there was no possibility of separating out the contribution of transport of ^{137}Cs in the gaseous phase, which accounted mainly for the contamination of areas outside the Union of Soviet Socialist Republics, and therefore the contribution of this mode is not taken into account in the line drawn for the temporal behaviour of intensity

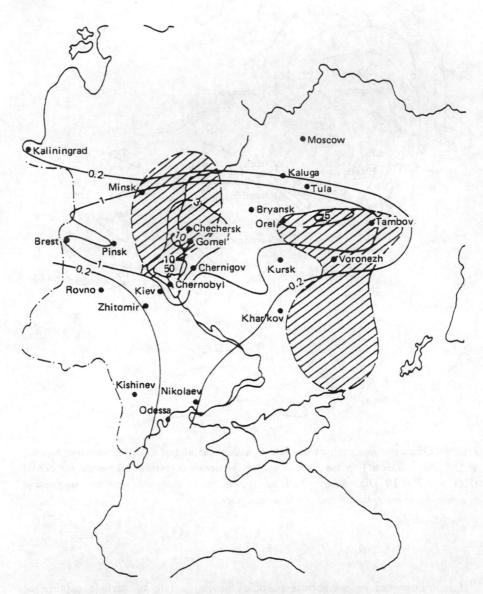

FIG. 5. Calculated field of ^{137}Cs fallout density (Ci/km^2) in the European territory of the USSR (1 Ci = 3.7 × 10^{10} Bq). Shading indicates regions where there was simultaneous passage of atmospheric precipitation and a radioactive flare.

TABLE I. AMOUNT OF ACTIVITY (Ci)[a] DISCHARGED FROM THE CHERNOBYL REACTOR BY 6 MAY 1986

Radionuclide	Source			Remarks
	[8]	[11]	Present authors	
I-131	0.9×10^7	—	1.3×10^7	
Cs-137 (near zone)	—	2.8×10^5	2.5×10^5	
Cs-137 (near zone, without entrainment abroad)	10^6	—	2.3×10^6	Without washout
			10^6	With washout
Sr-90	2.2×10^5	8.5×10^4	6×10^4	
Pu-239, Pu-240	1.7×10^3	—	10^3	

[a] $1 \text{ Ci} = 3.7 \times 10^{10} \text{ Bq}$.

as shown in Fig. 3. Figures 4 and 5 present maps showing the contamination of the far zone with ^{131}I (without allowance for decay) and ^{137}Cs. Figure 4 also presents data on total fallout of ^{131}I from topographical maps (without allowance for decay). In addition to the central region of elevated contamination, Fig. 5 shows 'spots' in the Byelorussian Soviet Socialist Republic and the Russian Soviet Federated Socialist Republic which are due to the combined effect of weather conditions and an abnormally high discharge of ^{137}Cs from the reactor on 27 April 1986. The locations of these spots differ somewhat from their actual whereabouts, owing to errors in the wind field determined by objective analysis. The calculation of caesium fallout was performed without taking precipitation into account. Subsequent calculation of washout did not alter the qualitative picture of the fallout field but reduced the estimated intensity of the discharge on 27 April by the amount shown in Fig. 3.

The ^{131}I source obtained was also used for performing calculations with the STOCH model for the discharge of this radionuclide beyond the boundaries of the USSR. According to the estimates made, about half of the ^{131}I in gaseous form was carried beyond the border.

In conclusion, we summarize in Table I the total discharge from the reactor in the gaseous–aerosol phase, according to the results of calculations making use of the proposed procedure for solving inverse problems.

REFERENCES

[1] BORZILOV, V.A., et al., Regional model of atmospheric transport of a polydisperse contaminant, Meteorol. Gidrol. No. 4 (1988) 57–65 (in Russian).

[2] YANENKO, N.N., Fractional Step Method for Solving Multidimensional Problems of Mathematical Physics, Nauka, Novosibirsk (1987) (in Russian).

[3] D'YAKONOV, E.G., Economic difference methods, based on the splitting of the difference operator for some systems of equations in partial derivatives, Vychislitel'nye Metody i Programmirovanie No. 6 (1987) (in Russian).

[4] TARNOPOL'SKIJ, A.G., SHNAJDMAN, V.A., Improved model of the planetary boundary layer, Meteorol. Gidrol. No. 10 (1979) 14–21 (in Russian).

[5] WIPPERMANN, F., Properties of the thermal boundary layer of the atmosphere obtained with a PBL model, Beitr. Phys. Atmos. 48 (1975) 30–45.

[6] TIKHONOV, A.N., ARSENIN, V.Ya., Methods of Solving Incorrect Problems, Nauka, Moscow (1986) (in Russian).

[7] VASIL'EV, F.P., Numerical Methods of Solving Extreme Problems, Nauka, Moscow (1988) (in Russian).

[8] ABAGYAN, A.A., et al., Information on the accident at the Chernobyl nuclear power plant and its consequences, prepared for the IAEA, At. Ehnerg. 61 5 (1986) 301–320 (in Russian).

[9] IZRAEHL', Yu.A., et al., Radioactive contamination of the environment in the area of the accident at the Chernobyl nuclear power plant, Meteorol. Gidrol. No. 2 (1987) 5–18 (in Russian).

[10] SKORER, R., Aerohydrodynamics of the Environment, Mir, Moscow (1980) 395–426 (in Russian).

[11] IZRAEHL', Yu.A., PETROV, V.N., SEVEROV, D.A., Modelling of radioactive fallout in the near zone of the Chernobyl nuclear power plant accident, Meteorol. Gidrol. No. 7 (1987) 5–12 (in Russian).

Invited Paper

COMPARISON OF NUCLEAR ACCIDENT
AND NUCLEAR TEST DEBRIS

R.W. PERKINS, D.E. ROBERTSON,
C.W. THOMAS, J.A. YOUNG
Battelle Pacific Northwest Laboratory,
Richland, Washington,
United States of America

Abstract

COMPARISON OF NUCLEAR ACCIDENT AND NUCLEAR TEST DEBRIS.

The first thermonuclear test, which was conducted in the Pacific Ocean in 1952, resulted in a worldwide distribution of fallout radionuclides. It was not until the United States and Soviet test series in 1961–1962, however, that very high concentrations of radionuclides became distributed throughout the Northern Hemisphere. Subsequent above ground nuclear tests by China and France maintained the levels of airborne radionuclides through the 1960s and 1970s and into 1980. At the US Department of Energy's Pacific Northwest Laboratory at Richland, Washington, the concentrations of some thirty airborne radionuclides have been measured for the past three decades. These radionuclides include those produced in the weapons testing programmes and cosmic ray bombardment of the atmosphere, radon emissions from the soil, natural radionuclides in dust and accidental releases from nuclear operations. The different sources result in different distribution patterns, atmospheric residence times and annual variations in concentrations. By maintaining monitoring programmes in which records are compiled of the atmospheric concentrations of radionuclides from the different sources, one can predict the behaviour of radionuclides entering the atmosphere from such sources and maintain a baseline against which those radionuclides entering the atmosphere from accidental releases can be compared. The contribution of both short and long lived radionuclides from the Chernobyl accident was easily observable at the monitoring station at Richland, Washington. However, the radionuclides did not represent a long term contribution to airborne radionuclide concentrations. The sources of the thirty radionuclides which have been measured for the past three decades are discussed and the factors affecting their annual and long term concentrations are considered.

1. INTRODUCTION

The worldwide distribution of man-made radionuclides became significant following the first thermonuclear test in the Pacific Ocean in 1952. However, it was during the 1961–1962 United States and Soviet test series that very substantial amounts of radionuclides were injected into the troposphere and stratosphere. These injections into the stratosphere, followed by interhemispheric mixing, resulted in deposition of nuclear weapons debris over the entire world with the major portion being in the Northern Hemisphere.

During the 1960s and 1970s, China and France conducted a series of above ground nuclear tests that contributed further to the stratospheric radionuclide inventory. Because of the long residence time of radionuclides in the stratosphere, the stratospheric debris is a continuous source of radionuclide movement into the troposphere.

At the Pacific Northwest Laboratory (PNL) of the US Department of Energy (USDOE), located in Richland, Washington, the concentrations of some thirty airborne radionuclides have been measured continuously for the past three decades and a record of their concentrations has been maintained. These records serve as a basis for evaluating suspected or accidental releases of radioactivity to the atmosphere. Variations in ground level atmospheric concentrations from past stratospheric injections follow an annual trend, with the spring maximum being caused by the so-called tropopause gap which allows major mixing of tropospheric air at mid-latitudes. This process alone accounts for variations in radionuclide concentrations of ground level air of up to an order of magnitude.

Radionuclides produced by cosmic ray bombardment are largely produced in the stratosphere and their tropospheric concentration shows annual variations similar to those of nuclear weapons debris. The change in concentration of other airborne radionuclides with time reflects their source and their time dependent release rates.

If one is to observe the contribution of an accidental radionuclide release to ground level air concentrations, it is very important to know the ambient concentrations of airborne radionuclides and their approximate variations in time. The Chernobyl accident did produce a substantial increase in tropospheric radionuclide concentrations over much of the Northern Hemisphere and, in many areas, its contribution to both airborne and deposited radionuclides greatly exceeded that from all previous nuclear weapons tests.

2. AIRBORNE RADIONUCLIDE CONCENTRATIONS DURING THE PAST THREE DECADES

During the past three decades, continuous large volume air sampling has been conducted at the PNL. Radiochemical analysis of the sampling filters has provided data for documentation of the airborne radionuclide concentrations which have resulted from nuclear weapons testing, cosmic ray bombardment of the atmosphere, naturally occurring radionuclides from the Earth's surface and reactor produced radionuclides. These measurements are graphically presented in Figs 1–12.

Text cont. on p. 124.

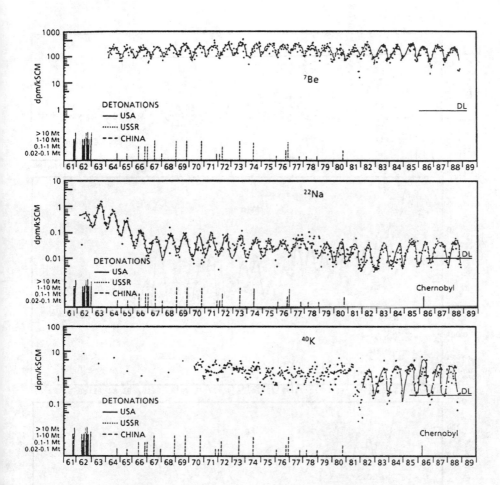

FIG. 1. Temporal distribution of airborne ^7Be, ^{22}Na and ^{40}K at Richland, Washington, USA (kSCM: 10^3 standard cubic metres; DL: detection level).

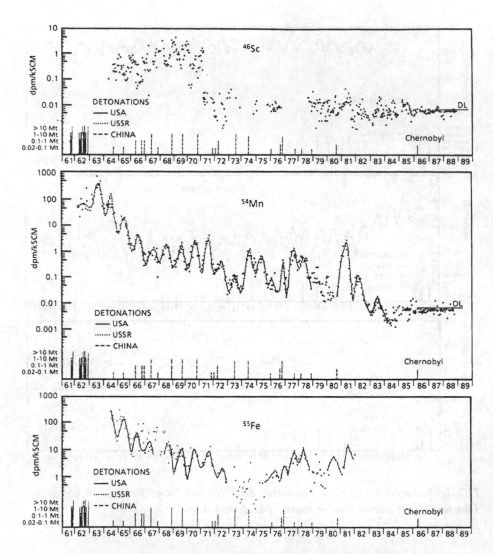

FIG. 2. *Temporal distribution of airborne* ^{46}Sc, ^{54}Mn *and* ^{55}Fe.

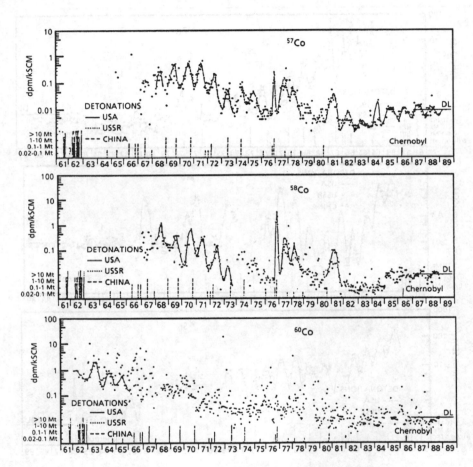

FIG. 3. Temporal distribution of airborne ^{57}Co, ^{58}Co and ^{60}Co.

PERKINS et al.

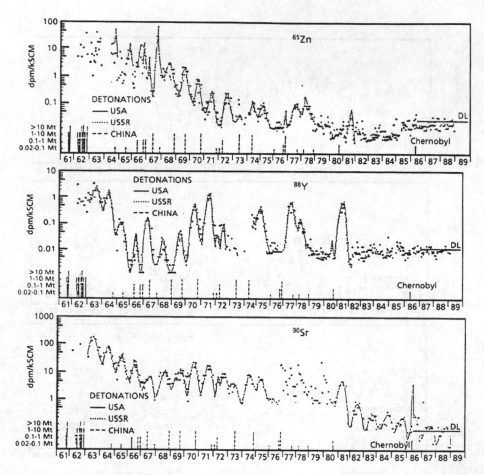

FIG. 4. *Temporal distribution of airborne* ^{65}Zn, ^{88}Y *and* ^{90}Sr.

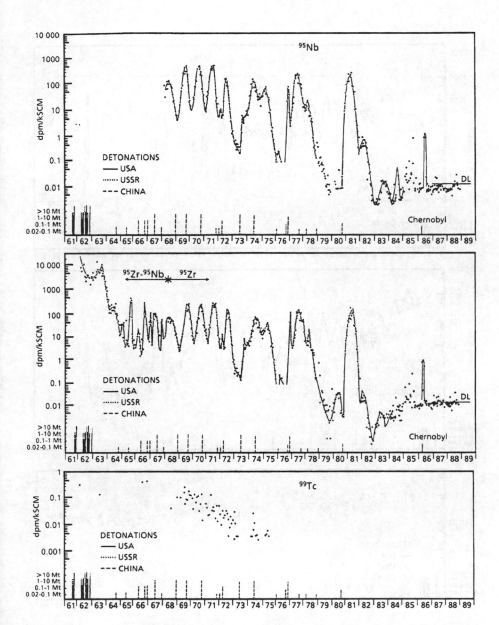

FIG. 5. Temporal distribution of airborne ^{95}Nb, ^{95}Zr-^{95}Nb, ^{95}Zr and ^{99}Tc.

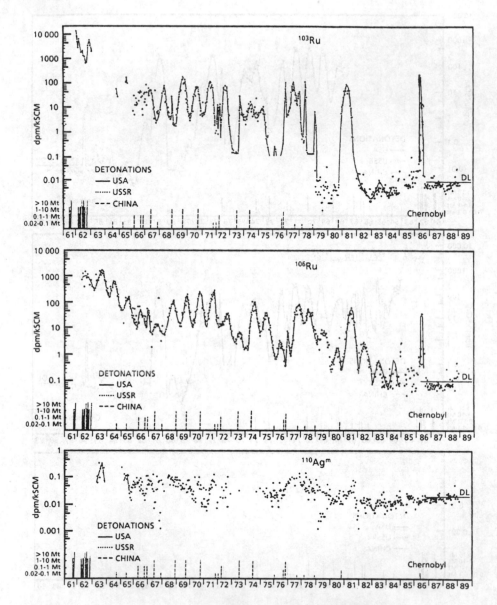

FIG. 6. Temporal distribution of airborne ^{103}Ru, ^{106}Ru and $^{110}Ag^m$.

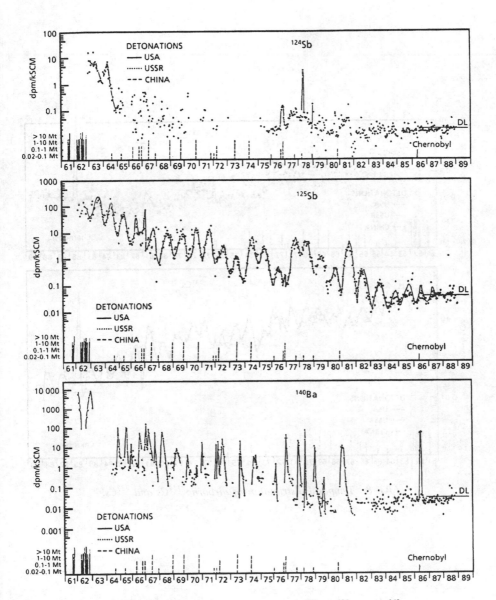

FIG. 7. *Temporal distribution of airborne* ^{124}*Sb,* ^{125}*Sb and* ^{140}*Ba.*

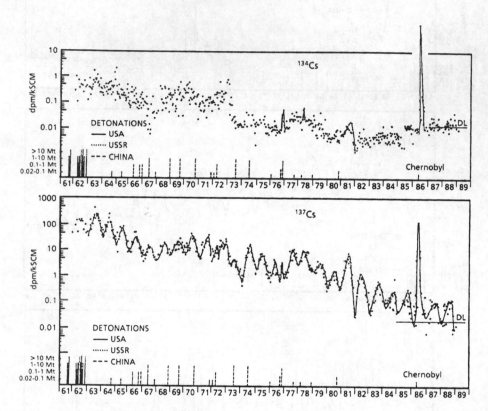

FIG. 8. *Temporal distribution of airborne* ^{134}Cs *and* $^{137}Cs.$

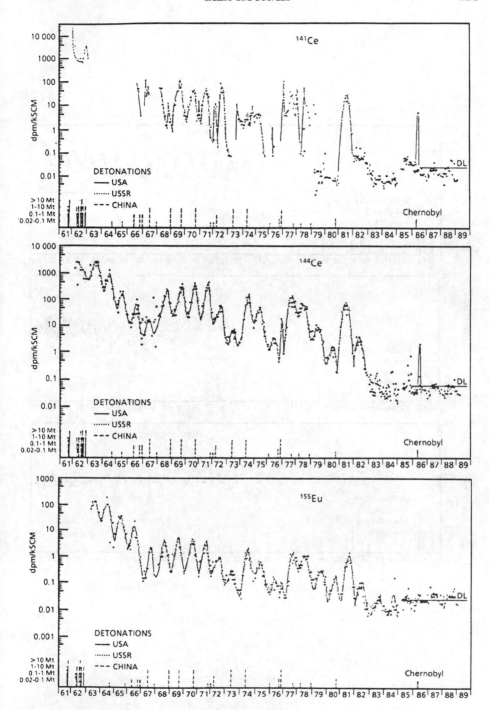

FIG. 9. Temporal distribution of airborne ^{141}Ce, ^{144}Ce and ^{155}Eu.

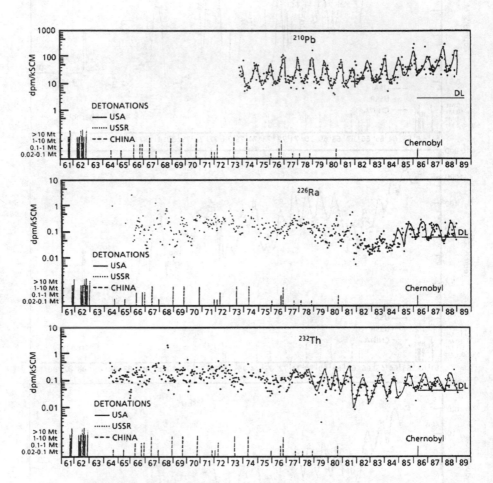

FIG. 10. *Temporal distribution of airborne* ^{210}Pb, ^{226}Ra *and* ^{232}Th.

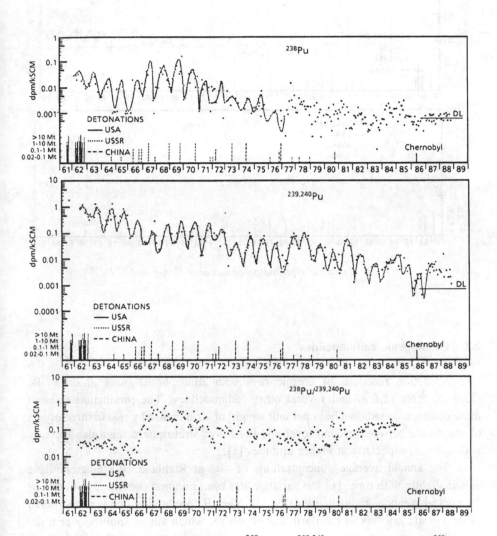

FIG. 11. Temporal distribution of airborne ^{238}Pu and $^{239,240}Pu$ and of the ratio of ^{238}Pu to $(^{239}Pu + {}^{240}Pu)$.

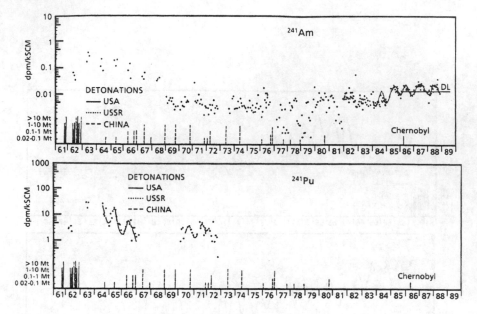

FIG. 12. Temporal distribution of airborne ^{241}Am and ^{241}Pu.

2.1. Cosmogenic radionuclides

Spallation reactions of cosmic rays with atmospheric gases produce ^7Be (53.3 d), ^{22}Na (2.6 a) and several other radionuclides. The production rates of these cosmogenic radionuclides per unit weight of air increase by two to three orders of magnitude between ground level and the lower stratosphere, and also increase with latitude, especially at higher altitudes [1].

The annual average concentrations of ^7Be at Richland, Washington, have varied slightly with time, but the variation has been minimal compared with that of the nuclear weapons radionuclides (Fig. 1). The ^7Be production rate is proportional to the cosmic ray flux in the Earth's atmosphere, which varies slightly over time.

These measurements have shown that the rate of transfer of radionuclides from the stratosphere into the troposphere reaches a yearly maximum in the spring, and a systematic annual cyclic effect in concentration is observed each year.

Sodium-22 concentration variations in the atmosphere are similar to those observed for ^7Be, except that the US and Soviet weapons testing in the early 1960s produced relatively large quantities of ^{22}Na, much of which was injected into the stratosphere (Fig. 1).

The airborne ^{22}Na concentrations at our sampling site at Richland reached a maximum in 1963 and then decreased rapidly through 1967, when they averaged about 5% of the 1963 concentrations. Between 1963 and 1966, the decay corrected concentrations of nuclear weapons produced ^{22}Na at the sampling site decreased with a residence half-life of 11.8 months. In making this calculation, it was assumed that the ^{22}Na concentrations in 1979 represented the steady state concentration due to cosmogenic ^{22}Na, and this concentration was subtracted from the measured annual peak concentrations in calculating the stratospheric residence time. The actual residence half-life in the stratosphere is probably less than 11.8 months since some ^{22}Na was produced in the Chinese nuclear tests.

In 1968 the ^{22}Na concentration increased substantially and remained at an elevated level for several years, evidently being supported by production in the Chinese nuclear tests. Over this period, the concentration continued to decrease, but at a much slower rate, perhaps because of some residual ^{22}Na in the upper stratosphere where the residence time would have been longer, and because of periodic input from the Chinese tests.

2.2. Nuclear weapons produced radionuclides

Radionuclides entering the atmosphere from nuclear weapons testing include fission products, activation products and transuranic radionuclides (Figs 2–12). The energy released in megatonnes and the time of detonation of the US, Soviet and Chinese atmospheric nuclear tests are indicated on the baseline of the concentration plots (Figs 2–12). The numerous atmospheric tests conducted in the 1961–1962 US and Soviet series resulted in the highest concentrations of radionuclides observed in the Northern Hemisphere.

The radionuclides introduced into the lower stratosphere at middle or high latitudes are generally transferred into the troposphere in maximum quantities the following spring. Therefore, the concentrations of the longer lived radionuclides introduced into the atmosphere by the 1961–1962 nuclear test series did not reach a maximum at the Richland ground level site until the spring of 1963. However, the ground level concentrations of the shorter lived radionuclides ^{95}Zr–^{95}Nb (64 d), ^{103}Ru (39.3 d), ^{140}Ba (12.7 d) and ^{141}Ce (32.5 d) decreased after 1961 because of radioactive decay, but always showed abrupt changes in concentration following nuclear weapons tests.

The concentrations of all the nuclear weapons produced radionuclides decreased rapidly until after the first Chinese high yield nuclear test in 1967 at Lop Nor (40° N, 90° E). Following this test, the concentrations of the shorter lived radionuclides decreased faster than did the concentrations of the longer lived radionuclides, as would be expected. In 1968 the concentrations of ^{137}Cs (30.2 a) at Richland averaged about 8% of the maximum concentrations measured in 1963, but the ^{103}Ru concentrations averaged only 0.2% of the maximum concentrations measured in 1961. After 1968, the concentrations of some of the long lived radionuclides increased somewhat (reaching another maximum in 1971) because of four 3 Mt tests conducted by China during 1967–1970 [2, 3]. In 1971 the ^{137}Cs concentrations at Richland averaged 11% of the 1963 concentrations.

By 1976, the annual average ^{137}Cs concentrations at Richland had decreased to 1.2% of the 1963 concentrations. China conducted two high yield tests in late 1976, causing the radionuclide concentrations to increase again in 1977. It conducted only three small tests in 1977 and 1978 and none in 1979, so there was a decrease in radionuclide concentrations again, continuing through 1979. In 1979 the ^{137}Cs concentrations at Richland averaged only 1.3% of the 1963 concentrations and 17% of the 1967 concentrations. However, in October 1980 China conducted a 0.2–1 Mt test, so concentrations increased again in the spring of 1981. That was the last Chinese atmospheric test and radionuclide concentrations have continued to decrease since that time. In 1966 the ^{137}Cs concentration at Richland was about 0.1% of the 1963 maximum concentration.

In early May 1986 a sharp peak in fission product concentrations at Richland resulted from atmospheric transport of Chernobyl debris following the reactor accident on 26 April. Because the Chernobyl radionuclide releases occurred at ground level and resulted in wide scale tropospheric transport, the residence time in the atmosphere was much shorter than in the case of the atmospheric weapons tests, which injected radionuclides into the stratosphere. Therefore, the Chernobyl fallout observed at Richland occurred over a relatively narrow time frame, and did not show the characteristic annual cyclic effects observed with weapons tests.

The Chernobyl debris arrived at the Richland air sampling station about 11 days after the accident, and radionuclides with half-lives of 2 d and greater were easily observable. The longer lived radionuclides which were measured as part of our air sampling and analysis effort and which produced major increases in atmospheric concentrations included ^{90}Sr, ^{95}Zr, ^{95}Nb, ^{103}Ru, ^{106}Ru, ^{140}Ba, ^{141}Ce and ^{144}Ce. These radionuclides produced only a short term increase in atmospheric concentrations. The Chernobyl debris is discussed further in Section 3.

Following the Chernobyl maximum, the airborne radionuclide concentrations continued to decrease at Richland, present concentrations being the lowest since before the 1961–1962 US and Soviet test series.

2.3. Stratospheric residence time

While low yield nuclear weapons tests may deposit a large fraction of the radionuclides in the troposphere, most of the radionuclides produced by high yield tests rise into the stratosphere [4, 5]. The residence time of radionuclides in the troposphere before they are deposited on the Earth's surface is around a month or less [6–9]. Therefore, except for the first few months following a nuclear test, the primary source of nuclear weapons produced radionuclides in the troposphere is from radionuclides that were originally injected into the stratosphere. As a result, the rate of decrease in the average annual radionuclide concentrations in the troposphere during periods when there is no nuclear testing should be equal to the rate of decrease in the concentrations in the lower stratosphere.

During the period from 1963 to 1966, the decay corrected concentrations of several long lived radionuclides decreased with an average residence half-life of about 11 months at Richland, indicating that the concentrations in the lower stratosphere were decreasing at this rate. During this period there was little atmospheric testing. France and China conducted a few small tests, mostly in late 1966, but the yield of these tests was less than 0.5% of the yield of the 1961–1962 test series [2, 3]. Therefore, the primary sources of airborne radionuclides were the tests conducted by the United States of America at 2° N, 157° W and 17° N, 169° E in 1961 and the Union of Soviet Socialist Republics at 75° N, 55° E in 1961 and 1962. However, the total yield of the Soviet tests was about eight times that of the US tests [5]. Also, about 60% of the total yield of the Soviet tests was contributed by tests in 1962 [2, 3], and much of the debris from the 1961 tests was transferred to the troposphere in 1962. Therefore, since almost all of the Soviet debris stabilized below 20 km [4], it may be considered that radionuclides measured at Richland during 1963–1966 were mostly injected into the lower polar (75° N) stratosphere in 1962. These results indicate that radionuclides introduced into the lower polar stratosphere will produce maximum ground level concentrations the following spring, and that the decay corrected stratospheric and tropospheric concentrations will decrease from then on with about an 11 month residence half-life.

The decay corrected concentrations of long lived radionuclides injected into the lower stratosphere at mid-latitudes (40° N) by Chinese tests also reached a maximum the following year at Richland and then decreased with an 11 month residence half-life. Following the 3 Mt Chinese test of 14 October 1970, the decay corrected concentrations of the long lived radionuclides reached a maximum in the spring of 1971 and decreased with an 11 month residence half-life during 1971–1973. China then tested a 1–3 Mt device on 26 June 1973, so the decay corrected concentrations again increased to a maximum in the spring of 1974 and then decreased with an 11 month residence half-life from 1974 through 1976, when China conducted two more tests.

2.4. Plutonium-238

Plutonium-238 is released in small quantities by nuclear weapons tests; however, it contributes less than 1% of the total disintegration rate of ^{238}Pu plus ^{239,240}Pu. In April 1964, a navigational satellite containing an electric power generator (SNAP-9A) which employed ^{238}Pu as a heat source burned up at an altitude of about 50 km at 11° S over the Indian Ocean and released 17 kCi[1] of submicrometre sized ^{238}Pu particles [10, 11].

---

[1] 1 Ci = 3.7×10^{10} Bq.

At the beginning of 1965, the ^{238}Pu was still above 25 km in the stratosphere [12]. However, by September–November 1965 layers of maximum concentration had developed in the lower stratosphere south of 30° S at an altitude of about 20 km and north of 40° N at an altitude of about 30 km. By June–August 1966, the concentration maxima were at about 20 km at middle and high latitudes of both hemispheres. Measurable concentrations of ^{238}Pu first appeared at Tokyo in late 1966 [13]. At Richland, ^{238}Pu concentrations began to increase in the spring of 1966 (Fig. 11). The ^{238}Pu/^{239}Pu ratio also increased, indicating that the ^{238}Pu originated from SNAP-9A rather than nuclear weapons tests (Fig. 11). Even though the ^{238}Pu concentrations had increased substantially and had developed the characteristic concentration maxima in the lower troposphere by the middle of 1966, ^{238}Pu concentrations at Richland increased to a new maximum in 1969, indicating that ^{238}Pu was still being transported downwards from the high stratosphere and perhaps across the Equator to the lower stratosphere in considerable quantities. During 1969–1976, however, the concentrations of SNAP-9A ^{238}Pu decreased with a residence half-life of about 12 months, indicating that the majority of the ^{238}Pu had been transported to the lower stratosphere. Beyond 1976 the concentrations of SNAP-9A ^{238}Pu became lower than the concentrations of nuclear weapons produced ^{238}Pu, so it was no longer possible to observe the decrease in the concentrations of SNAP-9A ^{238}Pu.

About 80% of the SNAP-9A ^{238}Pu was in the southern stratosphere during January–March 1966, and only 20% was in the northern stratosphere [11]. Plutonium-238 concentrations remained significantly higher in the southern stratosphere than in the northern stratosphere at least through 1972 [14]. Therefore, exchange between the hemispheres in the stratosphere should have slowed the rate of decrease of ^{238}Pu in the northern stratosphere. However, the measured rate of decrease in ^{238}Pu concentrations in the Northern Hemisphere was probably not significantly slower than the measured rates of decrease in the concentrations of radionuclides introduced into the northern stratosphere by the US, Soviet and Chinese nuclear tests. It does not appear that long term exchange between the northern and southern stratospheres was sufficient to change significantly the rate of decrease of the Northern Hemisphere radionuclide concentrations. It has been estimated that only about 16% of the air in the stratosphere of one hemisphere is exchanged with the stratosphere of the other hemisphere annually [15].

2.5. Lead-210

Lead-210 is a long lived (22.3 a) daughter of ^{222}Rn. Radon is a radioactive gas which is produced by the decay of radium in the Earth's crustal material. After its formation, radon diffuses into the atmosphere where it decays through a chain of daughter radionuclides until stable lead is produced. The radon daughters quickly become attached to atmospheric aerosols, and are therefore collected on air filters.

Since ^{222}Rn has a half-life of 3.8 d, it is able to mix upwards to a certain extent in the atmosphere, but its concentration decreases rapidly with altitude. The concentrations of ^{210}Pb also decrease with altitude in the lower troposphere over continental areas, but the decrease is less since its longer half-life allows for greater upward transport.

The airborne concentrations of ^{210}Pb at Richland from late 1973 through 1988 are plotted in Fig. 10. The concentrations showed pronounced seasonal variations, but the variations were out of phase with those of the nuclear weapons and cosmogenic radionuclides, with maxima occurring in the winter and minima in the summer.

Other investigators have observed similar concentration variations for ^{210}Pb and/or radon at continental stations [16–19]. It is believed by many investigators that a primary cause of the decrease in the concentrations of radon and radon daughters in the spring and summer is the lesser stability of the lower atmosphere at these times, leading to increased vertical mixing, reduced ground level concentrations and increased concentrations at higher altitudes [15, 16, 20, 21].

Ground level ^{210}Pb concentrations are, of course, also affected by the rate of emanation of radon from the soil and the time an air mass has spent over land during its recent history. The rate of radon emanation from ocean surfaces is about 1% of that from land surfaces [22]. However, the rate of emanation of radon from the soil should be greatest in the summer when the soil is driest, since soil moisture inhibits the escape of radon. The prevalence of marine air at the Richland sampling site is actually lower in the summer than in the winter, which should tend to increase the summer ^{210}Pb concentrations. Therefore, the observed low summer ^{210}Pb concentrations at Richland suggest that increased vertical mixing in the summer overshadows these other effects in reducing ^{210}Pb concentrations.

2.6. Nuclear reactor produced radionuclides (^{46}Sc, ^{55}Fe, ^{60}Co, ^{65}Zn, ^{134}Cs)

The concentration ratios of the various nuclear weapons produced radionuclides seem to have been internally consistent at the Richland sampling site. However, there are five radionuclides which exhibit concentration variations that suggest that they could have resulted, at least partially, from local sources. For example, ^{46}Sc shows an elevated concentration during the period of about 1964 through 1970 and then an abrupt drop in 1971, which is not what one might predict on the basis of nuclear weapons test frequency. The source of ^{46}Sc, and perhaps that of some of the other radionuclides, appears to be operation of the Hanford reactors, which were all shut down by early 1971.

The eight Hanford plutonium production reactors employed Columbia River water as their coolant on a once-through basis. Water from the Columbia River was first pumped into purification basins where water treatment processes somewhat similar to those employed on municipal waters were conducted. Following this, the water was pumped through the reactors to serve as a coolant, stored for a period of time in retention basins, and then allowed to flow back into the Columbia River. During this circulation, some trace elements were neutron activated in the reactor and subsequently entered the river. Processes that could lead to small amounts of these radionuclides entering the atmosphere include the entrainment of radionuclides in steam escaping from the retention basins and the areal suspension of Columbia River dried sediments which were exposed during normal fluctuations in the river level associated with the production of hydroelectric power at McNary Dam, some 50 km downstream from Richland.

While the hundredfold reduction in airborne ^{46}Sc coincides very well with cessation of the Hanford operations, the decreases in ^{55}Fe, ^{60}Co, ^{65}Zn and ^{134}Cs are not nearly so obvious, perhaps because of their much longer half-lives. In the case of ^{55}Fe, there is a significant reduction in the airborne concentrations after 1971, which suggests that a local source may be contributing. However, the increases in ^{55}Fe concentration during the period 1976–1978 and later must certainly be associated with nuclear weapons testing.

The concentrations of ^{60}Co during 1961–1971 and beyond seem to be almost random. Perhaps some order can be seen as suggested by the line drawn through the ^{60}Co points in Fig. 3. This area may possibly be the nuclear weapons contribution while additions from reactor effluent water sources may also contribute. This seems to be a likely source for at least part of the airborne ^{60}Co.

Zinc-65 (Fig. 4) also shows a rather random pattern during the period 1961–1968, which suggests that part of the zinc may be from a local source, but beyond that time the pattern appears to be consistent with nuclear weapons testing.

Caesium-134 concentrations (Fig. 8) also seem to be rather random and may therefore be explained by local as well as nuclear weapons sources.

2.7. Yttrium-88

The airborne concentrations of ^{88}Y (Fig. 4) at our sampling station showed clear annual variations, which suggests that nuclear weapons testing was responsible for production of this radionuclide. Its concentrations, however, do not show a consistent ratio with other fission products and therefore it appears that the target for its production in the nuclear weapons was a variable component. Its most likely mode of production is by an (n,2n) reaction on ^{89}Y.

3.   FREQUENT AIRBORNE RADIONUCLIDE MEASUREMENTS AT RICHLAND DURING THE ARRIVAL OF CHERNOBYL DEBRIS

The Chernobyl reactor accident on 26 April 1986 injected radioactive material into the atmosphere for about a ten day period until remedial actions at the reactor site terminated the releases. The airborne radionuclides were transported by the prevailing winds in a complex pattern over Europe before moving across Asia and the North Pacific Ocean to North America [23]. The first arrival of Chernobyl debris at the Richland air sampling site was detected at 07:00 on 7 May 1986. Once the debris had been detected, the air sampling frequency was increased and samples were changed every 12 h, or more frequently. The resulting ground level airborne distributions of ^{131}I and ^{137}Cs at Richland are shown in Figs 13 and 14. Also plotted in the figures for comparison purposes are the airborne concentrations of these radionuclides reported at Stockholm by the Swedish Defence Research Establishment. At Richland the airborne concentrations of ^{131}I and ^{137}Cs increased rapidly by nearly three orders of magnitude between 7 and 9 May 1986. Then their concentrations decreased by about an order of magnitude over the next two days, followed by two other concentration maxima which peaked on 12 and 17 May.

It should be emphasized that the ^{131}I concentrations plotted in Fig. 13 represent only particulate species collected on the air filters. Concurrent measurements of the various chemical forms of the ^{131}I showed that the particulate forms accounted for about 50% of the total ^{131}I, with the remaining forms being distributed approximately as 25% inorganic gases (i.e. I_2) and 25% organic gases (i.e. methyl iodide). Therefore, the total airborne ^{131}I concentration would be approximately twice that shown in Fig. 13. Also detectable in many of these samples were ^{95}Nb, ^{103}Ru, ^{106}Ru, ^{132}Te, ^{134}Cs, ^{141}Ce, ^{144}Ce and ^{140}Ba. On the basis of measurements with a six stage cascade impactor air sampler, it was shown that about 60% of the airborne Chernobyl radionuclides at Richland were associated with particle sizes of less than 0.5 μm.

It was evident that the Chernobyl debris was much different from that observed from earlier nuclear weapons test fallout. For example, the ^{137}Cs/^{131}I activity ratio for the Chernobyl debris at Richland was some 34 times higher than that observed for weapons test fallout. This was due to the buildup of ^{137}Cs in the Chernobyl fuel relative to ^{131}I during exposure in the reactor, and to the fact that a relatively large fraction (10–40%) of the ^{137}Cs in the reactor core was volatilized during the accident.

Figure 15 presents the activity ratios of some of the short lived fission products to ^{137}Cs observed at Richland during the period of peak Chernobyl fallout. These ratios varied by up to three- to fourfold over the five day period shown in the figure. These variations can be explained by the very nature of the accident. The initial breach of containment excursion involved an explosion which injected fragments of fuel, fission products and activation products into the atmosphere [23]. The explosion was followed by a fire which continued to burn until at least 5 May 1986, emitting additional quantities of radionuclides. It appears that the composition of emitted radionuclides varied considerably during the course of the release. At first, the emissions of volatile radionuclides (e.g. ^{131}I) dominated, but as the fuel temperature rose, the escape of more refractory radionuclides (e.g. ^{103}Ru and ^{140}Ba) increased [24]. Thus, the composition of the emissions over the ten day release

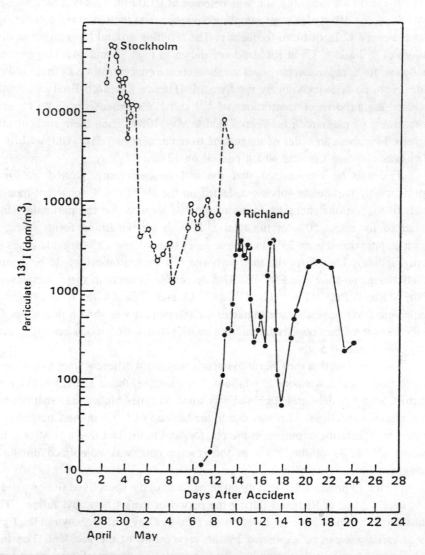

FIG. 13. Particulate airborne radioiodine concentrations at Stockholm and Richland measured at ground level following the Chernobyl accident.

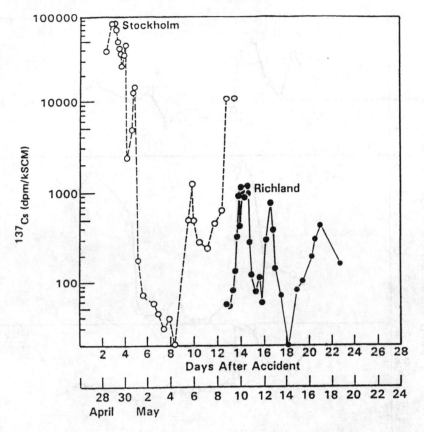

FIG. 14. Airborne ^{137}Cs concentrations at Stockholm and Richland measured at ground level following the Chernobyl accident.

period changed substantially, and these changes are reflected in the radionuclide ratios observed at Richland.

4. NATURE OF 'HOT PARTICLES' EMITTED DURING THE CHERNOBYL ACCIDENT

The Chernobyl accident injected a unique mixture of vaporized and particulate radioactive material into the atmosphere. The physical and chemical processes to which the uranium fuel was subjected during the initial explosion and subsequent fire influenced the nature of the debris which was released and deposited over Europe and the Northern Hemisphere.

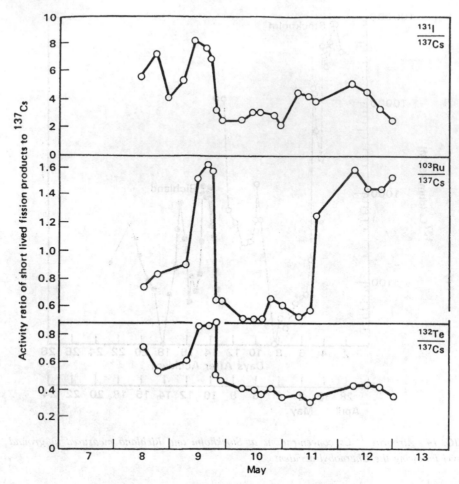

FIG. 15. *Activity ratio of short lived fission products to* ^{137}Cs *at Richland from 7 to 12 May 1986.*

One of the authors (D.E.R.) visited Sweden to evaluate the radiological conditions there a few days after the Chernobyl accident. With the co-operation of the Swedish National Institute of Radiation Protection (NIRP), environmental sampling was conducted along the eastern coast of Sweden from Gävle to Öland. During this sampling effort, it became apparent that much of the Chernobyl fallout in this area was in the form of discrete 'hot particles'. This was especially evident in areas where it had not rained, since the precipitation scavenged much of the vaporized radionuclides and those associated with small aerosols from the atmosphere. With such particles there is the possibility of local high irradiation exposure should a particle become attached to the skin or be inhaled. The composition of the particles is of interest since this information would improve our understanding of the behaviour of the fuel and fission products during the accident. In addition, the physical–chemical and radionuclide composition of the particles will affect their environmental behaviour and the long term fate of the Chernobyl debris.

Using a Geiger–Müller survey meter, 'hot particles' were located on various surfaces (grass, soil, wood, leaves), isolated, and packaged for laboratory analyses. The population density of the particles ranged from about 6 particles/m² at Stockholm to about 1 particle/m² at Öland. The particles were analysed by gamma ray spectrometry using a germanium diode detector at the PNL. Selected particles were subjected to further detailed measurements, including scanning electron microscopy, microprobe X ray fluorescence, mass spectrometry, alpha energy analyses and materials identification.

The results of the gamma ray spectrometric measurements of a few representative particles are given in Table I. It was obvious from the fission product composition of these particles that they consisted essentially of refractory fission products (^{103}Ru, ^{106}Ru, ^{141}Ce, ^{144}Ce, ^{95}Zr–^{95}Nb), suggesting that they were actually tiny particles of reactor fuel. The radionuclide composition of the particles was extremely variable. Some particles appeared to be composed mainly of ^{103,106}Ru whereas others appeared to consist mainly of ^{141,144}Ce, ^{95}Zr–^{95}Nb or ^{140}Ba. Most particles, however, were a mixture of these radionuclides. The volatile radionuclides (^{131}I, ^{132}Te, ^{134}Cs, ^{137}Cs) were usually very depleted compared with the refractory fission products in these particles. Attempts were made to correlate the concentration distributions of the various fission products in the particles, but it became obvious that no systematic correlations existed. It is apparent that the physical and chemical processes affecting the formation of these particles during the various stages of the reactor accident were extremely complex.

An examination of four selected particles by scanning electron microscopy indicated that at least two types of particles were released to the atmosphere: (1) micrometre sized fragments of microfractured fuel particles which probably were generated and released during the initial explosion, and (2) spherical particles of once-molten mixtures of fuel, fuel cladding and other structural materials such as concrete. What were initially judged to be single particles on the basis of surveys and visual observations were actually shown in several cases to consist of several discrete materials. The results of further examination of these and other particles by microprobe X ray fluorescence are given in Table II. The particles which were analysed ranged from 2.2 μm to 21 μm in diameter. Most of the particles consisted primarily of uranium oxide containing various fission product and elemental impurities. Some particles appeared to be fuel fragments (uranium oxide), some appeared to be mixtures of fuel and zirconium cladding, and others appeared to be mixtures of fuel plus cladding plus structural materials such as concrete and steel. One of the particles, a silicate, had an unusual composition and had become molten during its

TABLE I. RADIONUCLIDE CONCENTRATIONS IN HOT PARTICLES COLLECTED IN SWEDEN FOLLOWING THE CHERNOBYL ACCIDENT (pCi/particle)[a]

Sample No.	Zr-95	Nb-95	Ru-103	Ru-106	Te-132	I-131	Cs-134	Cs-136	Cs-137	Ba-140	Ce-141	Ce-144	Np-239	Mo-99
HP-1	4590 ±1.04%	6220 ±1.11%	1030 ±3.54%	219 ±18.2%	10 200 ±4.07%	1570 ±2.11%	1240 ±2.06%	62.5 ±13.2%	2510 ±1.18%	4890 ±3.68%	3310 ±1.09%	2480 ±1.47%	29 200 ±30.1%	
HP-2	4270 ±0.98%	4420 ±1.18%	788 ±1.74%	178 ±17.6%	521 ±55.4%	245 ±8.08%	27.4 ±10.3%	<27	57.6 ±7.61%	1800 ±3.79%	3230 ±1.38%	2320 ±1.59%	21 100 ±20.1%	983 ±21.4%
HP-8	4830 ±1.03%	5270 ±1.13%	3700 ±1.95%	630 ±11.6%	2490 ±11.0%	3030 ±3.58%	89.5 ±4.44%	<59	201 ±3.54%	3580 ±5.21%	4010 ±1.08%	2430 ±1.67%	54 700 ±18.0%	3670 ±13.5%
HP-20	2830 ±0.97%	3780 1.16%	781 ±1.70%	246 ±10.1%	197 ±22.5%	1040 ±2.89%	56.4 ±6.10%	<29	125 ±4.27%	1440 ±4.46%	2000 ±1.15%	1950 ±1.62%		
HP-34	154 ±4.51%	191 ±3.71%	56.7 ±7.36%	20.5 ±89.6%	<448	681 ±3.62%	29.0 ±7.62%	<29	105 ±4.10%	257 ±12.5%	7630 ±1.06%	4050 ±1.29%		

[a] Decay corrected to 26 April 1986 at 05:00 Chernobyl time (1 Ci = 3.7 × 10^10 Bq).

TABLE II. RANGES OF COMPOSITION OF HOT PARTICLES COLLECTED IN SWEDEN FOLLOWING THE CHERNOBYL ACCIDENT

Materials identification	Size (μm)	Elemental composition (wt%)	U ± 1σ			Pu ± 1σ		
			234/238	235/238	236/238	240/239	241/239	242/239
Multiple oxides	12.3	U 42–66, Zr 18–39, O 16–19	0.000 158 ±0.000 001	0.009 89 ±0.000 02	0.001 881 ±0.000 003	0.383 9 ±0.001 0	0.101 2 ±0.000 3	0.022 4 ±0.000 1
Uranium oxides	2.2–6.9	U 76–88, O 12–15, Zr 8–10, Si 3	0.000 139 ±0.000 001 to 0.000 159 ±0.000 001	0.007 63 ±0.000 02 to 0.011 95 ±0.000 01	0.001 606 ±0.000 004 to 0.002 165 ±0.000 006	0.338 3 ±0.000 7 to 0.546 4 ±0.003 0	0.098 0 ±0.000 2 to 0.147 2 ±0.001 2	0.016 10 ±0.000 04 to 0.047 65 ±0.000 4
Silicates	3.8	U 10, O 28, Pu 20, Si 16, Fe 12, Zr 6, Mn 4, Ce 2, Na 2	0.000 153 ±0.000 003	0.008 90 ±0.000 03	0.002 02 ±0.000 01	0.447 0 ±0.000 3	0.122 0 ±0.000 2	0.031 08 ±0.000 09
Quartz	—	O 53, Si 47	Mass spectrometric analysis not possible owing to the high carbon content of this particle					
Organic	21.0	C, O, U 1, Si 11, Zr 3, Fe 1, K 1, Al 1						

Note: In some cases data are presented for one particle only.

formation. This particle contained 20 wt% of plutonium, suggesting possible selective volatilization of plutonium from the fuel and condensation to form the particle. It also contained (in wt%) 16 Si, 12 Fe, 6 Zr, 4 Mn, 2 Ce and 2 Nd, indicating the presence of structural materials in the once-molten particle. One particle consisted mainly of organic material with small percentages of Si, Zr, Fe, K and Al.

The same selected particles were subjected to mass spectrometric analyses. The analytical results, consisting of elemental analyses, are summarized in Table II. As shown also by the atomic ratios in Table II, the fuel was slightly enriched uranium. The plutonium isotopic composition of the fuel particles showed a ^{240}Pu/^{239}Pu ratio which varied from 0.3383 to 0.5464 and a ^{241}Pu/^{239}Pu ratio which varied from 0.0980 to 0.1472.

5.   CONCLUSIONS

On examination of the airborne radionuclide concentrations over the past three decades, it is clear that background information of this type is essential in detecting any contribution of possible nuclear accidents. In the case of a serious accident, the debris may entirely override this background in the immediate vicinity of the release but at much greater distances the contribution will certainly blend into the 'ambient background', and unless one knows what this is, there will be a point where it will simply not be possible to distinguish new debris from that which is residual from past nuclear weapons testing and other radionuclide releases. It is therefore clear that monitoring programmes which measure the ambient airborne concentrations of radionuclides are critically important if we are to be in a position to determine what contribution subsequent releases may add to ambient levels.

REFERENCES

[1]   LAL, D., PETERS, B., "Cosmic ray produced isotopes and their application to problems in geophysics", Progress in Elementary Particle and Cosmic Ray Physics, North-Holland, Amsterdam (1962).

[2]   CARTER, M.W., MOGHISSI, A.A., Three decades of nuclear testing, Health Phys. **33** (1977) 55.

[3]   PERKINS, R.W., THOMAS, C.W., "World-wide fallout", Transuranium Elements in the Environment, DOE/OHER Rep. TID-22800, US Dept. of Energy, Washington, DC (1980).

[4]   FERBER, G.J., "Distribution of radioactivity with height in nuclear clouds", Radioactive Fallout from Nuclear Weapons Tests, AEC Symp. Series No. 5, US Atomic Energy Commission, Washington, DC (1964) 629.

[5]   PETERSON, K.R., An empirical model for estimating world-wide deposition from atmospheric nuclear detonations, Health Phys. **18** (1970) 357.

[6]   JUNGE, C.E., Air Chemistry and Radioactivity, Int. Geophys. Ser. No. 4, Academic Press, New York (1963) 227.

[7] ENHALT, D.H., Turnover times of Cs-137 and HTO in the troposphere and removal rates of natural aerosol particles and water vapor, J. Geophys. Res. **78** (1973) 7076.

[8] MARTELL, E.A., MOORE, H.E., Tropospheric aerosol residence times: A critical review, J. Rech. Atmos. **8** (1974) 903.

[9] BLEICHRODT, J.F., Mean residence time of cosmic-ray-produced Be-7 at north temperate latitudes, J. Geophys. Res. **83** (1978) 3058.

[10] HANSEN, H.E., CLARK, A.J., BENTZ, A.E., "Final report on re-entry flight demonstration No. 2", USAEC Rep. SC-RR-65-43, Sandia Corp., Albuquerque, NM (1965).

[11] KREY, P.W., Atmospheric burnup of a plutonium-238 generator, Science **158** (1967) 769.

[12] LIST, R.J., TELEGADAS, K., Using radioactive tracers to develop a model of the circulation of the atmosphere, J. Atmos. Sci. **26** (1969) 1128.

[13] MIYAKE, Y., KATSURAGI, Y., SUGIMURA, Y., A study on plutonium fallout, J. Geophys. Res. **75** (1970) 2329.

[14] KREY, P.W., SCHONBERG, M., TOONKEL, L., "Updating stratospheric inventories to March 1972", Fallout Program Quarterly Summary Report, December 1, 1972–March 1, 1973, USAEC Rep. HASL-273, Health and Safety Lab., New York (1973).

[15] REITER, E.R., Atmospheric Transport Processes, Part 4: Radioactive Tracers, Technical Information Center, US Dept. of Energy, Washington, DC (1978).

[16] JOSHI, L.U., RANGARAJAN, C., GOPALAKRISHNAN, S., Measurements of lead-210 in surface air and precipitation, Tellus **21** (1969) 107.

[17] JOSHI, L.U., MAHADEVAN, T.N., Seasonal variations of radium-D (lead-210) in ground level air in India, Health Phys. **15** (1968) 67.

[18] PEIRSON, D.H., CAMBRAY, R.S., SPICER, G.S., Lead-210 and polonium-210 in the atmosphere, Tellus **18** (1966) 427.

[19] LOCKHART, L.B., Natural radioactive isotopes in the atmosphere at Kodiak and Wales, Alaska, Tellus **14** (1962) 350.

[20] GALE, H.J., PEAPLE, L.H.J., Measurements on the near-ground radon concentrations on the A.E.R.E. airfield, Rep. AERE-HP/R, Atomic Energy Research Establishment, Harwell, UK (1958) 2381.

[21] BECK, H.L., GOGOLAK, C.V., Time dependent calculations of the vertical distribution of Rn-222 and its decay products in the atmosphere, J. Geophys. Res. **84** (1979) 319.

[22] WILKENING, M.H., CLEMENTS, W.E., STANLEY, D., "Radon-222 flux measurements in widely separated regions", Natural Radiation Environment 2 (ADAMS, J.A.S., Ed.), US Energy Research and Development Administration, Oak Ridge, TN (1975).

[23] GOLDMAN, M., CATLIN, R.J., ANSPAUGH, L., Health and Environmental Consequences of the Chernobyl Nuclear Power Plant Accident, Rep. DOE/ER-0332, US Dept. of Energy, Washington, DC (1987).

[24] USSR STATE COMMITTEE ON THE UTILIZATION OF ATOMIC ENERGY, The Accident at the Chernobyl Nuclear Power Plant and Its Consequences, information compiled for IAEA Experts Meeting, Vienna, 1986.

SHORT AND LONG TERM EFFECTS OF CHERNOBYL RADIOACTIVITY ON DEPOSITION AND AIR CONCENTRATIONS IN JAPAN

K. HIROSE, M. AOYAMA, Y. SUGIMURA
Geochemical Laboratory,
Meteorological Research Institute,
Tsukuba, Ibaraki,
Japan

Abstract

SHORT AND LONG TERM EFFECTS OF CHERNOBYL RADIOACTIVITY ON DEPOSITION AND AIR CONCENTRATIONS IN JAPAN.

Deposition and surface air concentrations of long lived radionuclides were measured continuously at the Meteorological Research Institute in Tsukuba, Ibaraki, Japan. After the Chernobyl accident, Chernobyl released ^{90}Sr and plutonium isotopes were detected in deposition and filter samples at Tsukuba, but these levels were much lower than those of volatile radionuclides, i.e. ^{137}Cs and ^{131}I, in contrast to the radioactivity composition in the fallout from atmospheric nuclear weapons tests. A typical feature of the Chernobyl activity observed in Japan is that the particle size distributions of ^{90}Sr and plutonium isotopes differed from the particle size distributions of the volatile radionuclides. The results of these measurements suggest that one of the important factors for controlling air concentrations of radionuclides is the particle size of radionuclide bearing particles. The Chernobyl fallout of long lived radionuclides, however, continued for more than two years. The results of the Japanese research indicate strongly that a part of the Chernobyl activity was injected into the stratosphere.

1. INTRODUCTION

Since 1957, research programmes on atmospheric behaviour and deposition of radioactivity derived from the atmospheric nuclear tests and nuclear accidents have been conducted by the Geochemical Laboratory of the Meteorological Research Institute in Tsukuba, Ibaraki, Japan [1, 2]. Surface air concentrations of ^{90}Sr and plutonium isotopes have been measured since 1978 [3, 4]. These studies revealed that the long term trend and typical seasonal variation of radioactive deposition and surface air concentrations are controlled by the stratospheric fallout; radioactive debris injected into the stratosphere was deposited on the Earth's surface with a stratospheric residence time of 1.4 a and its seasonal variation is characterized by a maximum from March to June and a minimum from September to November [5].

141

After the Chernobyl accident on 26 April 1986, an abrupt rise in radioactivity due to volatile fission products in surface air and rain water was observed in Europe. On 3 May 1986, radioactivity originating from the Chernobyl accident was detected in both surface air and rain water in Japan [6]. The composition of radionuclides in deposition and air concentrations differed from that of atmospheric nuclear tests [7, 8]. After markedly high deposition and air concentrations of the Chernobyl radioactivity were recorded in May 1986, they decreased rapidly in June 1986. The Chernobyl fallout, however, continued through March to June 1988; this is attributed to the stratospheric fallout of the Chernobyl radioactivity transported into the stratosphere.

In this paper, measurements of particle size distributions of gamma emitting nuclides, ^{90}Sr and plutonium isotopes as well as their surface air concentrations are described to clarify factors controlling the surface air concentrations of radioactivity and long term deposition of the Chernobyl released long lived radionuclides in Japan.

2.   SAMPLING AND METHODS

A high volume air sampler that was installed in a precipitation shelter was used to collect samples of airborne particulate materials on a glass fibre filter at a flow rate of 1000 L/min. To determine the size distribution of radionuclide bearing particles in surface air, a cascade impactor with five stages (the effective cut-off aerodynamic diameter of each stage being: I >7 μm; II 3.3–7.0 μm; III 2.0–3.3 μm; IV 1.1–2.0 μm; V <1.1 μm) was used at a flow rate of 600 L/min.

Total deposition samples were collected at Tsukuba on a monthly basis by using plastic open surface fallout collectors with surface areas of 1 m^2 and 4 m^2. Monthly deposition samples were evaporated to a dry state and residues were subjected to radioactivity measurements.

Filter and deposition samples were examined by gamma spectrometry followed by radiochemical extractions to permit the measurement of plutonium isotopes by alpha spectrometry and the measurement of ^{90}Sr by low background beta counting as described in detail elsewhere [1, 3].

All data on radionuclide concentrations released from the Chernobyl reactor are adjusted to the activity of 26 April 1986 as regards the radioactive decay.

3.   RESULTS AND DISCUSSION

3.1. Short term effect (surface air concentrations)

On 3 May 1986, an abrupt increase in concentrations of volatile radionuclides containing ^{131}I, ^{103}Ru, ^{106}Ru, ^{137}Cs, ^{134}Cs, etc., was observed in the surface air at

TABLE I. ACTIVITY MEDIAN AERODYNAMIC DIAMETERS (μm) OF THE CHERNOBYL RELEASED RADIONUCLIDES OBSERVED AT TSUKUBA

Sampling periods in 1986	Cs-137	Ru-103	Sr-90	Pu-239, Pu-240
6–10 May[a]	0.4_0	0.3_5	—[b]	—
13–17	0.8	0.8	1.7	3.8
17–24	0.8	0.7	—	—
24–31	0.7	0.7	1.3	4.8

[a] A low volume Andersen cascade sampler with 9 stages was used. Size distributions of Sr-90, Pu-239 and Pu-240 cannot be measured because of a low sample volume.

[b] Not measured.

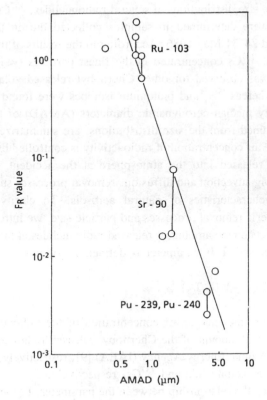

FIG. 1. Relationship between the F_R value and the AMAD of the Chernobyl radioactivity.

Tsukuba. The ^{137}Cs concentration in the surface air at Tsukuba increased to about four orders of magnitude higher than that of the previous month and, on 9 May, it reached a maximum of 61.2 mBq/m^3. On 1 June, the ^{137}Cs air concentration decreased to 0.59 mBq/m^3.

The monthly mean plutonium concentrations in the surface air in March and April 1986 were 0.012 and 0.009 μBq/m^3, respectively. From 3 to 6 May, the plutonium concentration in the surface air increased to 0.12 μBq/m^3. After 3 May, the ^{238}Pu/(^{239}Pu + ^{240}Pu) activity ratio in the surface air increased from 1–5% to 8–33%. In June, the plutonium level and the ^{238}Pu/(^{239}Pu + ^{240}Pu) activity ratio fell to the pre-Chernobyl levels. A significant part of the airborne plutonium at Tsukuba in May 1986 was due to the Chernobyl released plutonium isotopes, which are characterized by a higher plutonium concentration and a high ^{238}Pu/(^{239}Pu + ^{240}Pu) ratio [8].

The monthly mean ^{90}Sr concentration in the surface air at Tsukuba in April 1986 was 0.41 μBq/m^3. From 3 to 6 May, the ^{90}Sr concentration increased to 32 μBq/m^3, which is about two orders of magnitude higher than that of the previous months. In June, the ^{90}Sr level in air decreased rapidly to the level before the Chernobyl fallout.

The particle size distributions of several radionuclides, ^{137}Cs, ^{103}Ru, ^{90}Sr and ^{239}Pu + ^{240}Pu, were determined in samples collected during the periods 6–10, 13–17, 17–24 and 24–31 May 1986. According to the results of the activity particle size distributions, ^{137}Cs concentrated in the finest particles (< 1.1 μm), which is similar to what was observed for other Chernobyl released volatile radionuclides such as ^{103}Ru, whereas ^{90}Sr and plutonium isotopes were found in larger particle sizes. The activity median aerodynamic diameters (AMAD) of the radionuclides, which are determined from the size distributions, are summarized in Table I.

The surface air concentration of radioactivity is controlled by the total amount of radionuclides released into the atmosphere at the accident site, by transport processes including advection and diffusion, removal processes such as wet and dry deposition, and characteristics of natural aerosols. To clarify the relationship between atmospheric removal processes and particle size, we introduce a parameter to reduce the effects on the amount of released radionuclides at the accident site and on transport processes. This parameter is defined as follows:

$$F_R = (C_{A,R}/C_{A,Cs}) \cdot (I_R/I_{Cs})^{-1} \tag{1}$$

where C_A and I are the surface air concentration of the radionuclide observed at Tsukuba and the total amount of the Chernobyl released radionuclides reported by the International Atomic Energy Agency (IAEA) [9], respectively, and subscripts R and Cs represent a radionuclide and ^{137}Cs, respectively.

We examined the relationship between the parameter, F_R, and the AMAD for the Chernobyl radioactivity. The plot of logarithmic F_R value vs. logarithmic

AMAD was carried out. The result as shown in Fig. 1 indicated that the logarithmic F_R value is approximately linearly related with the logarithmic AMAD, whose slope is -3 within the range of these particle sizes; the F_R value, which reflects the effect of atmospheric removal processes, decreased with the increase of the particle size at a power of 3. The pattern of the size distribution of the F_R value is consistent with Junge's distribution of airborne particles. The result suggests that one of the important factors for removing radionuclides from the atmosphere is the particle size of radionuclide bearing particles.

3.2. Long term effect (deposition)

The recent results of the analyses of ^{90}Sr and ^{137}Cs in the fallout samples are shown in Table II. The annual deposition of ^{137}Cs in 1981 was 24 Bq/m^2 and that of ^{90}Sr was 19 Bq/m^2, which is a maximum during the period from 1980 to 1985. Using the $^{89}Sr/^{90}Sr$ activity ratio in the fallout, Katsuragi and Aoyama [10] estimated that about 90% of the ^{137}Cs and ^{90}Sr deposition in 1981 derived from the atmospheric nuclear weapons test conducted by China in October 1980. During the period from 1982 to 1985, the annual ^{137}Cs and ^{90}Sr deposition decreased by the stratospheric residence time of 1.4 a. The annual deposition of both radionuclides was minimal during the period from 1957 to 1985.

TABLE II. ANNUAL DEPOSITION OF Cs-137 AND Sr-90 (Bq/m^2) AND Cs-137/Sr-90 ACTIVITY RATIOS AT TSUKUBA FROM 1980 TO 1988

Year	Cs-137	Sr-90	Cs-137/Sr-90
1980	7.4	4.3	1.7
1981	24.0	19.0	1.3
1982	4.9	2.6	1.9
1983	2.2	1.3	1.7
1984	1.2	0.54	2.3
1985	0.68	0.34	2.0
1986	135	1.8	74
1987	0.95	0.15	6.4
1988	0.57		

TABLE III. CUMULATIVE DEPOSITION OF Cs-137, Cs-134 AND Sr-90 (Bq/m²) AS WELL AS Cs-137/Cs-134 AND
Cs-137/Sr-90 ACTIVITY RATIOS AT TSUKUBA FROM JANUARY 1986 TO DECEMBER 1988

Period		Cs-137	Cs-134	Sr-90	Cs-137/Cs-134	Cs-137/Sr-90
1986	Jan.–Apr.	0.30	ND[a]	0.17	—[b]	1.8
	May–Aug.	135	71.3	1.6	1.9	85
	Sep.–Dec.	0.35	0.19	0.062	1.8	5.6
1987	Jan.–June	0.68	0.30	0.098	2.3	7.0
	July–Dec.	0.30	0.097	0.055	3.0	5.4
1988	Jan.–June	0.39	0.13		3.0	
	July–Dec.	0.19				

[a] ND — not determined.
[b] Not measured.

The annual deposition of ^{137}Cs in 1986 increased up to 135 Bq/m^2, which is about 200 times higher than that in the previous year. In 1987, the annual ^{137}Cs deposition (0.95 Bq/m^2) decreased rapidly to 0.7% of that in 1986, but remained at a slightly higher level than that in 1985. On the other hand, the annual ^{90}Sr deposition in 1986 (1.8 Bq/m^2) was only about five times that in the previous year. In 1987, the annual ^{90}Sr deposition decreased to 0.15 Bq/m^2, which is slightly higher than that predicted from the stratospheric residence time. The excess ^{137}Cs and ^{90}Sr deposition in 1987 cannot be explained by resuspension of radionuclides deposited on the Earth's surface [11].

On the basis of the stratospheric residence time of 1.4 a and the ^{137}Cs/^{134}Cs activity ratio of 1.9 for the Chernobyl activity, Aoyama [11] reported that about 80% of the ^{137}Cs fallout at Tsukuba in 1987 was derived from the Chernobyl activity injected into the stratosphere, while 20% of the ^{137}Cs fallout originated from the previous atmospheric nuclear weapons tests. For ^{90}Sr, about 40% of the annual deposition in 1987 was estimated to originate from the stratosphere injected Chernobyl activity on the basis of the stratospheric residence time of 1.4 a.

The cumulative deposition of ^{137}Cs, ^{134}Cs and ^{90}Sr, as well as the ^{137}Cs/^{134}Cs and ^{137}Cs/^{90}Sr activity ratios are shown in Table III. The ^{137}Cs/^{134}Cs activity ratio from May to August 1986, the period of marked Chernobyl fallout in Japan, was 1.9, and was consistent with that in the Chernobyl released radioactivity. The ^{137}Cs/^{134}Cs activity ratio in the deposition increased systematically from January 1987 to June 1988. The increasing ^{137}Cs/^{134}Cs activity ratio indicates that the portion of the Chernobyl released ^{137}Cs against pre-existing nuclear weapons test derived ^{137}Cs decreased steadily in the stratospheric fallout. Therefore, it can be said that most of the Chernobyl radioactivity was injected into the stratosphere, and only a small portion of the total radioactivity release is present in its lower altitudes.

The ^{137}Cs/^{90}Sr activity ratios ranged from 1.3 to 2.3 from 1980 to 1985, which is the typical activity ratio for the fallout from the atmospheric nuclear weapons tests. After the Chernobyl accident, this ratio increased to 85. This result indicates that the Chernobyl radioactivity is characterized by higher ^{137}Cs/^{90}Sr ratios than those of the nuclear weapons tests.

We evaluated how much of the Chernobyl radioactivity has been transported into the stratosphere. According to the empirical method of Volchok [12] and Katsuragi [1], Aoyama [11] estimated that the global deposition of caesium isotopes in the Northern Hemisphere in 1987 was 1.6×10^{14} Bq for ^{137}Cs and 0.73×10^{14} Bq for ^{134}Cs with an uncertainty of 20%. Since 3.77×10^{16} Bq of ^{137}Cs (uncertainty of 50%) was released from the Chernobyl reactor [9], the global deposition of ^{137}Cs in 1987 corresponds to about 0.4% of the total released ^{137}Cs. The maximum stratospheric inventory of the Chernobyl released ^{137}Cs is estimated to be about 0.8% of the total released ^{137}Cs, considering the stratospheric residence time of 1.4 a. As well as Cs isotopes, the global deposition of ^{90}Sr in the Northern Hemisphere in 1987 can be estimated to be 1.3×10^{13} Bq. It was reported [9] that

8.9×10^{15} Bq of ^{90}Sr (uncertainty of 50%) was released from the Chernobyl reactor. Therefore, the global deposition of ^{90}Sr in 1987 corresponds to about 0.15% of the total released ^{90}Sr. The maximum stratospheric inventory of the Chernobyl released ^{90}Sr is estimated to be about 0.3% of the total released ^{90}Sr, considering the stratospheric residence time of 1.4 a. The results suggest that 0.15–0.3% of the total released ^{90}Sr, whose magnitude is lower than that of ^{137}Cs, was transported into the stratosphere.

The ^{137}Cs/^{90}Sr activity ratio of the Chernobyl radioactivity injected into the stratosphere is calculated to be 12 from their stratospheric inventories; this ratio differs substantially from that of the nuclear weapons tests (about 2).

4.   CONCLUSIONS

We detected in the surface air at Tsukuba gamma emitting nuclides, ^{90}Sr and plutonium isotopes which originated from the Chernobyl accident. The Chernobyl fallout decreased rapidly according to the apparent atmospheric residence time of about 25 d. One of the important factors for controlling air concentrations of radionuclides is the particle size of radionuclide bearing particles. The Chernobyl fallout of long lived radionuclides, however, has continued for more than two years. The results of our research indicate strongly that a part of the Chernobyl activity was injected into the stratosphere. The Chernobyl released ^{90}Sr, whose magnitude is lower than that of ^{137}Cs, injected into the stratosphere is estimated to be about 0.2% of the total strontium released.

REFERENCES

[1]   KATSURAGI, Y., A study of Sr-90 fallout in Japan, Pap. Meteorol. Geophys. (Tokyo) **33** (1983) 277–291.

[2]   HIROSE, K., et al., Annual deposition of Sr-90, Cs-137 and Pu-239, 240 from the 1961–1980 nuclear explosions: A simple model, J. Meteorol. Soc. Jpn. **65** (1987) 259–277.

[3]   HIROSE, K., SUGIMURA, Y., Plutonium in the surface air in Japan, Health Phys. **46** (1984).

[4]   HIROSE, K., et al., ^{90}Sr and $^{239+240}$Pu in the surface air in Japan: Their concentrations and size distribution, Pap. Meteorol. Geophys. (Tokyo) **37** (1986) 255–296.

[5]   EHHALT, D.H., HAUMACHER, G., The seasonal variation in the concentration of strontium 90 in rain and its dependence on latitude, J. Geophys. Res. **75** (1970) 3027–3031.

[6]   AOYAMA, M., et al., High level radioactive nuclides in Japan in May, Nature (London) **321** (1986) 819–820.

[7] AOYAMA, M., et al., Deposition of gamma-emitting nuclides in Japan after the reactor-IV accident at Chernobyl', J. Radioanal. Nucl. Chem. **116** (1987) 291–306.

[8] HIROSE, K., SUGIMURA, Y., Plutonium isotopes in the surface air in Japan: Effect of Chernobyl accident, J. Radioanal. Nucl. Chem. (in press).

[9] INTERNATIONAL NUCLEAR SAFETY ADVISORY GROUP, Summary Report on the Post-Accident Review Meeting on the Chernobyl Accident, Safety Series No. 75-INSAG-1, IAEA, Vienna (1986).

[10] KATSURAGI, Y., AOYAMA, M., Seasonal variation of Sr-90 fallout in Japan through the end of 1983, Pap. Meteorol. Geophys. (Tokyo) **37** (1986) 15–36.

[11] AOYAMA, M., Evidence of stratospheric fallout of caesium isotopes from the Chernobyl accident, Geophys. Res. Lett. **15** (1988) 327–330.

[12] VOLCHOK, H.L., Strontium 90: Estimation of worldwide deposition, Science (1964) 1451–1452.

LAVAGE DE L'IODE PAR LES PRECIPITATIONS
Expériences récentes in situ

C. CAPUT, H. CAMUS, D. GAUTHIER, Y. BELOT
Institut de protection et de sûreté nucléaire,
Commissariat à l'énergie atomique,
Fontenay-aux-Roses, France

Abstract–Résumé

PRECIPITATION WASHOUT OF IODINE: RECENT IN SITU EXPERIMENTS.

Precipitation washout of iodine is not a very well known subject: theoretical approaches differ depending whether or not one accepts the reversibility of gas absorption by raindrops. The only in situ experiments that have been carried out (by Engelmann in 1966) produce results which, depending on the method used, are either much lower or much higher than the theoretical values calculated on the assumption that absorption is irreversible. A series of experiments under varied conditions was therefore undertaken at the site of the Monts d'Arrée nuclear power plant. During each test, 100 g of stable iodine were released over a period of 10 minutes from a source at a height of 40 m above the site. A total of 80 precipitation collectors were placed roughly in two concentric arcs at average distances of 300 and 2000 m downwind from the source. The quantity of iodine in the rainwater samples was determined by ion chromatography together with electrochemical detection (Dionex system) which has a sensitivity better than 1 ppb (1 μg/L). As the total quantity of iodine released, the wind speed, the average magnitude of the deposits collected from a transverse section and the length of this section are known, it is possible to calculate the washout coefficient of the rain studied, this being characterized by various parameters (intensity, pH, temperature and dimensional spectrum of the drops). Ten experiments have been carried out so far: the washout coefficients ranged from 2.1×10^{-5} to 4×10^{-4} s^{-1}; in other words they were of the same order of magnitude as the theoretical values calculated assuming irreversible absorption of the gas by the drops. These initial results, which appear to be independent of rain intensity and distance from the source, were obtained with strong winds and a high level of turbulence. They will be complemented by measurements under different atmospheric conditions.

LAVAGE DE L'IODE PAR LES PRECIPITATIONS: EXPERIENCES RECENTES IN SITU.

Le lavage de l'iode par les précipitations est mal connu: les approches théoriques diffèrent selon que l'on admet ou non la réversibilité de la captation du gaz par les gouttes de pluie. Les seules expériences in situ, effectuées par Engelmann en 1966, conduisent, en fonction de la méthode utilisée, à des résultats tantôt beaucoup plus faibles, tantôt beaucoup plus forts que les valeurs théoriques calculées en supposant la captation irréversible. C'est pourquoi une série d'expérimentations en conditions variées a été entreprise sur le site de la centrale nucléaire des Monts d'Arrée. Lors de chaque essai, 100 grammes d'iode stable sont lâchés en 10 minutes à partir d'une source dominant le site d'une hauteur de 40 mètres. Des collecteurs de précipitations (80 en tout) sont disposés approximativement selon 2 arcs concentriques sous le vent de la source et à des distances moyennes de 300 et 2000 mètres. L'iode est dosé dans les échantillons d'eau de pluie par chromatographie d'ions associée à une détection électrochimique (système Dionex), ce qui autorise une sensibilité meilleure que 1 ppb (1 μg/L). Connaissant la quantité totale d'iode rejetée, la vitesse du vent, la valeur moyenne des dépôts collectés sur une section transversale et la longueur de cette section, on peut calculer le coefficient de lavage de la pluie considérée, cette dernière étant caractérisée par divers paramètres (intensité, pH, température et spectre dimensionnel des gouttes). Dix expériences ont été effectuées à ce jour: les coefficients de lavage sont compris entre 2,1 $\times$ 10^{-5} et 4 $\times$ 10^{-4} s^{-1}, c'est-à-dire de l'ordre de grandeur des valeurs théoriques calculées en supposant irréversible la captation du gaz par les gouttes. Ces premiers résultats, apparemment indépendants de l'intensité de la pluie et de la distance à la source, ont été obtenus en présence de vents forts et d'un niveau de turbulence élevé. Ils seront complétés par des mesures dans des conditions atmosphériques différentes.

1.   INTRODUCTION

Le lavage des polluants par la pluie est habituellement exprimé par un coefficient de lavage Λ, de dimension inverse de celle d'un temps qui est le rapport de la quantité de polluant collectée par mètre cube d'air et par seconde à la concentration dans l'air ($\Lambda = q/C$). On peut admettre dans le cas d'un gaz, soit que celui-ci est totalement soluble dans l'eau, ce qui permet d'appliquer à ce gaz les calculs de lavage irréversible effectués pour les aérosols [1], soit au contraire que les molécules peuvent se désorber de la goutte vers l'atmosphère [2, 3], cette dernière hypothèse conduisant à des valeurs plus faibles que celles qui résultent de la première. Indépendamment de ces considérations théoriques, Engelmann a effectué trois expériences en 1966 [4]: chacune d'elles consistait en un rejet d'iode et de brome stables à partir d'une tour de 12 m de hauteur. Des échantillons d'eau de pluie étaient prélevés dans des collecteurs disposés le long d'arcs de cercle centrés sur la source de manière à englober toute l'extension latérale du panache. Les deux éléments étaient mesurés par activation neutronique et les coefficients de lavage étaient calculés en utilisant la formule:

$$\Lambda = W(x)\,L\,u/Q \tag{1}$$

où W(x) est la quantité moyenne de polluant déposée par unité de surface sur la bande circulaire de longueur L située à la distance x de la source, u est la vitesse du vent et Q la quantité totale émise.

Les résultats expérimentaux, comparés aux valeurs théoriques calculées sur la base d'une entière solubilité de l'iode et du brome, sont présentés sur la figure 1. On observe que, dans le cas du brome, les valeurs expérimentales vérifient l'approche théorique à un facteur 2 près. Par contre, les coefficients de lavage de l'iode sont inférieurs de un à deux ordres de grandeur aux prévisions théoriques et

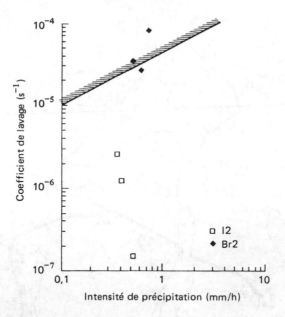

FIG. 1. Coefficients de lavage du brome et de l'iode. Comparaison entre valeurs théoriques (courbes continues), établies sur la base d'une entière solubilité des gaz dans l'eau, et valeurs expérimentales (points) obtenues par Engelmann en 1966.

indépendants de l'intensité de précipitation. Engelmann attribue cette différence de comportement à la différence de solubilité entre les deux corps. Lors de quelques essais complémentaires, il a utilisé comme source les traces d'iode radioactif rejetées par une installation industrielle, obtenant cette fois des coefficients de lavage très supérieurs aux valeurs théoriques.

Aucune autre expérimentation n'ayant à notre connaissance été effectuée, nous avons entrepris de déterminer ce paramètre en conditions météorologiques variées.

2.   METHODES

Les dix premières expériences, qui font l'objet de cette présentation, ont été effectuées en décembre 1988 et mars 1989 sur le site de la centrale des Monts d'Arrée (fig. 2). Le principe expérimental consiste à injecter dans l'atmosphère un panache d'iode élémentaire stable en présence de précipitations bien caractérisées et à échantillonner l'eau de pluie à différentes distances de la source tout au long de sections transversales complètes du panache. Le générateur est disposé au sommet d'un château d'eau dominant le site d'une hauteur de 40 mètres. Le gaz est produit

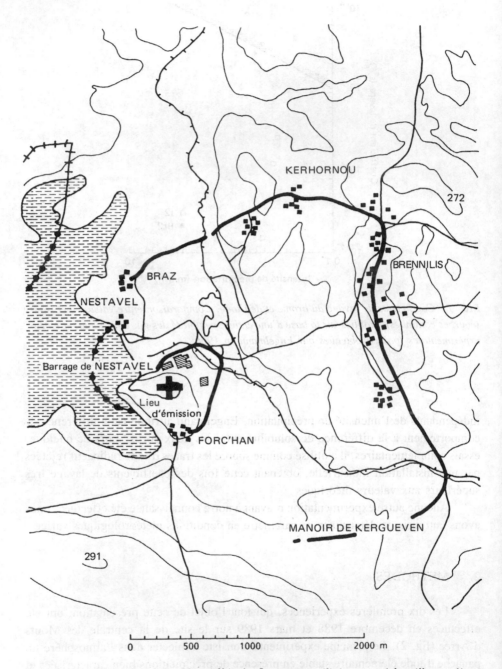

FIG. 2. *Site expérimental des Monts d'Arrée. Les routes utilisées pour échantillonner l'eau de pluie selon des sections transversales du panache d'iode sont tracées en trait gras.*

par sublimation en une dizaine de minutes de 100 g d'iode en paillettes et prédilution du gaz produit dans un flux de 3000 m³/h d'air chauffé à 80°C destiné à éviter une éventuelle condensation. L'échantillonnage s'effectue à l'aide de 80 collecteurs de précipitation de 400 cm² répartis selon deux sections transversales du panache situées respectivement à environ 300 et 2000 m de la source. Ces collecteurs sont mis en place sur chaque transect avant l'arrivée du panache et le prélèvement est effectué après la fin du passage de celui-ci de telle sorte que la totalité du gaz entraîné par la pluie au point considéré soit présente dans l'échantillon. La quantité d'eau prélevée par chaque collecteur est mesurée et la concentration en iode est dosée dans une aliquote de 5 ml par chromatographie d'ions associée à une détection électro-chimique (système Dionex), avec une sensibilité meilleure que 1 μg/L. La quantité d'iode W déposée au sol au point considéré est calculée à partir du volume et de la concentration du prélèvement, déduction faite de la teneur en iode de l'eau de pluie collectée hors panache. La moyenne W(x) des flux de dépôt déterminés sur toute la trace transversale du panache L à la distance x permet alors, connaissant la vitesse du vent u et la quantité d'iode émise Q, de calculer le coefficient de lavage Λ par la formule (1). Chaque pluie faisant l'objet d'une expérience est caractérisée par les paramètres suivants: intensité, pH, température et spectre dimensionnel des gouttes, ce dernier étant déterminé à l'aide d'un appareil mis à notre disposition par la Météorologie nationale, le *disdromètre*, dont le principe consiste à convertir l'énergie inertielle de chaque goutte lors de son impact sur le capteur en une impulsion électrique et à classer ensuite ces impulsions selon leur taille.

3. CONDITIONS EXPERIMENTALES ET RESULTATS

Le tableau I rassemble les informations suivantes: numéro de l'essai et indication de la distance à la source (P pour proche, soit environ 300 m, et L pour lointaine, soit environ 2000 m), vitesse du vent u mesurée au niveau de l'émission, intensité i de la pluie, diamètre médian des gouttes d, concentration moyenne de l'iode c dans l'eau collectée sous le panache et bruit de fond c_0 hors panache, valeur du coefficient de lavage Λ calculé par la relation (1).

Dans le cas de l'essai n° 5P, correspondant à une pluie extrêmement faible (3×10^{-2} mm/h), la détermination de Λ a été obtenue à partir du lavage des collec-teurs par une quantité connue d'eau distillée pour recueillir le prélèvement constitué par les très petites gouttelettes présentes sur les entonnoirs.

Sur l'ensemble des deux campagnes expérimentales, le pH de l'eau de pluie a varié entre 5,4 et 6,5 et sa température entre 4 et 8° C. Dans ces gammes de variabilité relativement faibles, surtout en ce qui concerne la température, l'influence de ces deux paramètres n'a pu être mise en évidence.

TABLEAU I.   COEFFICIENTS DE LAVAGE DE L'IODE MOLECULAIRE (Λ)
ET PARAMETRES ASSOCIES (SITE DES MONTS D'ARREE, 1988–1989)

N° essais	u (m/s)	i (mm/h)	d (mm)	c (μg/L)	c_0 (μg/L)	Λ (10^{-5} s^{-1})
1 P	5	1,1	1,2	5,7	0,6	3,5
L				0,9	0,6	2,1
2 P	12	1,4	1,3	3,7	0,3	6
L				1	0,4	5,3
3 P	18	2,4	1,3	3,5	0,3	21
4 L	12	5,7	1,9	0,5	0,5	19
5 P	5	0,03	0,4	13,3	0	7,2
6 P	14	5,9	1,9	4,7	0,6	42
L				0,7	0,3	8,7
7 P	14	0,4	0,8	22,7	1,1	40
L				3	1,1	24
8 P	10	0,5	0,6	5,3	0,7	11
L				1,4	0,7	6,4
9 P	15	0,5	0,7	1,9	0,6	8,6
L				2,4	0,6	11
10 P	14	1	0,7	2,2	0,6	7,5
L				1,5	0,6	6,6

u: vitesse moyenne du vent mesurée au niveau de l'émission

i: intensité moyenne de précipitation

d: diamètre moyen des gouttes de pluie

c: concentration moyenne de l'iode dans les échantillons d'un transect après déduction de la
concentration moyenne c_0 dans l'eau de pluie prélevée hors panache

Les résultats essentiels sont les suivants:

— les coefficients de lavage sont compris entre $2,1 \times 10^{-5}$ et 4×10^{-4} s^{-1},
c'est-à-dire de l'ordre de grandeur de ceux que l'on peut calculer théoriquement en
faisant l'hypothèse de l'irréversibilité de la captation du gaz par les gouttes;

— on ne constate pas de différence systématique et significative entre les coefficients de lavage mesurés à quelques centaines et quelques milliers de mètres de la

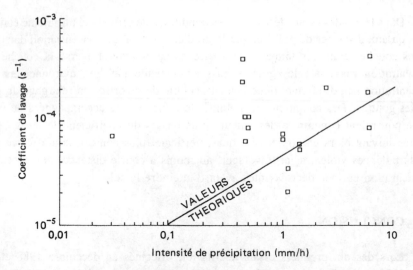

FIG. 3. Coefficients de lavage de l'iode moléculaire. Comparaison entre valeurs théoriques (courbe continue), établies sur la base d'une entière solubilité de l'iode dans l'eau, et valeurs expérimentales obtenues sur le site des Monts d'Arrée en 1988–1989.

source, peut-être en raison du niveau de turbulence très élevé régnant lors de ces dix premiers essais, lequel a probablement eu pour effet d'homogénéiser très rapidement le profil vertical de concentration dans l'air;

— la corrélation attendue entre le coefficient de lavage et l'intensité de précipitation n'apparaît pas (fig. 3); elle est probablement masquée par d'autres phénomènes plus déterminants, notamment l'effet de la trajectoire des gouttes, d'autant plus oblique, donc plus longue, que le vent est plus fort et aussi l'effet du niveau de turbulence, dont l'accroissement tend à augmenter les échanges entre phases liquide et gazeuse.

4.   DISCUSSION

Pour être captées par la pluie, les molécules de gaz doivent diffuser de l'atmosphère à la surface des gouttes, franchir l'interface vapeur–liquide et migrer à l'intérieur de la goutte où elles peuvent soit réagir chimiquement soit demeurer inchangées sous forme de molécules dissoutes. Conformément à la loi de Fick, le transfert diffusif vers la surface de la goutte dépend du gradient de concentration au voisinage de celle-ci. Si la goutte est saturée et ne peut absorber d'avantage de polluant, la diffusion s'annule. Si elle est en équilibre avec une certaine concentration dans l'air et si elle traverse une zone de concentration moindre, le gradient étant inversé, une partie du gaz capté sera réémise à l'atmosphère. Cette réversibilité de la captation de l'iode par les gouttes de pluie peut toutefois avoir une cinétique assez lente pour que tout se passe comme si la captation était irréversible.

Dans le cas des essais décrits ici, les conditions de diffusivité turbulente étaient telles qu'aux distances de prélèvement le gradient vertical de concentration dans les basses couches était très faible, c'est-à-dire qu'après avoir traversé la couche de concentration maximale les gouttes passaient ensuite par un environnement de concentration peu différente, donc peu susceptible de favoriser un relargage du gaz par les gouttes. Des conditions de stabilité de l'atmosphère accompagnées de vent faible pourraient conduire à des valeurs plus basses du coefficient de lavage, les gouttes suivant alors une trajectoire quasi verticale (plus courte), non affectée par des turbulences violentes, et traversant, au moins à courte distance, des zones de concentration en iode décroissantes avant d'atteindre le sol.

5.   CONCLUSION

Lors des dix premiers essais décrits ici, effectués en décembre 1988 et en mars 1989, si les intensités de pluie ont été variées, en revanche les conditions de vent ont été homogènes: vitesse et niveau de turbulence élevés. Les coefficients de lavage de l'iode moléculaire mesurés dans ces conditions sont compris entre $2,1 \times 10^{-5}$ et 4×10^{-4} s^{-1}, c'est-à-dire de l'ordre de grandeur de ceux que l'on peut calculer en faisant l'hypothèse d'irréversibilité de la captation du gaz par les gouttes. La corrélation théorique entre le coefficient de lavage et l'intensité de précipitation n'apparaît pas, probablement masquée par l'effet d'autres paramètres non pris en compte par les modèles.

REMERCIEMENTS

Ce travail a été réalisé en collaboration avec Electricité de France (Direction de l'équipement, Département Sites-environnement-information) dans le cadre d'un contrat de recherche à frais partagés.

REFERENCES

[1]   FUQUAY, J.J., "Scavenging in perspective", Precipitation Scavenging (Proc. Symp. Richland, WA, 1970) (ENGELMANN, R.J., SLINN, W.L., Eds), AEC Symp. Ser., US Dept. of Energy (1970).
[2]   POSTMA, A.K., "Effects of solubility of gases on their scavenging by raindrops", ibid.
[3]   HALES, J.M., Fundamentals of the theory of gases scavenging by rain, Atmos. Environ. 6 (1972) 635.
[4]   ENGELMANN, R.J., et al., "Washout coefficients for selected gases and particulates", Rep. BNWL-SA-657, Proc. 59th Annu. Mtg of Air Pollution Control Assoc., San Francisco, 1966.

ARTIFICIAL RADIONUCLIDE CONCENTRATIONS IN THE BULGARIAN POPULATION OVER THE PERIOD 1986–1988: AN ASSESSMENT OF THE IRRADIATION OF THE BULGARIAN POPULATION AS A RESULT OF THE GLOBAL RADIOACTIVE CONTAMINATION OF THE COUNTRY AFTER 1 MAY 1986

G. VASILEV*, V. BOSEVSKI**, A. BAJRAKOVA*, I. BELOKONSKI**,
Ts. BONCHEV[+], S. BONCHEV*, V. VELIKOV*, B. DONCHEVA*,
L. MINEV[+], B. MANUSHEV[+], Z. PASKALEV*, S. RADEV**,
S. SLAVOV*, S. STOILOVA*, G. FILEV*, M. KHRISTOVA*,
T. TSENOVA*, Z. PECHENIKOVA[++]

* Institute of Nuclear Medicine, Radiobiology and Radiation Hygiene

** Higher Military Medical Institute

[+] Department of Atomic Physics,
Sofia University

[++] Committee on the Use of Atomic Energy for Peaceful Purposes

Sofia, Bulgaria

Abstract

ARTIFICIAL RADIONUCLIDE CONCENTRATIONS IN THE BULGARIAN POPULA-
TION OVER THE PERIOD 1986–1988: AN ASSESSMENT OF THE IRRADIATION OF
THE BULGARIAN POPULATION AS A RESULT OF THE GLOBAL RADIOACTIVE
CONTAMINATION OF THE COUNTRY AFTER 1 MAY 1986.

During the time of radioactive contamination of the Bulgarian territory due to the Chernobyl accident after 1 May 1986 groups of the population were examined for determination
of artificial radionuclide concentrations in the body. From 1 May to 15 June tests were done
on about 6000 people, both male and female, of different ages and from different regions of
the country to measure the concentrations of ^{131}I in the thyroid glands. The highest concentrations were observed in the period 20–30 May. The highest mean values were observed to
be about 300 Bq, although in some people levels of the order of 6000 Bq were detected.
Between 1986 and 1988 about 4000 people were examined for artificial radionuclides with
fixed and portable whole body counters. During May–July 1986 about 15 artificial radionuclides were found. After the summer of 1986 mainly ^{137}Cs and ^{134}Cs were detected. Their
concentrations gradually increased and reached a maximum in April 1987 for ^{137}Cs and in
January 1987 for ^{134}Cs. Then they gradually decreased. It was found that the concentrations
of both radionuclides in the body depend on the age, sex, physiological characteristics and diet
of the individual and on their concentration in food, which was quite different for the different
regions of the country. The highest concentrations were detected in people living in southern
Bulgaria — these reached 300 Bq/kg (total for ^{134}Cs and ^{137}Cs).

1.   INTRODUCTION

Most of the fallout over the Bulgarian territory due to the Chernobyl nuclear power plant accident occurred during the first few days of May 1986. The National Radiation Monitoring Network made a vast number of measurements beginning on 30 April, when a sharp rise in the gamma radiation background, caused by fresh radioactive fallout, was detected. Much of the mass radiation measurement programme carried out in the country during the period 1986–1989 was devoted to direct measurements of artificial radionuclide concentrations in the human body.

Within the national radiation protection system much investigatory work was done on the situation in the country and a series of protective measures were implemented. Over 10 000 measurements of the gamma background were taken (during May, June and July 1986); over 20 000 spectrometric and radiometric analyses were performed (1986–1988) to determine levels of artificial radionuclides in liquid and solid substances (water and food products); hundreds of determinations of atmospheric radioactivity were performed; over 10 000 people were examined to determine ^{131}I concentrations in the thyroid gland and to detect other artificial radionuclides in the body; and a series of other specialized investigations were carried out. All the information collected was used to evaluate the irradiation of the Bulgarian population during the first and subsequent years after the accident.

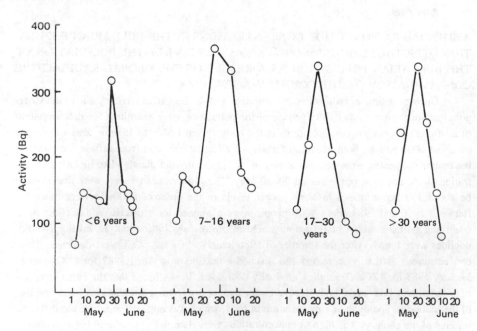

FIG. 1.   Dynamics of ^{131}I in the thyroid glands of a group of people in Sofia.

2.   DETERMINATION OF ^{131}I CONCENTRATIONS IN THE THYROID GLAND

From 1 May to 15 June 1986, direct measurements were taken of ^{131}I concentrations in the thyroid glands of around 6000 people of both sexes and various ages living in different areas of the country. Portable scintillation detectors and spectrometric apparatus were used for this work. The minimum detectable activity level in the thyroid gland was around 50 $\pm$ 20 Bq. Figure 1 shows an example of ^{131}I dynamics for a group of people living in the capital (Sofia). One can see that the highest concentrations were recorded for the period 20–30 May. Mean values were of the order of 300 Bq, but in certain areas of the country, where fallout density was highest, thyroid values of the order of 6000 Bq were detected in some subjects.

3.   DETERMINATION OF ARTIFICIAL RADIONUCLIDES IN THE WHOLE BODY

Between 1986 and 1988 more than 4000 members of the Bulgarian population were examined for artificial radionuclides. Fixed and portable personal radiation counters with scintillation and semiconductor detectors were used to perform these measurements. During May–July 1986 more than 15 artificial radionuclides were detected. During the subsequent months, mainly ^{134}Cs and ^{137}Cs were detected. Figure 2 shows the dynamics of ^{134}Cs concentration in the body in terms of the mean-weighted values for a group of people living in one region of the country (Sofia). Figure 3 shows the same for ^{137}Cs. As can be seen, ^{134}Cs and ^{137}Cs concentrations gradually rose from May 1986 onwards and reached a peak in January 1987 for ^{134}Cs and in April 1987 for ^{137}Cs; then they gradually subsided. The maximum mean-weighted ^{134}Cs value was of the order of 1.7×10^3 Bq; for ^{137}Cs it was 3.2×10^3 Bq. This corresponds to approximately 70 Bq/kg (for both radionuclides taken together). However, in some regions of the country (principally in southern Bulgaria), significantly higher values were registered for ^{134}Cs and ^{137}Cs — as high as 300–400 Bq/kg. For comparison, when atmospheric tests of nuclear weapons were being conducted (1964) the mean concentration of ^{137}Cs in the bodies of Bulgarian citizens (individual measurements) was 20 Bq/kg (Fig. 4). The amounts of ^{134}Cs and ^{137}Cs concentrated in the body depend on age (Figs 2 and 3), sex (the mean value in men is approximately 2.0–2.5 times higher than in women), physiological characteristics, diet and so on. The half-life for elimination of caesium radionuclides fluctuates within fairly wide limits. For a large group of men between the ages of 19 and 30 it is 105 $\pm$ 21 d, and for women it is 82 $\pm$ 17 d. Caesium-137 concentrations in the body are still being studied throughout the country, and these measurements are being used to evaluate the internal irradiation of the Bulgarian population.

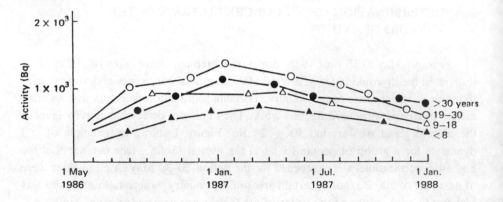

FIG. 2. Dynamics of ^{134}Cs in a group of people in Sofia.

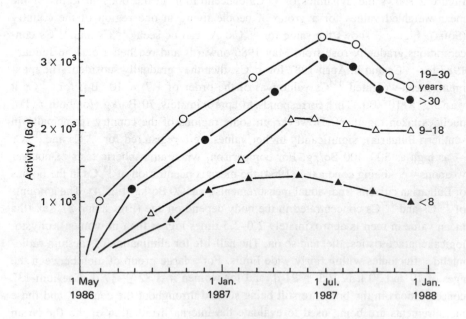

FIG. 3. Dynamics of ^{137}Cs in a group of people in Sofia.

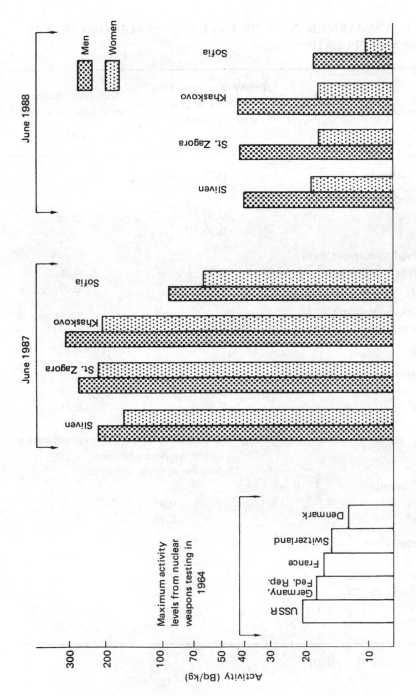

FIG. 4. Dynamics of specific ^{134}Cs and ^{137}Cs activity in the population of southern Bulgaria and Sofia over the period 1987–1988 (research performed by V. Bosevski and S. Radev).

TABLE I. SUMMARIZED DATA ON OVERALL IRRADIATION OF THE BULGARIAN POPULATION

Irradiation source	Effective dose equivalent over the first year after the accident (mSv)	Additional irradiation over subsequent years (mSv)
External gamma irradiation	0.30	—
Internal irradiation from I-131:[a]		
Children	0.30	—
Adults	0.10	—
Internal irradiation from Cs-137:		
Children	0.05	0.05
Adults	0.09	0.09
Internal irradiation from Cs-134:		
Children	0.04	0.04
Adults	0.09	0.09
Total irradiation:		
Children	0.69	0.09
Adults	0.58	0.18

	Background irradiation of the Bulgarian population: effective dose equivalent before the accident (mSv/year)
External irradiation:	
Gamma	0.45
Cosmic	0.28
Internal irradiation:	
K-40	0.16
Other radionuclides	0.17
Rn-222 and Th-220	1.22
Total	2.28

[a] Weighting factor W = 0.03.

4.   EXTERNAL GAMMA IRRADIATION

The gamma radiation background rose sharply on 30 April and 1 May 1986. Thus, for example, at the monitoring point in Sofia the gamma background rose from 63 nGy/h (on 27 April) to 1280 nGy/h (on 1 May). Subsequently, with a few fluctuations, it gradually returned almost to normal by the end of June and beginning of July 1986. Over this period, the mean-weighted individual dose equivalent calculated for a postulated average Bulgarian citizen comprised 0.30 mSv. It varied considerably, from 1.01 mSv at Kyrdzhali to 0.24 mSv at Burgas. (Here we are discussing a situation where there is virtually uniform irradiation of the whole body.)

5. ^{131}I IRRADIATION OF THE THYROID GLAND

The mean individual dose to the thyroid gland for the population group aged six and under was approximately 8–10 mSv, and for the rest of the population it was approximately 3–4 mSv. However, in certain regions of the country, where the contamination density was relatively high, the individual dose to the thyroid rose to as high as 100 mSv and above. This exposure occurred during May and June 1986.

6.   INTERNAL IRRADIATION OF THE BODY FROM ^{134}Cs AND ^{137}Cs

Two evaluations of internal irradiation from caesium radionuclides over the first year after the accident (May 1986–May 1987) were done. The first, based on calculated ingestion with water and food, produced values of 0.63 mSv for ^{137}Cs and 0.47 mSv for ^{134}Cs for people over 16, and 0.24 mSv and 0.24 mSv for children aged 16 and under. The second evaluation was based on an analysis of total caesium dynamics in the body. This yielded values for adults of 0.09 mSv and 0.09 mSv; and for children 0.04 mSv for ^{134}Cs and 0.05 mSv for ^{137}Cs. This means that, over the first year after the accident, children received an internal exposure dose from caesium of 0.09 mSv, and adults 0.18 mSv. In certain regions, however, individual doses were as high as 1.0 mSv. Over subsequent years, internal irradiation from total caesium will probably be found to have doubled.

Analyses have indicated that overall irradiation from other artificial radionuclides, including ^{90}Sr, is negligible.

Table I summarizes data on the overall irradiation of the Bulgarian population. For purposes of comparison, the table also gives typical natural background exposure dose levels for a period of one year prior to the accident.

POSTER PRESENTATIONS

IAEA-SM-306/101P

MESURE DE LA RADIOACTIVITE NATURELLE
A MADAGASCAR

RAOELINA ANDRIAMBOLOLONA, G. RAMONJY RABEDAORO,
E. RANDRIANASOLO, J.F. RATOVONJANAHARY
Laboratoire de physique nucléaire et
    de physique appliquée (LPNPA),
Antananarivo, Madagascar

Nos études portent sur les variations des taux et leurs corrélations avec les paramètres météorologiques (pression atmosphérique, température, hygrométrie, vitesse du vent, cyclone).

La multiplication du nombre des stations, le nombre d'années d'observation, l'importance de la reproductibilité des phénomènes nous ont amenés à informatiser l'étude de la radioactivité naturelle, en traitant les anciennes mesures et les nouvelles données obtenues.

Une partie de ce rapport a été déjà présentée par un de nos collègues du LPNPA à Nairobi (Kenya) lors d'un séminaire en 1988.

L'autre partie, complémentaire, consiste à la continuation des travaux et à l'amélioration des méthodes de surveillance de l'environnement. Dans ce cadre, nous avons pris les données de quatre stations.

Les mesures ont été faites à l'aide d'un détecteur NaI(Tl) pour des énergies 0,661 Mev < E < 3 Mev. A Antananarivo, on dispose en plus d'un détecteur NaI(Tl) pour des énergies E > 4 Mev. Le traitement des données a été programmé en langage PRO/BASIC et on a utilisé en partie le logiciel RS/1. Nous présentons ci-dessous les différents résultats de l'étude des taux correspondants aux spectres entiers.

a) *Influence de la pression:* Nos stations ont observé des coefficients de corrélation positifs entre la pression et l'intensité horaire. Mais entre mai et août 85 et entre fin juillet et début septembre 86, le centre d'Antananarivo a montré une corrélation négative. Ce qui montre la complexité de l'influence de la pression sur la radioactivité naturelle.

b) *Influence de la température:* Nous avons observé des coefficients de corrélation négatifs, mais des coefficients de corrélation positifs ont été obtenus pour la station d'Antananarivo de mai à septembre 85.

c) *Influence de l'hygrométrie:* Des coefficients de corrélation positifs dominent dans les différentes stations.

167

d)   *Influence de la vitesse du vent:* Des coefficients de corrélation négatifs ont été observés. Cependant, à Antananarivo, des résultats contraires sont apparus entre mai et août 85.

e)   *Effets des cyclones:* Mentionnons l'effet du cyclone Honorinina qui est passé à Antananarivo du 15 au 16 mars 86 (accompagné de pluie du 12 au 19 mars 86): deux jours avant le passage du cyclone, le taux de radioactivité a diminué et a atteint son minimum le jour suivant.

f)   *Horaire des maxima et des minima:* Les résultats des différentes stations présentent des régularités. Nous tenons à remarquer que la fluctuation statistique est presque constante dans les différentes stations sauf à Antananarivo.

g)   *Variations journalières:* Pendant la saison sèche, le taux de radioactivité est plus important que pendant la saison des pluies. Mais à Anatananarivo, de mai à septembre (saisons sèches de 85 à 87), une chute du taux a été observée.

h)   *Calcul des intégrales des différents pics du spectre de la radioactivité naturelle:* La connaissance de la distribution spectrale de la radioactivité naturelle par le traitement sur microordinateur facilite l'analyse et permet de garder les résultats. On peut ainsi comparer deux spectres différents.

Ce traitement facilite la surveillance de l'environnement.

IAEA-SM-306/99P

RADIOBIOLOGICAL STUDIES
AND RISK ESTIMATION
IN THE EVENT OF INCORPORATION
OF HOT PARTICLES

I. BELOKONSKI*, Ts. BONCHEV**, V. BOSEVSKI*,
T. KOLUSHEVA*, L. MINEV**, S. RADEV*, V. RADEVA*,
G. RUSSEV*, I. UZUNOV**

* Higher Military Medical Institute

** Department of Atomic Physics,
   Sofia University

Sofia, Bulgaria

1.   BACKGROUND

The small number of humans who inhaled high concentrations of beta emitting hot particles (HPs) and the low concentrations of such HPs during the period of nuclear weapons tests in the atmosphere did not provide sufficient data to estimate the carcinogenic and/or early effects by means of epidemiological studies.

Therefore after the Chernobyl accident, risk assessment for cancer induction could be carried out only by modelling. The single appropriate model available was developed by Mayneord and Clarke [1]. We used this model and to take into account cell proliferation, we used the model of Tsanev and Sendov [2].

In Bulgaria the highest concentrations of HPs were recorded during the period 2–8 May 1986. The number and activity of inhaled HPs were estimated from their deposition on the filters routinely exposed at the meteorological monitoring stations. It was initially estimated that, for instance, a farmer working 8 h/d on the field during the period 2–8 May could inhale about 50 HPs and after 1 month, 8 particles may have been retained.

From the dose distributions around HPs obtained by Sommermeyer in the late 1950s and early 1960s [3] for an HP of 260 Bq (an activity frequently measured in Bulgaria during the first two weeks in May 1986) and 2 μm diameter, the lung cells adjacent to the particle surface could receive a dose rate of 9000 Gy/h; at 10 μm from the particle, a rate of 60 Gy/h and at 50 μm a dose rate of 1.5 Gy/h. Therefore at some distance from the particle, a zone of sublethal damage exists where cancer induction has a high probability.

From the curves calculated by Mayneord and Clarke [1], it can be seen that for a low level homogeneous contamination of the lungs (which was the case for the Bulgarian population), the probability for cancer induction is lower by orders of magnitude as compared to inhomogeneous irradiation by an HP of the same activity. On the contrary, if the homogeneous contamination is high, it can be more carcinogenic than irradiation by an HP, because a major portion of the dose is 'wasted' in the production of necrotic tissue.

In contrast to the inhalation of alpha emitting HPs (e.g. Pu), the encapsulation of the particle by necrotic tissue cannot stop most of the beta radiation and at some distance from the particle surface a zone exists where the doses are highly carcinogenic. Furthermore computer experiments [2, 4] clearly show that cell proliferation stimuli could amplify significantly the radiation induced carcinogenic effect. Therefore the stimulating properties of the necrohormones and the irritating effect of the hitherto unknown substance contained in the radioactive plume which affected the tracheobronchial tract of people in many countries (skin lesions, inflammation of lymph nodes and diarrhoea were observed as well) must be taken into account according to the formula used in Ref. [4].

In view of these considerations and using the theoretical curves of Mayneord and Clarke [1], the carcinogenic risk for persons exposed to high levels of dust during the period 2–8 May 1986 may be as high as 0.1–1% [5].

2.   INITIAL ESTIMATION

In order to confirm or reject the initial estimation of the number and activity of HPs retained in the lungs, we used the autoradiographic technique for detection of HPs in the lung ash of persons who died in accidents approximately one year after May 1986. The ash was dispersed evenly and fixed between two X ray films and screens (to discriminate artefacts).

After 40 days of exposure, 8–11 HPs were detected in three of the five lungs investigated. This confirmation and the observation by Reubel and Muss [6] on enhanced cell proliferation of human lung fibroblasts exposed in vitro to HPs show that the initially estimated carcinogenic risk was not unrealistic.

3. INVESTIGATION OF HOT PARTICLE EXPOSURE

To investigate early radiobiological effects, studies of embryonic neural tissue of *Triturus cristatus* exposed to HPs were carried out. During the ten days after the implantation of HPs, changes in the myoblasts and emptied synapsical buttons were seen; in the neuroblasts, destruction of mitochondria, splitting of contacts between cells, encircling of mitochondria with ring shaped cysterns were apparent and in the nucleus — gigantic polymorph perichromatin granules were observed.

It is obvious that HPs induced ultrastructural changes on the level of the whole cellular organization, which is characteristic for the effects of embryoactive substances.

The results of all these studies justify, in retrospect, the emphasis placed on the countermeasures recommended initially for the population, i.e. precautionary actions against dust exposure.

REFERENCES

[1] MAYNEORD, W., CLARKE, R., Quantitative assessment of carcinogenic risks associated with "hot particles", Nature (London) **259** (1976) 535.

[2] TSANEV, R., SENDOV, B., An epigenetic mechanism for carcinogenesis, Z. Krebsforsch. **76** (1971) 299.

[3] SOMMERMEYER, K., "Die Dosisverteilung in der Umgebung der heißen Partikel", paper presented at Colloq. Bad Schwalbach, 1959.

[4] UZUNOV, I., Das Risiko/Nutzen-Verhältnis bei der Radontherapie, Z. Angew. Bäder- und Klimaheilkunde **4** (1979) 425.

[5] UZUNOV, I., "Measurements of hot particles in Bulgaria and carcinogenic risk assessment", Hot Particles (Proc. Workshop, Theuern, 1988), Vol. 12, Bergbau- und Industriemuseum (1988) 141.

[6] REUBEL, B., MUSS, M., "Effect of a beta-hot particle on biophysical parameters of human lung cells", ibid., p. 111.

IAEA-SM-306/35P

DETERMINATION OF PHYSICOCHEMICAL
FORMS OF RADIONUCLIDES DEPOSITED
AFTER THE CHERNOBYL ACCIDENT

B. SALBU, H.E. BJØRNSTAD, T. KREKLING,
H. LIEN, G. RIISE, G. ØSTBY
Isotope and Electron Microscopy Laboratories,
Agricultural University of Norway,
Ås, Norway

Information on the physicochemical forms of radionuclides deposited after a
nuclear accident is essential for modelling their transport, distribution and biological
uptake. Low molecular weight species are believed to be mobile and available for
active biological uptake. Particles and colloids play, however, an important role for
passive uptake.

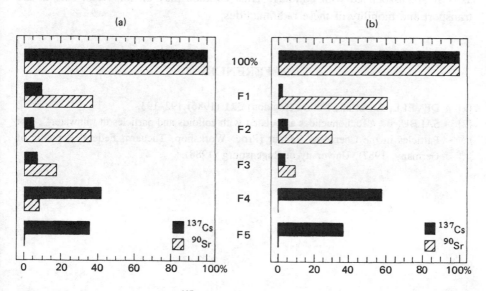

FIG. 1. Sequential extraction of ^{137}Cs and ^{90}Sr in soil (0–2 cm) from central Norway. Relative distribution (%) in different extracting media (F1: 1M NH$_4$Ac; F2: 0.04M NH$_2$OH HCl
in HAc; F3: H$_2$O$_2$ in HNO$_3$; F4: 7M HNO$_3$; F5: residue). The $^{134}Cs/^{137}Cs$ ratio was
0.49–0.55 as of April 1986. Two areas of central Norway were measured: (a) Heimdalen in
1989: 38.4 kBq/kg ^{137}Cs; 0.31 kBq/kg ^{90}Sr, and (b) Vestre Slidre in 1988: 265 kBq/kg ^{137}Cs;
1.2 kBq/kg ^{90}Sr.

Hot particles have been identified in several countries in Europe [1]. The presence of high molecular weight species (colloids) containing radionuclides has also been identified in rain water during the fallout episode [2]. The transformation of these inert species into mobile species depends on the carrying matrix as well as on weathering processes.

In the work described in this presentation, scanning electron microscopy has been used for identifying hot particles. Large (several micrometres) particles deposited in central Norway during the first episode (28–30 April 1986) and in southern Norway during the second episode (5 May and after) have been identified. Leaching experiments have demonstrated that the hot particles and colloids in the rain water are relatively inert.

The major fraction (>90%) of caesium isotopes is still present in the upper soil layer (0–2 cm). On the basis of sequential extraction (Fig. 1), the caesium isotopes are strongly associated with organic and mineral components (fractions F3 and F4) in the soil. To a large extent, isotopic exchange with stable caesium seems to have taken place three years after the deposition.

Strontium-90 is easily leachable from the soil and, compared to the caesium isotopes, a higher fraction of ^{90}Sr should be considered mobile and available for root uptake. Fractionation of runoff waters illustrates that caesium isotopes as well as ^{90}Sr are associated with colloids. Thus colloids play an important role in the transport and mobility of these radionuclides.

REFERENCES

[1]   DEVELL, L., et al., Nature (London) **321** (1986) 192–193.
[2]   SALBU, B., "Radionuclides associated with colloids and particles in rainwater", Hot Particles in the Chernobyl Fallout (Proc. Workshop, Theuern, Federal Republic of Germany, 1987), University of Regensburg (1988).

IAEA-SM-306/54P

RELATION BETWEEN PHYSICOCHEMICAL STATES OF ^{103}Ru, ^{131}I, ^{134}Cs AND ^{137}Cs IN CHERNOBYL FALLOUT AND THEIR BEHAVIOUR DURING AND AFTER DEPOSITION ON PLANTS AND SOILS

V. JANSTA, Š. WIRDZEK
Institute of Radioecology and Applied Nuclear Techniques,
Košice, Czechoslovakia

An attempt to generalize selected results gained in the short and long term after the Chernobyl accident is presented. Interpretation of radionuclide deposition rates, removal from the foliar surfaces of four plant species, and migration into two soil profiles is confronted with the physicochemical state of the radionuclides in the wet and dry fallout from April to May of 1986. The accident fallout was analysed for the contents of undissolved forms (more than 0.4 μm) as well as for the so-called dissolved forms (cationic, anionic, colloidal and neutral) [1].

In the case of rain, the deposition rates increased by more than two orders of magnitude for all radionuclides, proving that rain is decisive for radionuclide deposition from clouds [2, 3]. Analysis of deposition rate values together with knowledge of radionuclide forms in the accident fallout led to an estimation of ^{131}I in gaseous form — a total of about 85%, without been determined. In wet and dry fallout, most of the radioiodine was in dissolved form (8 Gy); the dominant forms were anionic (81%). The same analysis for ^{103}Ru led to an estimate of its partial existence in the atmosphere (about 20%) on particles smaller than 0.1 μm; simultaneously, 21% of its activity was found in anionic form. As much as 77% of ^{103}Ru was in undissolved form and still more — up to 92% for ^{134}Cs and ^{137}Cs — was determined.

The study of desorption of all radionuclides from clover, alfalfa and lettuce foliage under tap water resulted in an unambiguous conclusion that retention of the radionuclide increases with exposure time. The greatest gradient of activity retention was ascertained within the first 5–10 days. The possibility of radionuclide removal from the foliage after 1–3 days was typically 60–70% for ^{103}Ru or ^{137}Cs, and 40–50% for ^{131}I. Generally more significant retention of ^{131}I can be attributed to its higher availability in the water phase. The foliar uptake of ^{103}Ru and ^{137}Cs has mainly the same value in the range of counting errors; however, uptake depends also on the plant species (for grass more ^{137}Cs and for lettuce more ^{103}Ru were desorbed).

The migration of all radionuclides into undisturbed meadow profiles was minimal and results had the character of a systematic sampling error. Even after two years, the penetration of ^{134}Cs was maximal at 3 cm. At the same time the decrease of its gross activity was by more than a factor of 15. The observed facts can be explained as follows: determined undissolved forms of ^{134}Cs and ^{137}Cs in the accident fallout were still the most significant ones on soil surfaces during the first one to two years and were effectively rinsed out by rain water. On the other hand when arable soil is considered, the highest short term migration into its profiles for ^{131}I was found (6 cm), i.e. for the radionuclide with the dominant amount of dissolved forms. The penetration of ^{103}Ru, ^{134}Cs and ^{137}Cs was similar in this case.

Results presented here indicate that the physicochemical state of radionuclides in accident fallout probably plays the important role for the behaviour and fate of radionuclides in the environment. Because of the possibility of having radionuclides of different chemical backgrounds in individual localities, it seems that the determination of forms of radionuclides in selected wet and dry fallout samples, and consequently in the environmental samples, should be the integral part of first analyses carried out after a major nuclear accident.

REFERENCES

[1] JANSTA, V., J. Radioanal. Nucl. Chem. **121** 2 (1988) 295.

[2] WIRDZEK, Š., JANSTA, V., NAVARČIK, I., Rep. IRANT 23-805, Institute of Radioecology and Applied Nuclear Techniques, Košice, Czechoslovakia (1988).

[3] WIRDZEK, Š., JANSTA, V., NAVARČIK, I., Rep. IRANT-12-805, Institute of Radioecology and Applied Nuclear Techniques, Košice, Czechoslovakia (1986).

IAEA-SM-306/9P

APPLICATION OF THE WEERIE CODE
FOR ESTIMATING POPULATION EXPOSURE
RESULTING FROM AN ACCIDENT
AT THE EWA AND MARIA RESEARCH
REACTORS IN POLAND

K. NOWICKI
Institute of Atomic Energy,
Otwock-Świerk, Poland

1. INTRODUCTION

The WEERIE program has been developed to facilitate comprehensive analyses of the consequences of radioactive releases discharged from nuclear installations to the atmosphere. A detailed description of the physical models employed in the program is given in Ref. [1] and only a brief outline of the various stages of the calculation is included in this presentation. The original version of the code used on the IBM 370/165 computer was adapted to the CDC-6400 computer installed in the Institute of Atomic Energy. The program was applied earlier to predict the radiological consequences of various accidents at the EWA and MARIA research reactors for internal safety reports [2, 3].

2. DOSE AND CONTAMINATION ESTIMATES

This presentation summarizes the application of the code for estimating the population exposure resulting from a single fuel element melting in the MARIA and EWA reactors. All doses and contamination levels were estimated for people living in the downwind direction at a distance of from 100 m to 100 km under the following conditions:

(1) Stability category A (most unstable weather conditions): wind speed (u) = 1 m/s, height of mixing layer (L) = 1500 m;
(2) Stability category D (moderate weather conditions): u = 3 m/s, L = 500 m;
(3) Stability category F (most stable weather conditions): u = 2 m/s, L = 200 m.

Two air pathways are considered for 22 isotopes of the noble gases (xenon and krypton) and for many isotopes of iodine, strontium, caesium and ruthenium; the first pathway is from external radiation during passage of the radioactive cloud, and the second from the inhalation of radionuclides in the cloud. Doses are estimated for the total body and the thyroid.

Doses from gamma radiation of ground contamination are calculated for people exposed for 2 h, 24 h and 50 a. The results of the estimation were compared with values of the intervention levels used in Poland.

3.   CONCLUSIONS

The main conclusion is that for the EWA reactor the predicted radiation doses are much below values of the intervention levels of doses for evacuation, for sheltering and for the administration of stable iodine. However, it may be necessary to apply restrictions on the consumption of milk in the vicinity of the Świerk Nuclear Centre.

For the MARIA reactor the results are as follows:

(a)   It is necessary to consider the possibility of implementing protective measures such as: sheltering people during passage of the radioactive cloud up to the downwind distance of 800 m for stability categories A and D and up to 2 km for category F; administration of stable iodine for a downwind distance of up to 1 km for category A and 5 km for category D;

(b)   The emergency reference level for ground contamination of ^{131}I, from the viewpoint of milk consumption, would be exceeded up to 10 km for category A, up to 50 km for category D and up to 100 km for category F;

(c)   Emergency reference levels for ground contamination of ^{137}Cs and ^{90}Sr would not be exceeded.

REFERENCES

[1]   CLARKE, R.A., Physical Aspects of the Effects of Nuclear Reactors in Working and Public Environments, PhD Thesis, Central Electricity Generating Board, London (1973).

[2]   NOWICKI, K., Operational Safety Report of the MARIA Reactor, Internal Rep. No. 0-27/ORiPI/86, Institute of Atomic Energy, Otwock-Świerk, Poland, 1986.

[3]   NOWICKI, K., The Analyses of Nuclear Safety and Radiation Protection for Research Reactor EWA, Internal Rep. No. 0-2/ORiPI/87, Institute of Atomic Energy, Otwock-Świerk, Poland, 1987.

Part I

RADIOACTIVE CONTAMINATION
OF THE ENVIRONMENT

(c) Soil

RADIATION AND MIGRATION CHARACTERISTICS OF THE CHERNOBYL RADIONUCLIDES CAUSING ENVIRONMENTAL CONTAMINATION IN VARIOUS PROVINCES IN THE EUROPEAN SECTOR OF THE USSR

V.A. VETROV, G.A. ANDRIANOVA, A.A. KASIMOVSKIJ,
A.L. POSLOVIN, A.V. TOLOKONNIKOV, S.I. YAMSHCHIKOV
Laboratory for Environmental and Climatic Monitoring,
USSR State Committee for Hydrometeorology and
    USSR Academy of Sciences,
Moscow,
Union of Soviet Socialist Republics

Abstract

RADIATION AND MIGRATION CHARACTERISTICS OF THE CHERNOBYL RADIO-NUCLIDES CAUSING ENVIRONMENTAL CONTAMINATION IN VARIOUS PROVINCES IN THE EUROPEAN SECTOR OF THE USSR.
The radionuclide composition of the Chernobyl fallout has been studied over practically the whole European sector of the Union of Soviet Socialist Republics using soil sampling methods, the upper turf layer of the soil being treated as a natural planchet. Regular observation of the migration of radionuclides in the soil–plant system has been in progress since July 1986 at a stationary network of landscape–geochemical testing sites and sampling grounds; two types of natural ecosystem have been looked at — meadowland and forest. It has been found that the radionuclide composition varies greatly, principally in relation to two spatial factors: distance and direction from the source. The behaviour of individual radionuclides is evaluated using the coefficient of fractionation in relation to ^{95}Zr or ^{144}Ce. The migration characteristics of individual radionuclides were revealed to depend both on soil conditions and on the type of radionuclide composition. The main features of the vertical profile of radionuclide concentrations in soils were clear 2–3 months after the contamination occurred. During subsequent years the migrational profiles changed depending on the individual characteristics of the radionuclides and soil conditions. Information on the radionuclide composition and evolution of the vertical distribution of radionuclides in soils was used to calculate and predict exposure gamma radiation dose from the Chernobyl fallout in various provinces in the European sector of the Soviet Union. When monitoring forest sampling grounds, samples were taken of the soil (litter, humus, mineral layer), the phytomasses of ligneous species (pine needles, leaves, bark, wood), moss and lichens. The main aim of monitoring surveys from 1986 to 1989 was to evaluate the distribution of Chernobyl radionuclides in the compartments of the forest ecosystem. In particular, 60–80% of the total amount of ^{137}Cs was bound with the humus layer of the forest soil. The investigations showed that the Chernobyl fallout opened up great prospects for the solution of many problems regarding biogeochemical migration of artificial radionuclides in the environment in highly varied soil, climate and landscape conditions.

1.   INTRODUCTION

Owing to fallout of radioactive products from the accident at the Chernobyl nuclear power plant in April and May of 1986, the radiation background in a significant portion of the European sector of the Union of Soviet Socialist Republics showed a noticeable rise. In certain regions of the Ukrainian and Byelorussian Soviet Socialist Republics as well as the Russian Soviet Federated Socialist Republic, radioactive contamination of the environment reached such a level that the contents of long lived Chernobyl radionuclides in the soil, water and atmosphere are still to this day the determining factors as to whether people can be allowed to live in those regions [1]. Therefore, immediately after the accident, a large scale programme of radioecological field research had to be implemented to study, evaluate and produce a prognosis of the radiation and migration characteristics of the Chernobyl radionuclides in natural and artificial (mainly agricultural) landscapes and ecosystems.

This radioecological research covered two main groups of problems:

(1)   A detailed study of the radiation characteristics of contamination of the Earth's surface or underlying surface soil, i.e. the spread, composition and density of radionuclides deposited over the region, and a forecast of the change in the radiation characteristics over the course of time, including dose levels from external gamma radiation;

(2)   A study of radioactive contamination of compartments of the natural and agricultural ecosystems, and study and prediction of the migration characteristics of individual radionuclides in the soil–plant system.

Problems relating to the uptake of Chernobyl radionuclides in agricultural plants are reviewed in Ref. [2]. The present report gives a brief account of the approaches used and some of the results of the research into the radiation and migration characteristics of the Chernobyl radionuclides in natural ecosystems in the European sector of the Soviet Union.

2.   RADIOACTIVE CONTAMINATION OF THE ENVIRONMENT AND RADIATION CHARACTERISTICS OF CHERNOBYL RADIONUCLIDES

The two main characterizing features of the radioactive contamination of natural landscapes are the density of the individual Chernobyl radionuclides deposited, σ_i (Ci/km^2)[1], and the radionuclide composition of the fallout. The latter is expressed in terms of a set of ratios of the activity levels of individual Chernobyl

---

[1] $1 \text{ Ci} = 3.7 \times 10^{10} \text{ Bq}$.

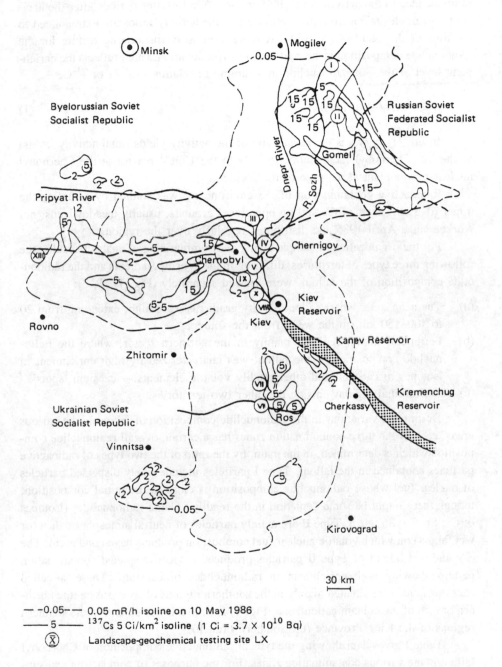

FIG. 1. Map showing the positions of landscape–geochemical testing sites.

radionuclides to the activity level of ^{95}Zr (or ^{144}Ce for later periods after the accident) in samples taken from the environment; the activity ratios are extrapolated to the time of the accident $(A_i/A_{95})_0$. A more general characterizing feature for the radionuclide composition is the coefficient of fractionation which reflects the enrichment level of the i-th radionuclide in a sample in relation to ^{95}Zr or ^{144}Ce:

$$f_{i,95} = (A_i/A_{95})_0/(y_i/y_{95}) \tag{1}$$

In Eq. (1), y_i/y_{95} is the initial ratio of the activity yields (total activity levels) of the i-th Chernobyl radionuclide to ^{95}Zr in the Unit 4 reactor at the Chernobyl nuclear power plant at the time of the accident.

Radioactive contamination of the environment was studied by sampling the upper (0–3 cm) layer of the soil on open level grounds, usually meadowlands, not worked since April 1986 and distributed regularly over the region.

For the organization of sample taking and summarizing of analysis results, the following three types of territories, differing in the density gradients and the radionuclide composition of the fallout, were viewed separately (Fig. 1) [1, 3]:

(a)   The near zone, where fallout density gradients were high, extending from 20 to 100–150 km (to the west) from the source;
(b)   Peripheral fallout regions, mainly in the northern sector, where the radionuclide composition of the fallout was characterized by high enrichment in isotopes of caesium and other readily volatile elements — caesium 'spots';
(c)   Background regions outside the other two territories.

Despite the variations in the radionuclide composition of fallout in the various areas, each of the three contamination zones has a certain overall radionuclide composition which is determined, in the main, by the ratio of the two types of radioactive particles contained in the fallout. Type I particles include finely dispersed particles of nuclear fuel whose radionuclide composition is close to the initial composition, though there might be some depletion in the readily volatile radionuclides (isotopes of Cs, Sr, Te, Sb, etc.). Type II are mainly particles of neutral atmospheric dust (or wet fallout) on which volatile nuclear fuel combustion products have condensed. The dry and wet fallout of Type II particles produced a specific spotted contamination pattern showing high enrichment in radionuclides of caesium. These so-called caesium spots were situated mainly in the southern regions of Byelorussia (the northern branch of the contamination zone, Fig. 1) and, to a lesser extent, in the southern regions of the Kiev Province (the southern branch).

Table I gives data showing the typical radionuclide composition of Chernobyl fallout in the various contamination zones. For the purposes of comparison and subsequent fractionation evaluations, the second column of Table I shows the initial radionuclide composition of the nuclear fuel in the reactor. This was calculated from the preliminary evaluations of the composition of the radioactive discharge during

TABLE I. OVERALL RADIONUCLIDE COMPOSITION OF THE PRODUCTS FROM THE CHERNOBYL NUCLEAR POWER PLANT ACCIDENT IN TYPICAL CONTAMINATION ZONES $(A_i/A_{95})_0$

Radio-nuclide	Initial radio-nuclide composition	Close zone (up to 100 km)			Caesium spots		Background
		North	South	West	Byelorussia	Southern part of Kiev Province	North Caucasus
Np-239	10	2.0	14	5.0	(10)[a]	(10)	(10)
Mo-99	1.0	1.0	2.5	1.5	(2.0)	(1.5)	(1.5)
Te-132	0.9	5.0	1.3	3.6	(230)	(3.0)	(3.0)
I-131	0.7	5.0	3.0	3.0	180	(3.0)	(3.0)
Ba-140	1.0	1.0	2.0	1.4	12	(1.5)	(1.5)
Ce-141	1.0	1.1	1.0	0.9	2.0	1.5	1.8
Ru-103	0.8	0.8	1.2	0.8	35	8.0	6.0
Sr-89	0.4	0.2	1.2	0.7	2.5	1.0	1.0
Y-91	0.3	0.8	0.8	0.9	(1.0)	0.5	0.6
Zr-95	1.0	1.0	1.0	1.0	1.0	1.0	1.0
Ag-110m	0.001	0.003	0.001	0.001	0.15	(0.03)	(0.01)
Ce-144	0.66	0.7	0.6	0.6	1.2	1.0	1.2
Ru-106	0.4	0.3	0.5	0.3	13	3.0	2.3
Cs-134	0.03	0.15	0.05	0.1	9.0	1.5	0.5
Sb-125	0.004	0.006	0.01	0.01	0.9	0.3	0.1
Sr-90	0.04	0.04	0.15	0.1	0.25	0.1	0.1
Cs-137	0.06[b]	0.3	0.1	0.2	18	3.0	1.0
Pu	0.001	0.001	0.001	0.001	—[c]	—	—

[a] The numbers in brackets are evaluations based on indirect data.
[b] Sum of the α emitting isotopes ^{238}Pu, ^{239}Pu, ^{240}Pu.
[c] — indicates no data.

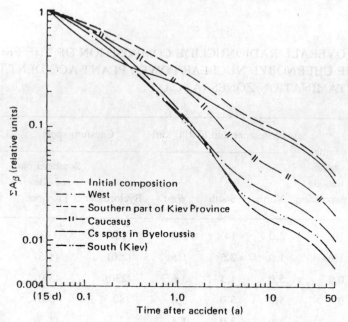

FIG. 2. *Total beta activity (ΣA_β) of the Chernobyl fallout in various radioactively contaminated regions.*

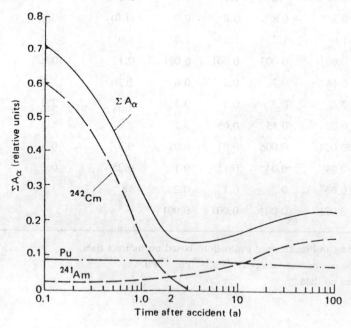

FIG. 3. *Change over time in the total alpha activity (ΣA_α) of the Chernobyl radionuclides.*

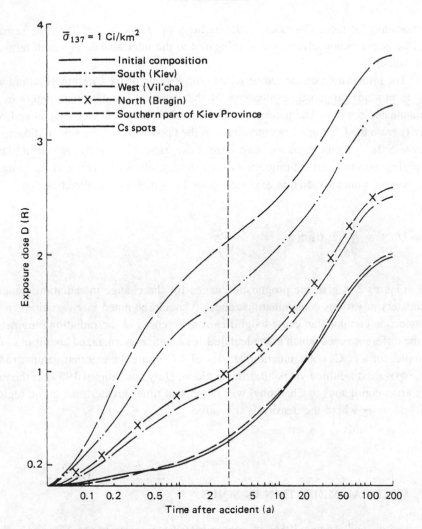

FIG. 4. *Cumulative external gamma radiation exposure dose from the Chernobyl fallout in various radioactively contaminated regions for a ^{137}Cs contamination level of 1 Ci/km^2 at the time of the fallout (1 Ci = 3.7 × 10^{10} Bq).*

the accident [4], taking into account more detailed data on the radionuclide composition of fallout in the various regions of the near zone obtained through our monitoring in 1986 [3].

As may be seen from the data in Table I, the group of refractory Chernobyl radionuclides (for which $f_{i,95} \simeq 1$–2 in practically all the contamination zones) included radionuclides of Np, Mo, Ce and Pu. Radionuclides from the group of readily volatile fission products — Te, I, Ru, Sb, Cs — showed a high degree of fractionation in relation to ^{95}Zr; in the caesium spots at the sites, the coefficients of fractionation for these Chernobyl radionuclides were $f_{i,95} > 10$–100. The remaining Chernobyl radionuclides can be relegated to the intermediate group in terms of their volatility.

The information on the radionuclide composition of the Chernobyl fallout was used to produce prognostic evaluations of the radiation situation parameters in the contaminated regions. The main parameters used included: the total alpha and beta activity from the Chernobyl radionuclides in the upper layer of the soil (radioactivity reserve); the gamma radiation exposure dose rate P_γ (mR/h) on virgin land, taking into account vertical migration down through the soil layer; and the cumulative external gamma radiation exposure dose D_γ which is calculated as:

$$D_\gamma(t) = \int_0^t P_\gamma(t)\, dt \qquad\qquad (2)$$

Figures 2–4 give the prognostic curves for the change in radiation situation parameters in various contamination zones. It should be noted that variations in the radionuclide composition caused significant differences in the radiation parameters for the different zones which are calculated for identical normalized conditions. For example, for a ^{137}Cs contamination density of 1 Ci/km^2, the external gamma radiation exposure dose three years after the accident (July and August 1989) in the near zone areas contiguous to Chernobyl will be several times greater than in the regions of Byelorussia where the density is the same.

3.   VERTICAL MIGRATION IN SOILS

In July and August 1986, in contaminated territories in the Ukraine and in Byelorussia, a network of landscape–geochemical testing sites was set up (see Fig. 1). In each of these testing sites there were several sampling grounds exhibiting typical landscape ecology conditions (forest, meadowlands, river flood plains, etc.). From July 1986 at these testing sites, systematic observation began of the migration of radionuclides of Ce, Ru, Cs, Sb, Ag, Sr and Pu through the soil profile and out of the soil into ligneous and herbaceous plants. The aim of this monitoring is to obtain an overview of the migration characteristics of the Chernobyl radionuclides in the soil–plant system in natural plant ecosystems in relation to environmental factors and to the time after the accident. These data are, in turn, used as real parameters for the models for predicting radioactive contamination of parts of natural ecosystem compartments, and to predict the radiation situation in other contaminated territories (Figs 5, 6). A satisfactory approximation of the specific

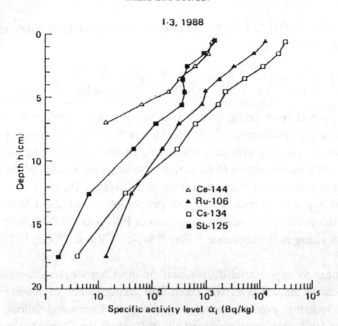

FIG. 5. *Vertical distribution of the Chernobyl radionuclides in the soil of a meadowland sampling ground at the LI landscape-geochemical testing site in 1988.*

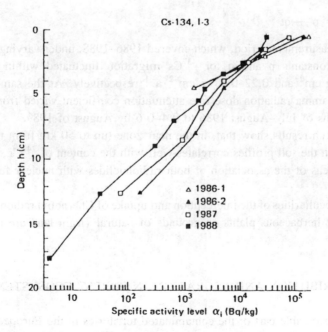

FIG. 6. *Vertical distribution of ^{134}Cs in the soil of a meadowland sampling ground at the LI landscape-geochemical testing site from 1986 to 1988.*

activity profiles α_i (Bq^{-1}) may be obtained by superposing the exponent and the Gaussian distribution:

$$\alpha_i(x,t) = \frac{a_1}{L} \exp\left(-\frac{x}{L}\right) + \frac{a_2}{\beta \sqrt{D_i}} \exp\left(-\frac{(x-vt)^2}{4\,Dt}\right) \qquad (3)$$

where x is the soil layer depth, g/cm^2; L is the migration depth parameter, g/cm^2; D is the diffusion coefficient, $g^2 \cdot cm^{-4} \cdot s^{-1}$; v is the diffusion speed, $g \cdot cm^{-2} \cdot s^{-1}$; β is the standardization coefficient; and a_1, a_2 are the weighting parameters in the sense of the relative proportion of the activity participating in each of the two migration processes ('slow' and 'fast' migration respectively), Bq/cm^2.

Depending on soil conditions, the proportion of all the Chernobyl radionuclides in the soil involved in slow migration in 1987 was 75–99%, with the migration velocity rising in the sequence [125]Sb, [90]Sr < [106]Ru < [144]Ce, [134]Cs, ^{137}Cs, as a rule.

The near to exponential distribution of the Chernobyl radionuclides in the upper 4–5 cm of the soil layer was taken into account when evaluating changes in the gamma radiation exposure dose rate due to the Chernobyl fallout, $P_\gamma(t)$. (By July–August 1988, this layer contained 83–99% of all the Chernobyl radionuclides present in virgin plots of subsoil.) The empirical migration depth L(t) was used in the model for calculating the dose rate attenuation coefficient; L(t) was approximated as a linear function:

$$L(t) = p + qt \qquad (4)$$

Over the measurement period, which covered 1986–1988, under varying soil conditions, the constants p and q for ^{137}Cs migration fluctuated within the ranges 0.14–0.77 g/cm^2 and 0.22–0.47 $g \cdot cm^{-2} \cdot a^{-1}$ respectively. At the same time, the calculated gamma radiation dose rate attenuation coefficient varied from 0.55–0.7 in the months of July–August 1986 to 0.4–0.6 by August of 1989.

Research results show that, in the near zone (up to 50 km from the source), Pu content in the soil profiles correlates well with the content of [144]Ce, which supports our ideas of the association of both radionuclides with nuclear fuel particles (Type I).

The peculiarities of the interception and uptake of Chernobyl radionuclides into meadowland herbaceous plants on grounds of natural virgin land are reviewed in Ref. [2].

4.   DISTRIBUTION AND MIGRATION IN FOREST ECOSYSTEMS

A considerable part of the contaminated territories in the European sector of the Soviet Union is covered with conifer (mainly pine) and mixed forest. Consequently the study of the distribution and migration of the Chernobyl radionuclides

TABLE II. DISTRIBUTION OF CHERNOBYL ^{137}Cs IN COMPARTMENTS OF FOREST ECOSYSTEMS OF CONTAMINATED REGIONS IN BYELORUSSIA FROM 1986 TO 1988 *(percentage of the total level per unit of area)*

Compartment samples	Testing sites (see also Fig. 1)	Percentage of total level		
		July 1986	July 1987	August 1988
	LI	1–2		0.6
Pine needles	LII	10	0.2	0.2
	LV	23	15	6
	LI			1.8
Bark plus wood	LII	Not measured	Not measured	0.4
	LV			0.6
	LI	73	65	62
Surface deposits plus humus layer	LII	70	74	83
	LV	50	62	63
	LI	26	35	35
Mineral layer of soil	LII	25	27	17
	LV	40	23	31

in forest ecosystems formed an important part of the radioecological monitoring programme.

In the process of monitoring forest sampling grounds in the landscape-geochemical testing sites, samples were taken of the soil (surface litter, the mulch–humus layer, mineral layer), and the phytomasses of ligneous species (pine needles, leaves, bark and wood). The main aim of the monitoring from 1986 through 1988 was to evaluate the distribution of all the individual Chernobyl radionuclides present at the sampling grounds in the various compartments and layers of the forest. One highly interesting feature of this process was the study of the contamination dynamics of ligneous plants with a view to predicting contamination levels in wood in order to evaluate its eventual industrial use. Radioactive contamination of lichens and mosses was also studied since they are sensitive indicators of radioactive fallout.

TABLE III. TRANSFER COEFFICIENTS (K_t, 10^{-9} km²/kg) OF THE CHERNOBYL RADIONUCLIDES TO MOSSES AND LICHENS IN CONTAMINATED AND BACKGROUND REGIONS IN 1987, K_t

Region	σ^a_{137} (Ci/km²)	Type of sample		Cs-137 (global)	Cs-134	Ru-106	Ce-144	Sb-125
Byelorussia	>40	Pine needles[b]	C	— [c]	7-14	2-8	2	1
			M	—	14-30	0.2-0.8	1-2	1-2
	<0.2	Fir needles	C	6	60	63	20	50
			M	12	10	5	10	8
Estonia	<0.2	Fir needles	C	17	70	160	50	60
			M	6.5	39	35	70	20
Caucasus	~1	Silver fir needles	C	—	2	1	1	1.5
			M	—	5	2	0.6	0.6
Byelorussia	<0.2	Sphagnum moss		160	820	1300	700	160
	<0.2	Sphagnum moss		320	700	270	310	140
Byelorussia	<0.1	Epiphytic lichen		4200	11 000	3500	2300	830
Transbaikalia	<0.1	Epiphytic lichen		180	1500	2200	—	400
Transbaikalia	<0.1	Terrestrial lichen reindeer moss		240	910	700	—	230

[a] σ — Chernobyl soil contamination density.

[b] For the pine needles, values are given both for 'young' (1987 — 'M') and 'old' (1986 and older — 'C') needles.

[c] — indicates not measured.

Table II gives data showing the dynamics of a pinewood ecosystem typical for the contaminated regions in Byelorussia. The typical forest sampling sites LI, LII and LV (Fig. 1) had the following characteristics:

LI — Pine forest with patches of oak and alder; age of the pine: 70–100 years; contamination density: $\sigma_{137} > 40$ Ci/km^2 (caesium spot);

LII — Artificial pinewood plantation; age: 30–40 years; $\sigma_{137} > 40$ Ci/km^2 (caesium spot);

LV — Artificial pinewood plantation with high crown density; age: 15–20 years; $\sigma_{137} = 1$ Ci/km^2.

Since the ^{137}Cs levels in the wood of the pine trees at these sampling sites were not determined during 1986–1987, the overall amount and the distribution of ^{137}Cs were evaluated on the assumption that no more than 2% of all the ^{137}Cs on the site during that period was in the wood and bark.

Despite the natural divergence of data, due both to error in the evaluation of ecosystem parameters (for instance, the quantity of pine needles per unit area) and to differences in the parameters themselves at different sites, the data in Table II show fairly clearly the distribution picture for ^{137}Cs in the forest ecosystem and the dynamics of that distribution from 1986 to 1988. It should be noted that the high uptake of radioactive fallout by the pine needles in sampling site LV, by comparison with sampling site LI, might be due to the differing capacity of the pine needles to retain Type I (characteristic of the fallout in the 30 km zone in which sampling site LV is situated) as opposed to Type II fuel particles (fallout in caesium spots). Vertical distribution of radioactive caesium and other Chernobyl radionuclides in the soils of forested areas is significantly different from the distribution in the soils in meadowlands. As may be seen from Table II, in 1987–1988, 60–80% of the total amount of ^{137}Cs was bound in the upper organic layer of the soil (humus).

The study of radioactive contamination of pine needles, mosses and lichens, whose role in the redistribution of radionuclides in the forest ecosystems is not in any way significant, is of great interest when it comes to their potential as indicators for monitoring atmospheric contamination, including radioactive fallout. Table III gives some data on the contamination of these items by global nuclear weapons testing fallout ^{137}Cs and certain Chernobyl radionuclides in contaminated and background regions in 1987. The following transfer coefficient was used to characterize the contamination:

$$K_t = \frac{\text{specific activity level of dry plant mass, } \alpha_i \text{ (Ci/kg)}}{\text{soil contamination density, } \sigma_i \text{ (Ci/km}^2)} \tag{5}$$

Careful analysis of the data in Table III provokes the following comments.

Firstly, the capacity of all the different plant indicators to adsorb recent Chernobyl fallout (using ^{134}Cs as a label), and to act as an indicator, turned out to be significantly greater than their capacity to adsorb global nuclear weapons testing fallout ^{137}Cs. Monitoring of the contamination of 'young' pine needles by Chernobyl radioactive caesium (using ^{134}Cs as a label) over a series of years could be one way of keeping track of exchange process dynamics in the soil–tree stand system. Such monitoring would also be a way of determining when a state of equilibrium has been reached between the Chernobyl radioactive caesium and the stable caesium in the soil. The existence of these processes is confirmed by the high values obtained for the transfer coefficient of ^{134}Cs for young pine needles in 1987, $K_t = (5–40) \times 10^{-9}$ km^2/kg; these figures reflect the high extent to which Chernobyl radioactive caesium is involved in the exchange processes of conifer species.

Secondly, the very high adsorption of the Chernobyl radionuclides by mosses and lichens is clear evidence of the exclusive capacity of these plants to concentrate and retain atmospheric fallout; mosses are ten times more efficient, and lichens hundreds of times more efficient, than pine needles in acting as indicators. In contrast to global radioactive fallout, we have specific times for the Chernobyl fallout — the end of April to the beginning of May 1986 — which means that the capacity of mosses and lichens to retain contamination which they have adsorbed from the air can be monitored over a period of time.

Lastly and more generally, the exact dating we have of the Chernobyl fallout opens up great prospects for the solution of many problems regarding biogeochemical migration of artificial radionuclides into the environment under highly varied soil, climate and landscape conditions.

REFERENCES

[1]   IZRAEHL', Yu.A., Chernobyl: The past and predictions for the future, Pravda **79** 25 797 (20 March 1989) (in Russian).

[2]   VETROV, V.A., ANDRIANOVA, G.A., OLEJNIK, R.N., IAEA-SM-306/116, these Proceedings, Vol. 2.

[3]   IZRAEHL', Yu.A., et al., Radioactive contamination of the environment in the Chernobyl nuclear power plant accident zone, Meteorol. Gidrol. No. 2 (1987) 5–18 (in Russian).

[4]   ABAGYAN, A.A., et al., Information on the accident at the Chernobyl nuclear power plant and its consequences prepared for the IAEA, At. Ehnerg. **61** 5 (1986) 301–320 (in Russian).

POST-CHERNOBYL DISTRIBUTION OF ^{137}Cs CONCENTRATION IN SOIL AND ENVIRONMENTAL SAMPLES IN MOUNTAINOUS AND PLAIN AREAS OF THE PROVINCE OF SALZBURG, AUSTRIA

H. LETTNER
Division of Biophysics,
University of Salzburg,
Salzburg, Austria

Abstract

POST-CHERNOBYL DISTRIBUTION OF ^{137}Cs CONCENTRATION IN SOIL AND ENVIRONMENTAL SAMPLES IN MOUNTAINOUS AND PLAIN AREAS OF THE PROVINCE OF SALZBURG, AUSTRIA.

The Province of Salzburg is one of the regions with the highest contamination due to the Chernobyl fallout in western Europe. The average value of the accumulated ^{137}Cs deposition has been determined to be 70 kBq/m^2 in flat land and on valley floors and 105 kBq/m^2 on Alpine pastures. The results of measurements continuing since the reactor accident have shown that the soil to vegetation transfer factors vary by at least one order of magnitude. The region with the highest transfer factors and consequently highest contamination in milk and dairy products is mainly restricted to pasture land of the mountain region along the southern border of the Province of Salzburg north of the highest elevations in the Alps. Taking into consideration geological and geographical conditions appears to lead to a better general estimation of the expected ^{137}Cs levels of locally produced food in contaminated areas.

1. INTRODUCTION

Since the accident at Chernobyl continuous measurements have been made of the content of ^{134}Cs and ^{137}Cs in samples of soil, vegetation and locally produced food in the Province of Salzburg (Fig. 1). The accumulated ^{137}Cs deposition according to the grid model of the World Health Organization (WHO) was estimated to be greater than 20 kBq/m^2 in most regions of the province and 10–20 kBq/m^2 in the southern mountainous areas [2]. According to this more or less uniform distribution, ^{137}Cs contamination values for agricultural products from different parts of the province were expected to be of the same order of magnitude. The Cs levels monitored in agricultural products from the first summer after the reactor accident conformed with these expectations, but ongoing measurements in the following years have shown that for longer periods reliable predictions of contamination levels can only be made if more detailed information about regional environments is available.

193

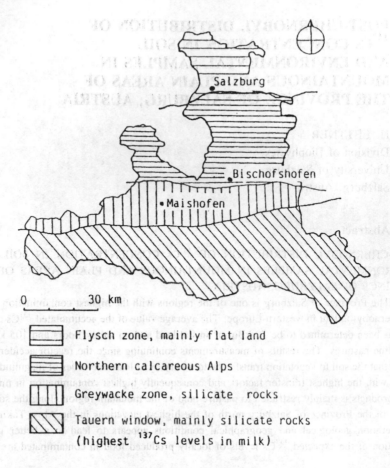

Flysch zone, mainly flat land

Northern calcareous Alps

Greywacke zone, silicate rocks

Tauern window, mainly silicate rocks
(highest ^{137}Cs levels in milk)

FIG. 1. The Province of Salzburg: locations of dairies investigated and geological background (adapted from Ref. [1]).

2.   MATERIALS AND METHODS

The data presented in this paper include ^{137}Cs contamination levels of soil, vegetation (green forage) and milk samples. Levels of ^{134}Cs were found to be 0.5 times those of ^{137}Cs at the time of the fallout. All samples were analysed by HPGe and Ge(Li) detectors with relative efficiencies of 18–35%. The soil samples, including the layers from 0 to 6–8 cm depth, were collected using a soil sampler of cylindrical form with a cross-sectional area of 28 cm^2 for calculation of the accumulated ^{137}Cs deposition. Soil and vegetation samples were dried at 105°C. Soil samples were homogenized and measured in 100 mL beakers. Vegetation samples were compacted and measured in 500 mL beakers. Beakers of 500 mL capacity

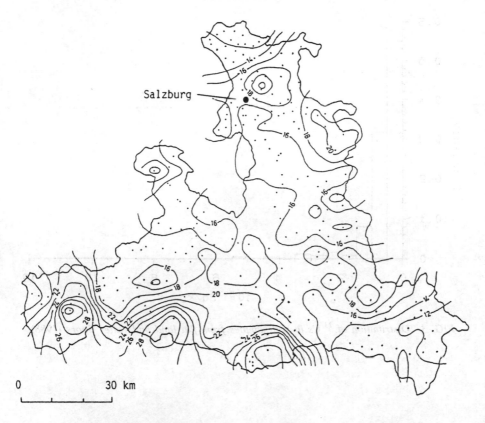

Salzburg

0   30 km

FIG. 2. External gamma exposure rates (μR/h) in 1988 in the Province of Salzburg after the Chernobyl fallout (100 μR/h ≈ 1 μSv/h).

were also used for milk samples. The calibration was carried out with standard nuclide solutions from Amersham International (United Kingdom) and the Physikalisch-Technische Bundesanstalt (Federal Republic of Germany). Most of the milk data are taken from continuous food control measurements by laboratories responsible for the surveillance of nutrition (Lebensmitteluntersuchungsanstalt Linz and Österreichisches Forschungszentrum Seibersdorf).

3.   RESULTS AND DISCUSSION

The accident at Chernobyl happened at the time when dairy farms had started with the feeding of the first green forage cut, which consequently was contaminated by the fallout. A main source of information for the estimation of the accumulated ^{137}Cs deposition is the distribution of external gamma exposure (Fig. 2) [3]. After

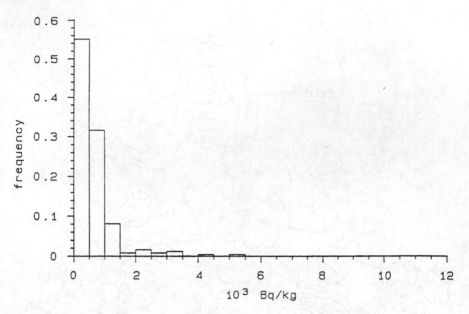

FIG. 3. Distribution of ^{137}Cs in soil samples from flat land and valley floors (1988).

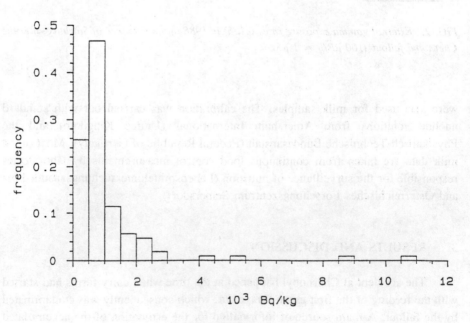

FIG. 4. Distribution of ^{137}Cs in soil samples from Alpine pastures (1988).

reduction of these values by the amount of the natural gamma background [4] an average increase of 0.10–0.12 μSv/h remains even after three years. The analyses of the soil samples from flat land and valley floors (Fig. 3) yield a mean accumulated deposition of 70 kBq/m^2; the corresponding mean value of soil samples from Alpine pastures (Fig. 4) is 105 kBq/m^2. These values are significantly higher than the deposition rates reported on the basis of the WHO grid model, with 10–20 kBq/m^2 in the southern parts of the province and above 20 kBq/m^2 in the northern parts [2]. The accumulated deposition in flat land and valley floors seems to be quite uniform throughout the province, while the higher levels at elevated altitudes can be correlated with higher rainfall at these altitudes and are partly due to the function of mountains as a blocking region for rainfall, which is certainly the case for the Tauern region, with the highest mountains in the eastern Alps.

After deposition of the fallout the supply with contaminated forage was most likely to remain elevated in most cases for a period of three to four weeks, owing to the feeding practices of the dairy farms. During this period the contamination in milk reached maximum values after a few days [5], decreasing slowly afterwards owing to the weathering half-life [6] of the fallout, and then quite drastically with the beginning of the feeding of the second green cut. The absolute values of ^{137}Cs contamination in milk in the period immediately after the fallout deposition from different regions of the province are of the same order of magnitude (Figs 5–7), which seems to be good evidence for similarly high deposition throughout the province. The contamination levels above 150 Bq/kg in milk samples from mountainous regions (Fig. 8) are due to the greater deposition at elevated altitudes in these regions. Though the deposition on Alpine pastures in the whole province is of the same order of magnitude, the contamination in milk samples from the southern region is significantly higher than from the regions to the north of the Tauern (greywacke zone and northern calcareous Alps).

The geological and geographical conditions (Fig. 1) are of great importance: the region in the south is dominated by the silicate rocks of the Tauern window, which have high resistance to weathering, high relief energy and less soil production compared with the greywacke zone in the north. The resulting differences, mainly in pH values (acid in the south) and clay mineral content, are controlling factors for Cs retention in soil [7, 8].

The higher values reported from the central dairy in the city of Salzburg (Fig. 5) are related to one specific region in the flat land northeast of the city.

About four to six weeks after the fallout the feeding period with the second, less contaminated green cut forage began. The contamination level in milk dropped drastically within a few weeks and reached an almost constant level during the summer (Figs 5–7). With the beginning of the hay feeding period in the autumn the Cs contamination of the milk samples quickly increased again owing to the high contamination of the hay used, and reached approximately the same levels as those reported in the period immediately after the fallout. The maximum values in each

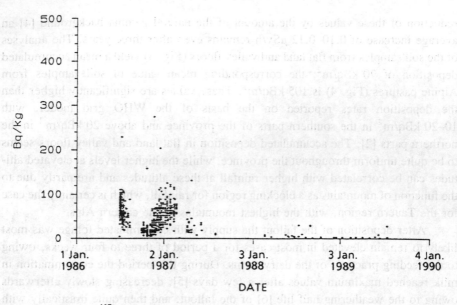

FIG. 5. Levels of ^{137}Cs in milk samples from the central dairy in the city of Salzburg, 1986–1989.

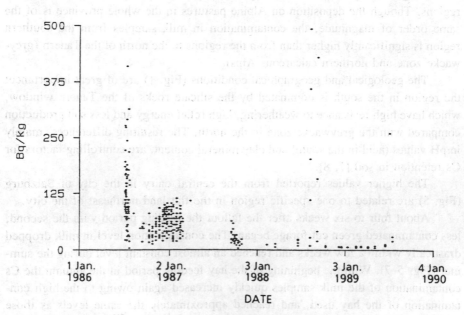

FIG. 6. Levels of ^{137}Cs in milk samples from the dairy in Bischofshofen, 1986–1989.

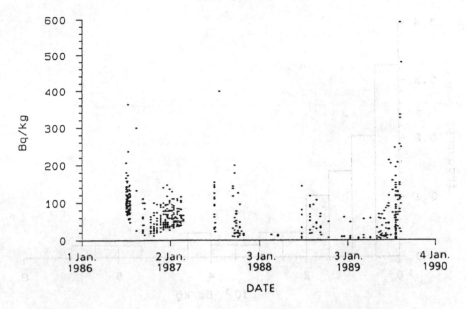

FIG. 7. Levels of ^{137}Cs in milk samples from the dairy in Maishofen, 1986–1989.

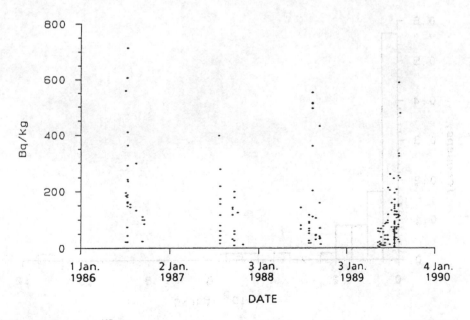

FIG. 8. Levels of ^{137}Cs in milk samples from cattle grazing on Alpine pastures, 1986–1989.

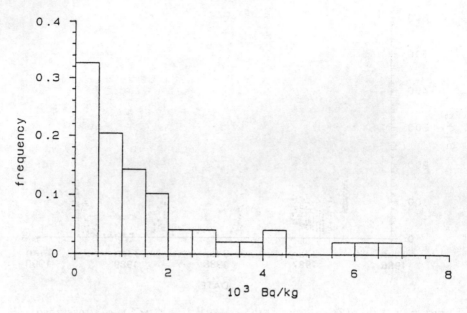

FIG. 9.   Distribution of ^{137}Cs in forage samples from Alpine pastures (1988).

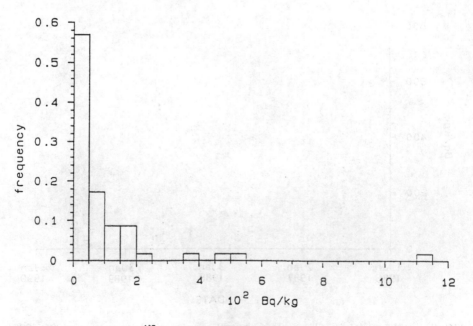

FIG. 10.   Distribution of ^{137}Cs in forage samples from flat land and valley floors (1988).

of the three dairies investigated were reached at the end of 1986, approximately two months after the beginning of the hay feeding period and four months before the period of the first green cut of the next vegetation period.

This time dependent behaviour is best demonstrated by the example of the central dairy in the city of Salzburg, which has the most homogeneous catchment area, mainly the surrounding flat land (Fig. 1). In the following years a divergent development of the contamination in milk samples from the different areas investigated can be seen. Milk samples from the central dairy in Salzburg were only slightly contaminated after the first winter following the fallout (Fig. 5), with a few high levels, which might be partly due to milk production on pastures at elevated altitudes.

The situation in the inner Alpine regions is entirely different, with an extreme maximum in contamination levels appearing in the summer period, coinciding with farming on Alpine pastures (Figs 6-8). The contamination level of ^{137}Cs in milk produced from these pastures has only decreased by 20-60% compared with the levels in the first summer after the fallout (Fig. 8).

The highest contamination levels in milk samples from the mountainous areas can be correlated with the highest levels in soil and vegetation samples (Figs 4 and 9; cf. Fig. 10). After the annual summer farming period the ^{137}Cs level in milk drops quickly owing to the ending of the supply with highly contaminated forage [5]. The percentage of production represented by milk from Alpine pastures during the period from June to September is quite high, for example 30% at the dairy in Maishofen.

TABLE I. TRANSFER FACTORS FOR SOIL TO GREEN FORAGE AND GREEN FORAGE TO MILK IN DIFFERENT ENVIRONMENTS OF THE PROVINCE OF SALZBURG

(data based on analyses of 1988, soil and green forage dry mass, milk fresh mass)

Environment	Mean contamination (Bq/kg)			Daily activity intake[a] (Bq/kg)	Transfer factor	
	Soil	Green forage	Milk		Green forage / Soil	Milk / DAI (d/kg)
Alpine pastures	1 070	1 410	89	21 150	1.3	0.004 2
Flat land and valley floors	620	90	4	1 350	0.14	0.002 9

[a] The daily activity intake (DAI) was estimated assuming an average feeding need of 15 kg dry mass of green forage per day per cow.

The average transfer factor for soil to green forage was calculated on the basis of the dry mass of green forage and soil and was defined as the mean ^{137}Cs activity per kilogram of green forage divided by the mean activity per kilogram of soil. The transfer factor for forage to milk was defined as the mean activity per kilogram of milk per daily activity intake. The daily activity intake per cow was estimated assuming an average feeding need of 15 kg dry mass of green forage per day (unpublished recommendations of the University of Agricultural Sciences, Vienna) for a milk production rate of 15 kg/d. The average transfer factor for soil to forage on Alpine pastures is 1.3 and the corresponding value for flat land and valley floors is 0.14 (Table I). Although the daily activity intake was estimated, which clearly reduces the accuracy of the green forage to milk transfer factor, the higher green forage to milk transfer factors from mountainous areas could be related to the different feeding habits of the cattle on the pastures ('bottom grazers' [9]).

4.   CONCLUSIONS

(a)   While in the three years since the Chernobyl fallout the ^{137}Cs concentration in milk from flat land and valley floors has decreased by at least one order of magnitude, the decrease in milk from Alpine regions in the south of the Province of Salzburg (Tauern window) has been in the range of 20–60%.

(b)   The long term contamination of agricultural products largely depends on the specific environment. In the worst case (e.g. Alpine regions) the concentration of ^{137}Cs in milk can remain almost constant for years after contamination by fallout.

(c)   Taking into consideration geological and geographical conditions offers a possibility for improving the prediction of the contamination levels in agricultural products related to dairy farming.

REFERENCES

[1]   DEL NEGRO, W., Geologie des Landes Salzburg, Schriftenreihe des Landespresse-büros, Salzburg (1983).

[2]   WORLD HEALTH ORGANIZATION, Working Group on Assessment of Radiation Dose Commitment in Europe during the Chernobyl Accident, WHO Regional Office for Europe, Copenhagen (1986).

[3]   STEINHÄUSLER, F., HOFMANN, W., DASCHIL, F., REUBEL, B., Chernobyl and its radiological consequences for the Province of Salzburg, Environ. Int. **14** (1988) 91–111.

[4]   TSCHIRF, E., BAUMANN, W., NIESNER, R., VICHYTIL, P., Strahlenkarte Österreichs, terrestrische und kosmische Strahlung, Ministry of Health and Environmental Protection, Vienna (1975).

[5]   BERTILSSON, J., ANDERSSON, I., JOHANSON, K.J., Feeding green cut forage contaminated by radioactive fallout to dairy cows, Health Phys. **55** (1988) 855–862.

[6]   MARTIN, C.J., HEATON, B., THOMPSON, J., Cs-137, Cs-134 and Ag-110m in lambs grazing pasture in NE Scotland contaminated by Chernobyl fallout, Health Phys. **56** (1989) 459–464.

[7]   SCHULLER, P., HANDL, J., TRUMPER, R.E., Dependence of the Cs-137 soil to plant transfer factor on soil parameters, Health Phys. **55** (1988) 575–577.

[8]   KÜHN, W., HANDL, F., SCHULLER, P., The influence of soil parameters on Cs-137 uptake by plants from long term fallout on forest clearings and grass land, Health Phys. **46** (1984) 1083–1093.

[9]   BURMANN, F.J., Influence of grazing intensity on Cs-137 levels in milk, J. Dairy Sci. **50** (1967) 1891–1896.

MONITORING METHOD BASED ON LAND CLASSIFICATION FOR ASSESSING THE DISTRIBUTION OF ENVIRONMENTAL CONTAMINATION

A.D. HORRILL, D.K. LINDLEY
Merlewood Research Station,
Institute of Terrestrial Ecology,
Grange-over-Sands, Cumbria,
United Kingdom

Abstract

MONITORING METHOD BASED ON LAND CLASSIFICATION FOR ASSESSING THE DISTRIBUTION OF ENVIRONMENTAL CONTAMINATION.

Following the Chernobyl accident, areas of western Great Britain were subject to contamination, primarily by rainfall. A rapid response was needed to assess the degree and distribution of the radionuclide deposition. Sample collection over Great Britain was made possible by the wide geographical spread of the research stations of the Institute of Terrestrial Ecology (ITE). Samples were required to be representative of the area from which they were collected and sample numbers were a limiting factor. A land classification developed by ITE was used as a basis of the sampling scheme. In all, 320 sample sites from 16 land classes were used and a map of deposition produced. The results agree well with estimated deposition patterns based on rainfall data. Using data from the survey combined with those from the land use database, predictive estimates of the fallout in Great Britain have been made. These estimates show that in the best case up to 88% of the variability in the ^{137}Cs values can be accounted for by variables used in the land classification. It is suggested, in view of the information available, that this approach could be extended to indicate vulnerable areas and ecosystems within Europe.

1. INTRODUCTION

Following the Chernobyl accident, areas of Great Britain were subject to contamination, primarily by rainfall [1]. It was important that a rapid response to the situation was mounted both for scientific and regulatory purposes. Representative vegetation samples needed to be obtained quickly before radionuclide concentrations were affected by weathering, plant uptake or removal by grazing animals. The wide geographical spread of the research stations of the Institute of Terrestrial Ecology (ITE) allowed samples to be collected simultaneously over Great Britain by staff familiar with field sampling.

2.   METHOD

With regard to the samples, a number of requirements had to be fulfilled. The samples needed to be similar morphologically, since other work [2] had shown that the trapping efficiency of aerial pollutants differs greatly according to vegetation type. The samples needed to be collected from areas typical of the region being sampled. Sample numbers needed to be reasonably low and were limited by the availability of counting equipment. A number of alternative sampling schemes were considered. These included random samples, the use of a regular sampling grid, stratification by vegetation type, and the collection of individual species. All these alternatives presented difficulties either in that too many samples would be collected or that there was no even distribution over the country of the sample type.

Morphological variability was overcome by concentrating on a single type, graminoid vegetation (grasses, sedges and rushes). This choice had the advantage that the land use of much of Great Britain consists of some form of grassland management resulting in meat or dairy products. An important food pathway back to man was therefore covered. In order to obtain representative sampling of areas from the whole of Great Britain, a land classification was used.

The land classification used has been developed at ITE [3] and is based on data from 1228 squares each measuring 15 km × 15 km. For the central 1 km × 1 km of each square, environmental data have been abstracted from maps. These data are of four types: climatic, topographic, geological and human artefacts (road type, building density, etc.). This data set was then subjected to Indicator Species Analysis [4] (a divisive polythetic method) which separates the data set into 2, 4, 8, 16 or 32 land classes. The technique does not divide the data set on a single attribute but on groups of attributes. As a result of this procedure, the land classes may be described in terms of the several attributes used in their definition.

In order to illustrate the type of result obtained from such a classification, 3 of the 16 land classes are described; these are classes 1, 7 and 15.

Class 1 is typical of central and southeastern England. The topography is of a gently rolling nature with moderate relief. Arable farming dominates the landscape and hedgerows and small woodlands are frequent.

Class 7, in contrast, is often coastal in nature. The landscape is flat with few trees and hedges. Arable farming is common but pasture is also extensive. This type is common in the northwest and midland plain of England.

Class 15 contains the rugged terrain of northwest Scotland. There are many rock exposures, streams and lakes. The land is often peaty and there is little or no arable land. The dominant land use is grazing either by domestic sheep or wild red deer.

Sampling over Great Britain was based upon such an analysis which produced 16 land classes. The rationale for this approach was that the samples obtained would be representative of the whole country. Each of the 16 land classes was subsampled by extracting twenty 1 km squares at random from the main data set; there were thus 320 sampling sites. Samples were collected by clipping the vegetation from a 1 metre square, with the restriction that sampling was not to take place near roadsides (to avoid splash), but was to be carried out away from overhanging vegetation (to avoid throughfall from trees) and not in an area that was obviously heavily grazed. The 1 m square sample was taken as near to the centre of the 1 km area as was practically possible. At a later date, soil samples were collected from the sites which enabled total deposition to be estimated. All samples were oven dried at 80°C, homogenized by grinding and analysed by high resolution gamma spectrometry using hyperpure germanium detectors with relative efficiencies of 20–25%.

3. RESULTS AND DISCUSSION

The data set obtained from the sample sites enabled the levels of ^{137}Cs to be expressed in three ways: as concentration on vegetation, concentration in soil, and deposition per unit area. As would be expected from such a wide ranging survey, the whole data set showed a large spread of values. Variability in ecological data sets is always large, even for a small area [5]. A range of interception factors for grass from 0.05 to 0.3 has been reported [1] and this could add to the variability of the overall picture. By using a computer package [6], the pattern of distribution was determined (Fig. 1) for the concentration, expressed as Bq/kg dry weight, on graminoid vegetation shortly after deposition.

The algorithm used constructs a regular grid over the area to be mapped and each grid point is then assigned a value based on a nearest neighbour analysis using a previously determined number of sample points (in this case 8). A contouring routine then overlays lines of equal activity on the final map. The map obtained shows the pattern of ^{137}Cs deposition over Great Britain with the highest concentrations in North Wales (highest recorded value 1790 Bq/m^2), western Cumbria (6670 Bq/m^2), the southwest and central highlands of Scotland (15 900 Bq/m^2) and the Shetland Isles. This map of surface deposition on grass agrees well with that of rainfall measured from 2 to 4 May 1986 compiled from 4000 rain gauge station measurements [7].

A relatively small number of sample sites thus produce a realistic distribution pattern of the Chernobyl fallout. Nevertheless, the collection and processing of such a set of samples takes a considerable amount of time and manpower, even for an area the size of Great Britain. It was therefore considered worth while to use the present data set to discover if a predictive approach could be developed using the information from which the land use classification was derived. If such a scheme could work, then areas in which intensive sampling might be needed could be easily and quickly defined.

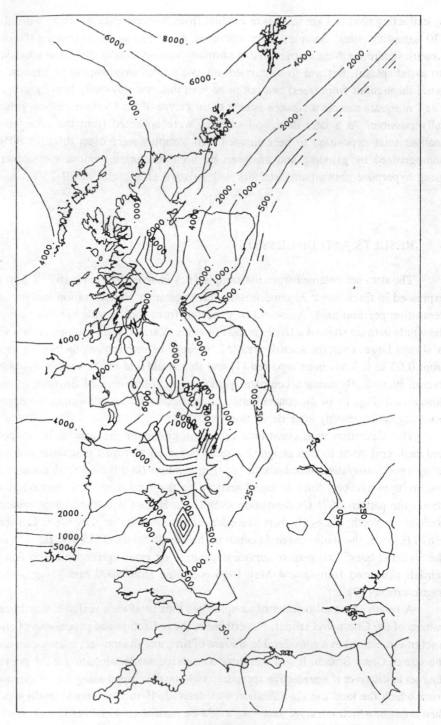

FIG. 1.   *Caesium-137 concentrations (Bq/kg, dry weight) measured in UK grassland vegetation sampled after the Chernobyl reactor accident. All samples collected in May 1986.*

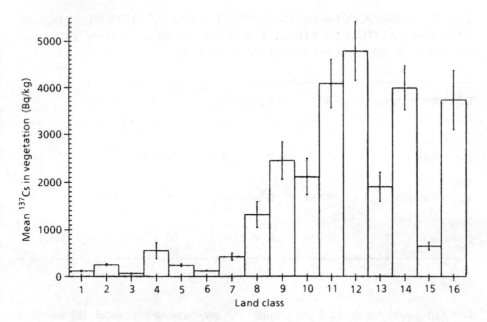

FIG. 2. Mean ^{137}*Cs values in graminoid vegetation (Bq/kg, dry weight) for each of 16 land classes. (Error bar is one standard error.)*

In the course of the multivariate analysis used to construct the land classification, sets of axis scores are produced. The axes are a result of the analysis being based on the ordination method of reciprocal averaging [8, 9]. The axes represent the main trends of variation in the data set, but are not correlated necessarily with any single variable. The approach used was to investigate the correlation between the axis scores for the land use groups and the radiocaesium values. A stepwise multiple regression analysis was then used to formulate a simple model. Values were predicted from the model and the residual values between these and the observed values were examined.

As a preliminary trial the mean values for ^{137}Cs from each land use type were plotted (Fig. 2). Despite the large standard errors, there is a considerable degree of difference between the land classes. Classes characteristic of upland Great Britain (classes 10–16) contain many of the larger values. Each land class described in the main analysis was derived from the scores on five axes. The mean ^{137}Cs value for each land class was then plotted against all five axis scores for all three types of Cs measurement. A visual examination of the plots indicated that the first axis always showed a relationship. In other cases there was little indication of any correlation.

After these preliminary investigations, the data were subjected to a stepwise multiple regression. All three forms of ^{137}Cs measurement were tested against the

TABLE I. SUMMARY OF RESULTS FROM STEPWISE MULTIPLE REGRESSION FOR CAESIUM-137 CONCENTRATION ON VEGETATION, CONCENTRATION IN SOILS, AND TOTAL DEPOSITION

Dependent variable	R^2	Significance	Equation
Vegetation concentration	0.753 84	$P > 0.0001$	$X = 79.65 \text{ Axis1} - 725.84$
Soil concentration	0.880 87	$P > 0.0001$	$X = 5.92 \text{ Axis1} - 48.8$
Total deposition	0.550 36	$P > 0.001$	$X = 70.46 \text{ Axis1} + 843.5$

five axis scores for the land use groups. For inclusion in the model, the variables had to meet a 0.05 significance criterion. Upon investigation it was revealed that in all cases a significant regression was only obtained with the first axis score. The results (Table I) show that the degree of variation accounted for by the regression varies between 55% and 88%. The result for the soil concentration accounts for the most variability. Using the regression equations obtained, predicted mean values for each land use were calculated. The percentage differences in actual and residual values are presented in Table II.

The first axis of the ordination has been attributed to a trend from the high rainfall uplands of the north with acidic soils to the low rainfall lowlands of the south with basic soils. The land classification contained a significant amount of climatic information and it is known that fallout radiocaesium is closely related to rainfall [10]. Both these facts agree with the finding that the first ordination axis has emerged as significant in the regression analyses. The distribution of land classes follows a north–south trend, those with the higher numbers occurring in the north. Examination of the residual percentages in Table II shows that, particularly with reference to Cs concentrations on vegetation, the models are overestimating for the southern classes. Actual mean values in these southern classes are low, often being an order of magnitude less than those of the uplands. Estimates of soil concentrations yield results most in agreement with measurements, this regression accounting for 88% of the variability in the data. The best estimates are obtained for land classes typical of the north and west. The interaction of Cs with soils has been widely investigated and it is possible to classify the degree of Cs mobility in terms of soil properties [11]. The soil represents an integration of many of the attributes in the land classification and this probably accounts for the best fitting regression.

TABLE II. SUMMARY OF PERCENTAGE DIFFERENCES IN ACTUAL AND RESIDUAL VALUES FOR CAESIUM-137 CONCENTRATION ON VEGETATION, CONCENTRATION IN SOILS, AND TOTAL DEPOSITION, USING 16 LAND CLASSES

Dependent variable	Cs-137 Concentration on vegetation	Cs-137 Concentration in soil	Total deposition
Independent variable	Axis score	Axis score	Axis score
R^2	0.753 84	0.880 87	0.550 36

LAND CLASS

1	−114.01	28.86	−84.14
2	61.51	24.17	−67.24
3	−731.29	−82.91	−122.18
4	−29.62	39.76	37.50
5	−179.44	−94.53	−138.54
6	−115.04	−33.03	−127.29
7	−77.70	−2.39	−2.53
8	28.45	1.67	15.29
9	−4.16	−8.80	14.68
10	−26.78	−18.00	−9.49
11	6.65	21.00	−0.05
12	−8.75	−13.92	−44.65
13	25.14	−55.46	2.60
14	52.94	23.45	44.29
15	−254.79	−37.73	32.27
16	29.74	25.16	8.21

The use of the land classification has thus provided a meaningful basis for sampling over Great Britain. Simple models using the axis scores from the classification tend to overpredict values for lowland areas. This is probably due to the high correlation found between fallout Cs and rainfall, there being relatively little variation in rainfall over the lowland areas of Great Britain compared with the highland areas. One drawback of the method is that where large areas of a single land type exist, the sampling density is low. It is thus possible that localized areas of deposition may not be detected. Nevertheless the approach has produced a reliable picture of the deposition, which agrees well with other work [5]. Such an approach can indicate where, in the case of future releases, limited resources should be deployed to be most effective. It is not impossible to consider such an exercise applied over the whole of Europe. Data have been published in map form for soils [12], vegetation [13] and climate [14]. There are also numerous data sets available from individual countries. Combined with the recorded Chernobyl deposition, this information could form the basis for predictive studies showing vulnerable areas and ecosystems in the event of future accidents.

ACKNOWLEDGEMENTS

R. Patel is thanked for performing most of the computational work. Much of this work was funded by the United Kingdom Department of the Environment as part of its Radioactive Waste Management Research Programme. The results will be used in the formulation of, but at this stage do not necessarily represent, Government policy.

REFERENCES

[1]   CLARK, M.J., SMITH, F.B., Wet and dry deposition of Chernobyl releases, Nature (London) 332 (1988) 245–249.

[2]   FOWLER, D., "Removal of sulphur and nitrogen compounds from the atmosphere in rain and dry deposition", Ecological Impact of Acid Precipitation (Proc. Symp. Oslo, 1980) 22–32.

[3]   BUNCE, R.G.H., LAST, F.T., How to Characterize the Habitats of Scotland, Annual Report, Edinburgh Centre for Rural Economics, Edinburgh (1980–1981) 1–14.

[4]   HILL, M.O., BUNCE, R.G.H., SHAW, M.W., Indicator Species Analysis, a divisive polythetic method of classification, and its application to a survey of native pinewoods in Scotland, J. Ecol. 63 2 (1975) 597–613.

[5]   LIVENS, F.R., QUARMBY, C., Sources of variation in environmental radiochemical analysis, Environ. Int. 14 (1988) 217–275.

[6]   NATURAL ENVIRONMENT RESEARCH COUNCIL COMPUTER SERVICES, GRAFIX/GKS, Swindon, United Kingdom (1984).

[7]   SMITH, F.B., CLARK, M.J., The Transport and Deposition of Airborne Debris from the Chernobyl Nuclear Power Plant Accident with Special Emphasis on the Consequences to the United Kingdom, Meteorological Office Paper No. 42, Her Majesty's Stationery Office, London (1989).

[8]   HILL, M.O., Reciprocal averaging: an eigenvector method of ordination, J. Ecol. 61 (1973) 237–249.

[9]   HILL, M.O., Correspondence analysis: a neglected multivariate method, Appl. Stat. 23 (1974) 340–354.

[10] PIERSON, D.H., CROOKS, R.N., FISHER, E.M.R., Radioactive Fallout in Air and
      Rain, Rep. AERE-R3358, Atomic Energy Research Establishment, Harwell, UK
      (1960).
[11] LIVENS, F.R., LOVELAND, P.J., The influence of soil properties on the environ-
      mental mobility of caesium in Cumbria, Soil Use Manage. **4** 3 (1988) 69–75.
[12] COMMISSION OF THE EUROPEAN COMMUNITIES, Soil Map of the European
      Communities, CEC, Luxembourg (1985).
[13] COMMISSION OF THE EUROPEAN COMMUNITIES, Council of Europe Map of
      Natural Vegetation, CEC, Luxembourg (1987).
[14] WORLD METEOROLOGICAL ORGANIZATION, UNITED NATIONS EDUCA-
      TIONAL, SCIENTIFIC AND CULTURAL ORGANIZATION, Climatic Atlas of
      Europe, WMO, Geneva (1970).

PREDICTIONS FOR THE DURATION
OF THE CHERNOBYL RADIOCAESIUM
PROBLEM IN NON-CULTIVATED AREAS
BASED ON A REASSESSMENT
OF THE BEHAVIOUR OF FALLOUT
FROM NUCLEAR WEAPONS TESTS

K. HOVE
Department of Animal Science,
Agricultural University of Norway,
Ås

P. STRAND
National Institute of Radiation Hygiene,
Østerås

Norway

Abstract

PREDICTIONS FOR THE DURATION OF THE CHERNOBYL RADIOCAESIUM
PROBLEM IN NON-CULTIVATED AREAS BASED ON A REASSESSMENT OF THE
BEHAVIOUR OF FALLOUT FROM NUCLEAR WEAPONS TESTS.

The presence of two radiocaesium isotopes in fallout from the Chernobyl accident allowed estimation of pre-Chernobyl radiocaesium contamination of animals and animal products in areas where the fallout from Chernobyl was small. Average concentrations of ^{137}Cs of weapons tests origin in lambs slaughtered in the month of September in the years 1986 to 1988 were 70–100 Bq/kg. Values during 1966–1972 ranged from 40 to 280 Bq/kg. The effective half-life for the 1965–1988 period was about 20 years. The level observed in lamb meat per kBq/m^2 of weapons tests ^{137}Cs in both the 1966–1972 and the 1986–1988 periods was 13–100 Bq/kg, indicating that the bioavailability was high after more than 20 years. A similar ratio between Chernobyl radiocaesium in lamb meat and in soil was observed in 1986–1988 in areas with Chernobyl fallout in the 20–200 kBq/m^2 range. Whey cheese from goat milk in the county of Troms contained 5–800 Bq/kg ^{137}Cs in the period 1963–1966 and 170–300 Bq/kg of pre-Chernobyl radiocaesium in 1988, giving an effective half-life of between 15 and 25 years. In conclusion, the binding of weapons tests caesium deposited in natural ecosystems in Norway has had little significance in making the radiocaesium unavailable. Similar ratios between radiocaesium originating from Chernobyl and that from the weapons tests indicate that radiocaesium from Chernobyl will be of major concern to agriculture for several decades.

1.   INTRODUCTION

Norway has had fallout of radioactive material from nuclear weapons tests and from the Chernobyl accident. In both cases, radioactive caesium was a major pollutant, giving radiation doses to man through external radiation, inhalation and food intake. Important pathways of contamination through food are from products from cultivated areas, but natural and seminatural ecosystems contribute significantly to man's total dose through meat (lamb, venison) and dairy products [1]. In Norway, radiocaesium levels from the nuclear weapons tests were easily detected in man by whole body counting until 1985 [2]. By comparison, levels of radiocaesium in Danes were very low after 1972 and undetectable after 1978 [3]. A major difference between agricultural practices in Norway and those in Denmark is the comparatively larger production of food from natural (uncultivated) lands in Norway.

The soil contamination of radiocaesium was typically 3–5 kBq/m^2 after the nuclear weapons tests. Following the Chernobyl accident, radiocaesium (^{134}Cs and ^{137}Cs) soil contamination levels in areas which received significant fallout were generally from 20 to over 200 kBq/m^2. A large fraction of the sheep production in Norway occurs in these polluted areas. About 30% of the sheep (0.3 million) were above the action limit of 600 Bq/kg in September of 1986 and of 1988, and in preparation for slaughter had to be fed uncontaminated feeds for a period of 2 to 12 weeks. These facts show how important it is to be able to predict the long term consequences of the Chernobyl fallout on our natural or seminatural ecosystems.

In Norway the measurement of radiocaesium from nuclear weapons tests has been extended to the years following the Chernobyl accident (1986–present) in an attempt to obtain information on transfer and availability which may be useful in predicting the longevity of the Chernobyl problem.

2.   METHODS

The Norwegian Defence Research Establishment's surveillance programme for fallout from the nuclear weapons tests included measurements of soil, milk and meat. The programme ended in 1972 with the exception of reindeer meat measurements. A new surveillance programme for radiocaesium in food was started after the Chernobyl accident. A ratio of ^{134}Cs to ^{137}Cs of 0.6 in the fallout from the Chernobyl accident was estimated from measurements of air [4, 5] and 400 samples of lamb meat with radiocaesium concentrations of 2–10 kBq/kg. The presence of only one radiocaesium isotope in the fallout from nuclear tests makes it possible to calculate nuclear tests fallout in food which also contains contamination from the Chernobyl accident. We have observed the decline of nuclear tests fallout over a period of 25 years by combining the results of the monitoring of lamb meat and goat milk from 1986 to 1988 with those of the Norwegian Defence Research Establishment's surveillance programme. The number of samples available for our study was restricted by the extensive use of caesium binders and the feeding of uncontaminated feeds in preparation for slaughter. Samples from 1986 to 1988 were analysed partly by the National Institute of Radiation Hygiene and partly by the local food authority laboratories. Goat milk was analysed by the National Control Laboratory for dairy products.

Local laboratories used NaI (2 and 3 in)[1] scintillation counters (Harshawe type) with lead shielding. The counter was connected with a multichannel analyser (Canberra type 10); the integration time was 0.1–24 hours. The limit of detection of ^{134}Cs in lamb meat was 8 Bq/kg. The standard deviation in measurements of nuclear tests ^{137}Cs can be estimated from counting statistics. Since the decay of ^{134}Cs was considerable during the period of study, the error in estimating nuclear tests ^{137}Cs was dependent not only on activity levels but also on the decay of ^{134}Cs. For example, with an activity level of 100 Bq/kg of ^{137}Cs and 10 Bq/kg of ^{134}Cs in a meat sample, the uncertainty in the estimation of radiocaesium of nuclear origin is 8–12%, 10–15% and 12–18% for the years 1986, 1987 and 1988, respectively. The coefficient of variation was normally < 10% in the range of activities studied and exceeded 25% only in cases with very low levels of activity.

The bioavailability of soil radiocaesium may be estimated by calculating the gross transfer factors, i.e. the ratios between radiocaesium activities of soil and those of milk and meat. Such transfer factors represent average values for the area grazed by the animals. We have used radiocaesium levels in the soil after the Chernobyl accident [6] to calculate average levels for two counties and for Norway as a whole. By comparing transfer factors in a county with low Chernobyl fallout (Rogaland), where nuclear tests fallout could be measured, with one county with high Chernobyl fallout (Oppland), the bioavailability of the radiocaesium from the two different sources could be compared.

3.   RESULTS

The radiocaesium activity of atmospheric air reached a peak in 1963 and declined to very low levels throughout the 1970s [5, 7]. On the basis of air activity as an indicator of the rate of fallout, it was found that about 85% of the fallout occurred in 1950–1965 (Fig. 1). This amount is probably an underestimate because of resuspension. Radiocaesium concentrations in lamb meat during the first series of measurements between 1958 and 1971 reached a maximum of 280 Bq/kg in 1966 and declined rapidly (Fig. 1). The values were particularly low in 1971 when measurements ceased (Fig. 1). During the years 1986–1988, average radiocaesium concentrations from the weapons tests were 70–100 Bq/kg. When data on nuclear

---

[1] 1 in = 2.54 cm.

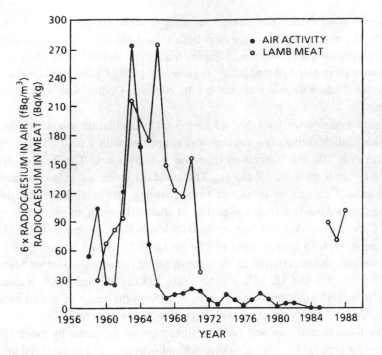

FIG. 1. Radiocaesium activity levels derived from the nuclear weapons tests in lamb meat in 1959–1971 and 1986–1988. Values for radiocaesium in atmospheric air (in fBq/m³) were multiplied by 6 to fit the scale of the meat.

weapons tests radiocaesium from the 1958–1971 and 1986–1988 periods were combined, an effective half-life of 22 years was calculated for the 1963–1988 period and a half-life of 27 years for the 1966–1988 period.

Table I compares the estimated transfer factors for radiocaesium from the nuclear weapons tests with those for fallout from the Chernobyl accident. Levels in lamb meat of 13–100 Bq/kg of weapons tests ^{137}Cs were observed per kBq/m² in both the 1966–1972 and 1986–1988 periods. Transfer factors from Chernobyl radiocaesium were of similar magnitude (24–136 Bq/kg per kBq/m²) in 1986–1988 shortly after the Chernobyl accident.

Radiocaesium is accumulated by fungi and may be present in high concentrations in fruiting bodies of many species. Experience from the 1988 grazing season showed that fungal radiocaesium could have a great impact on radiocaesium activities in food derived from domestic animals using natural ecosystems [8]. In 1988, there was a high density of fruiting bodies of *Rozites caperata* in the mountain birch forests and above the tree line in the whole of Norway. Transfer factors between soil and fungal fruiting bodies for radiocaesium from nuclear weapons tests and from the Chernobyl accident are very similar (Table II).

TABLE I. COMPARISON OF RADIOCAESIUM IN LAMB MEAT, GROUND DEPOSITS AND ESTIMATED TRANSFER FACTORS FOR THE NUCLEAR WEAPONS TESTS AND CHERNOBYL FALLOUT

(The county of Rogaland (southwest Norway) received two levels of fallout from the Chernobyl accident)

Caesium origin	Area	Year	Samples per year	Lamb meat activity[a] (Bq/kg)	Soil activity (kBq/m^2)	Transfer factor[b] (Bq/kg per kBq/m^2)
Nuclear	Norway	1966–1972	30	40–280	3	13–93
weapons	Rogaland	1966–1972	6	104–250	5	21–50
tests	Norway	1986–1988	100	70–100	1	70–100
	Rogaland	1986	44	80	1	80
Chernobyl	Norway	1986	7 000	800	11	72
	Norway	1987	30 000	700	11	63
	Norway	1988	30 000	1 300	11	136
	Oppland	1986	1 000	1 300	42	31
	Oppland	1987	3 000	1 000	42	24
	Oppland	1988	3 000	2 800	42	68

[a] Range of yearly averages.
[b] Calculated from measurements of lambs slaughtered at the end of the summer grazing period.

Remaining radioactivity from the nuclear tests as measured in the county of Troms in northern Norway was of the same magnitude as the fallout from the Chernobyl accident, i.e. 1–2 kBq/m^2. This provided a good basis for the comparison of radiocaesium transfer from the two fallout sources to goat milk (Fig. 2). Radiocaesium levels were low during the indoor feeding season but increased ten- to thirtyfold during the summer, when the goats mostly find their feed in birch forests and other uncultivated areas. The radiocaesium levels were especially high in August 1988, when fungi were present in quantity in the pastures. Whey cheese from goat milk contained 5–800 Bq/kg ^{137}Cs in 1963–1966, while cheese produced in late summer 1988 had 170–300 Bq/kg ^{137}Cs after the nuclear weapons tests. An effective half-life in the range of 15–25 years may thus be calculated for ^{137}Cs in this area as well.

TABLE II. GROSS TRANSFER FACTORS FOR RADIOCAESIUM FROM SOIL TO FUNGI AFTER THE NUCLEAR WEAPONS TESTS AND THE CHERNOBYL ACCIDENT

Caesium origin	Fungal species	Year	Fungal activity (Bq/kg) (fresh weight)	Soil activity (kBq/m^2)	Transfer factor (Bq/kg per kBq/m^2)
Nuclear	Mixed[a]	1987	65	1	65
weapons	Mixed[a]	1988	220	1	220
tests	*Rozites caperata*	1988	260	1	260
Chernobyl	Mixed[a]	1987	500	13	38
	Mixed[a]	1988	900	15	60
	Rozites caperata	1988	15 000– 25 000	100	150–250

[a] Averages of results from 15 samples in 1987 and 61 samples in 1988 of fungi of the genera *Leccinum*, *Boletus*, *Cantharellus*, *Russula*, *Lactarius* and *Hydnum*. The contents of fallout from nuclear weapons tests and Chernobyl were measured in the same samples, except in *Rozites caperata* where nuclear tests radiocaesium was measured in specimens from an area with low levels of Chernobyl fallout, while the transfer factor for Chernobyl radiocaesium was calculated for an area with high fallout.

Ground deposits of ^{137}Cs after the nuclear weapons tests were about 3.5 kBq/m^2 in 1965–1966 and 1.1 kBq/m^2 in 1988 [6]. The gross transfer factors from soil to whey cheese were 140–228 Bq/kg of radiocaesium in cheese per kBq/m^2 of soil in 1963–1966 and 150–270 Bq/kg per kBq/m^2 in 1988.

4. DISCUSSION

The similarity in transfer factors for lamb meat, goat milk, whey cheese and fungi from the 1960s until the end of the 1980s indicates that the bioavailability of radiocaesium in natural terrestrial ecosystems in Norway was maintained at a consistently high level during the period studied. It would therefore appear that immobilization of radiocaesium by natural binding agents such as clay minerals is of minor

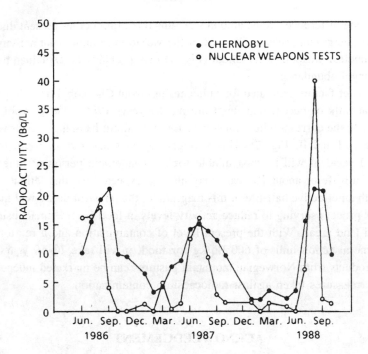

FIG. 2. Radiocaesium from the nuclear weapons tests and from the Chernobyl accident measured in goat milk during the period June 1986 to October 1988. The milk was collected from a dairy in the county of Troms where the fallout levels from the nuclear tests and Chernobyl accident were of the same magnitude.

importance in most natural Norwegian soils. On the basis of the radiocaesium concentrations in lamb meat, we calculated the half-lives to be between 22 and 27 years. Since the air concentrations indicate that about 15% of the fallout appeared after 1966, true half-lives are probably shorter. Thus, it appears reasonable to use a half-life of about 20 years for natural ecosystems producing both lamb meat and goat milk in Norway. This long half-life contrasts with the short half-lives (1–3 years) for the radiocaesium of the nuclear weapons tests reported in traditional agricultural production systems with crop rotation and frequent ploughing [9, 10]. Considerable uncertainty is involved in comparing the behaviour of radiocaesium accumulating over a series of years with an acute deposit following an accident such as in the Chernobyl reactor. Accurate assessment of the effective half-life of radiocaesium from the Chernobyl accident in natural ecosystems cannot be made for many years afterwards.

The level of radiocaesium in green vegetation in the Norwegian mountain pastures showed very little change during the years 1987–1989 [10, 11]. The major cause of variability in radiocaesium levels in goat milk and lamb meat during these years was the differences in the amounts of fungal fruiting bodies present during the last half of the grazing season [8]. Part of the variation in lamb meat activity levels after the nuclear weapons tests fallout (Fig. 1) may probably be explained by variation in fungal abundance. ·

Transfer factors estimated for radiocaesium from Chernobyl from soil to lamb meat, goat milk or cheese and fungi during the years 1986–1988 were of similar magnitude to the corresponding transfer factors for fallout from the nuclear weapons tests (Tables I and II, Fig. 2). This strongly suggests that radiocaesium from the Chernobyl accident will be bioavailable for a considerable period of time and an effective half-life of about 20 years may also be expected for this fallout.

With an effective half-life of this magnitude, the physical decay will in fact be the major process serving to reduce activity levels in food from animals using non-cultivated land areas. With the present level of contamination and a reinforcement of the general action limit of 600 Bq/kg for food, it will take 20–50 years before animal products from Norwegian mountain pastures can be marketed independently of countermeasures taken against radiocaesium contamination.

ACKNOWLEDGEMENT

The work was supported by grants from the Agricultural University of Norway and the Nordic Council of Scandinavian Governments.

REFERENCES

[1]   STRAND, T., STRAND, P., BAARLI, J., Radioactivity in foodstuffs and doses to the Norwegian population from the Chernobyl fallout, Radiat. Prot. Dosim. **20** (1987) 211–230.

[2]   WESTERLUND, E.A., LIND, B., Whole Body Measurements of Personnel in the Institute in 1984, Rep. No. 10, National Institute of Radiation Hygiene, Østerås (1985).

[3]   AARKROG, A., et al., Environmental Radioactivity in Denmark in 1986, Rep. Risø-R-549, Risø Natl Lab. (1988).

[4]   ANTILLA, M., Technical Report No. Tsherno-2/86, Technical Research Centre of Finland, Otaniemi (1986).

[5]   NATIONAL INSTITUTE OF RADIATION HYGIENE, Measurements of Radioactivity in Atmospheric Air, Østerås (unpublished data).

[6]   BACKE, S., BJERKE, H., RUDJORD, A.L., UGLETVEIT, F., Fallout pattern in Norway after the Chernobyl accident estimated from soil samples, Radiat. Prot. Dosim. **18** (1987) 105–107.

[7]   LILLEGRAVEN, A., HVINDEN, T., Measurement of Cesium 137 in Norwegian soil samples 1960–1979, Norwegian Defence Research Establishment, Kjeller (1982) (in Norwegian with English summary).

[8]   HOVE, K., PEDERSEN, Ø., GARMO, T.H., HANSEN, H.S., STAALAND, H.,
      Fungi: A Major Source of Radiocaesium for Grazing Ruminants in Norway, Health
      Phys. (in press).

[9]   AARKROG, A., Environmental Studies on Radioecological Sensitivity and Variability
      with Special Emphasis on the Fallout Nuclides ^{90}Sr and ^{137}Cs, Rep. Risø-R-437, Risø
      Natl Lab. (1979).

[10] HAUGEN, L.E., "Radiocaesium in natural pastures in Norway after the Chernobyl
      accident", paper presented at Scandinavian Organization of Agricultural Research
      Sem. on Deposition and Transfer of Radionuclides in Nordic Terrestrial Environment,
      Beitostølen, Norway, 1989, abstract appears in Nordisk Jordbrugsforskning 1990 1.

[11] GARMO, T.H., PEDERSEN, Ø., HOVE, H., STAALAND, H., "Radio-caesium in
      dairy goats and semi-domestic reindeer from a mountain area in southern Norway",
      ibid.

POSTER PRESENTATIONS

IAEA-SM-306/58P

RADIOLOGICAL PROTECTION CONSIDERATIONS IN THE USE OF ASH FROM CHERNOBYL CONTAMINATED PEAT FUEL

P. JÄRVINEN
Imatran Voima Oy,
Helsinki, Finland

In late April and early May 1986, southern and central parts of Finland received a considerable amount of radioactive fallout originating from the Chernobyl NPP. The deposition of ^{137}Cs ranged from 2 kBq/m^2 to 60 kBq/m^2. In Finland peat is used as fuel in a number of power plants, mainly for co-production of electricity and heat for district heating. The summer of that year was very favourable for peat production, so much so that in some power plants peat produced in 1986 was used up to early 1989.

Some peat production areas received quite a large amount of the Chernobyl fallout, resulting in the need for radiological protection considerations in the handling and utilization of peat ash. The aim has been to utilize ash to the largest possible degree as an element in concrete, in landfills, etc. After Chernobyl it was no longer possible to use the ash for concrete, as the ^{137}Cs concentration rose to as high as 60 kBq/kg.

The Rauhalahti power plant uses peat produced in the heaviest fallout area. Annual use is about 400 000 t, the equivalent production area being about 2000 hectares. The annual ash production is about 20 000 t. The Chernobyl related ^{137}Cs activity transported to the power plant is estimated to be of the order of 1 TBq.

Peat ash has been used as landfill material for landscape enhancement. Ash has been dumped on a separate site close to the power plant. The relatively high activity of the ash, however, made it necessary to take into account radiological considerations for the handling and dumping of the ash. Unnecessary restrictions on later use of the fill sites were to be avoided by careful planning for the finishing of the dump sites. Moreover, the ashfill sites are located within a few hundred metres of several houses, so precautions were deemed necessary.

Dose rates from the unfinished dump site as well as the dose from handling and transporting the ash were estimated. The ash is collected in a silo capable of storing one day's production. From the silo, ash is transported by truck to the nearby dumping site. In all, about 1000 truckloads were dumped annually.

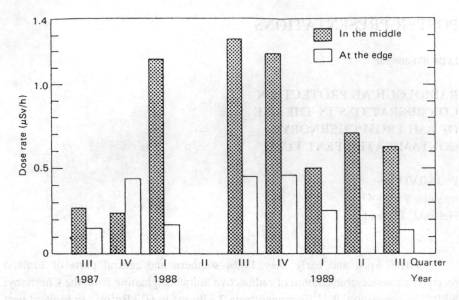

FIG. 1. External dose rates (μSv/h) at the edge and in the middle of an ashfill site at the Rauhalahti power plant.

The radiological dose estimated for the person handling and transporting the ash was approximately 2.5 mSv/a. The dose rate in the landfill area has been monitored with TL dosimeters. Surface water and groundwater samples have also been measured and the results show that dose estimates were conservative and that the dose rate at a finished dump site does not exceed natural background.

IAEA-SM-306/64P

THE PATTERN OF GROUND CONTAMINATION IN THE GERMAN DEMOCRATIC REPUBLIC AFTER THE CHERNOBYL ACCIDENT

E. ETTENHUBER, P. MARSCHNER, H.-U. SIEBERT
National Board for Atomic Safety
   and Radiation Protection,
Berlin

In the German Democratic Republic, the inflow of unevenly contaminated air masses and different weather conditions, in particular locally limited precipitation, gave rise to different degrees of surface contamination as a consequence of the Chernobyl accident. For the evaluation of the resulting radiation dose to man, especially of the external dose caused by surface contamination, it was necessary to have as thorough a knowledge as possible of the deposition of ^{134}Cs and ^{137}Cs.

The ^{134}Cs and ^{137}Cs deposition was calculated on the basis of the gamma spectrometric determination of ^{134}Cs concentration in soil samples. The sampling was performed in the following way:

(a)   The average distance between the sampling sites was 10 km.
(b)   The samples were taken from the upper layer (0–5 cm) of grassland soils.
(c) ^{134}Cs was measured without preparing the sample.

The ^{137}Cs deposition due to the Chernobyl accident was calculated from the values obtained, using the determined ^{134}Cs/^{137}Cs ratio of 1:2.

The ^{134}Cs activities per square metre were calculated from the ^{134}Cs concentrations, which, however, were not necessarily representative of the whole area (10 km × 10 km). As the results showed, on about 80% of the territory, the ^{134}Cs activity was smaller than 2000 Bq/m^2 and the ^{137}Cs activity smaller than 4000 Bq/m^2.

The effective dose equivalent from gamma ground exposure in the administrative districts was calculated on the basis of the contamination pattern as follows:

$$H_{eff} = A_{0,\,dist.} \times SF \times DF$$

where $A_{0,\,dist.}$ is the average surface contamination in the district (Bq/m^2), SF the shielding factor of 0.3 and DF the dose conversion factor devised by Jacob [1].

The results showed that for about 90% of the population the radiation dose to man was smaller than 20 μSv during the period May–December 1986. Only 0.4%

of the population lives in areas where, during the indicated period, effective dose equivalents from 50 to 60 μSv were possible.

The effective dose equivalent per caput due to the external radiation exposure as a result of the reactor accident was 12 μSv in 1986.

REFERENCE

[1] JACOB, P., PARETZKE, H.G., ROSENBAUM, H., ZANKL, M., Effective dose equivalents for photon exposures from plane sources on the ground, Radiat. Prot. Dosim. **14** (1986) 299–310.

Part I

RADIOACTIVE CONTAMINATION
OF THE ENVIRONMENT

(d) Water

RADIOACTIVE CONTAMINATION OF WATER SYSTEMS IN THE AREA AFFECTED BY RELEASES FROM THE CHERNOBYL NUCLEAR POWER PLANT ACCIDENT

S.M. VAKULOVSKIJ, O.V. VOJTSEKHOVICH,
I.Yu. KATRICH, V.I. MEDINETS,
A.I. NIKITIN, V.B. CHUMICHEV
'Typhoon' Scientific and Industrial Association,
USSR State Committee for Hydrometeorology,
Obninsk,
Union of Soviet Socialist Republics

Presented by V.A. Vetrov

Abstract

RADIOACTIVE CONTAMINATION OF WATER SYSTEMS IN THE AREA AFFECTED BY RELEASES FROM THE CHERNOBYL NUCLEAR POWER PLANT ACCIDENT.

The radioactive substances discharged into the atmosphere by the Chernobyl nuclear power plant accident caused contamination of water systems as a result of both aerosol fallout onto the water surface and washing down of contamination by rain and melted snow. During the first month after the accident, the main isotopes causing radioactive contamination of water systems were ^{131}I, ^{140}Ba, ^{95}Zr, ^{95}Nb, ^{103}Ru and ^{141}Ce. Later, the most important ones were ^{90}Sr and ^{137}Cs. The paper gives information on the distribution of the isotopes in the water–suspension–bottom deposit system in the cascade of the Dnepr reservoirs and in the Sea of Azov and the Black Sea; it also gives information on the integrated flux values for ^{90}Sr and ^{137}Cs from the Rivers Pripyat and Dnepr into the Kiev Reservoir and from the Dnepr into the Black Sea.

1. RADIOACTIVE CONTAMINATION OF RIVER WATER

The radioactive products released into the atmosphere by the Chernobyl nuclear power plant accident settled both directly on the water surface of rivers, reservoirs and seas and on the surface of catchment areas, which then became a secondary source of river contamination. Since the rate of entry of radioactive substances into rivers depends on the geographic location of the latter with respect to the source and on the rate of release from the source, the change in the concentration of radioactivity in rivers was not monotonic and reached maximum values on different days.

TABLE I. MAXIMUM RADIONUCLIDE CONCENTRATIONS (10^{-9} Ci/L)[a] IN RIVERS DURING MONITORING PERIOD

River	Date of monitoring (1986)	I-131	Ba-140	Zr-95	Ru-103	Ce-141
Pripyat	2 May	120	60	42	22	
Teterev	3 May	54	34	39	19	18
Irpen'	6 May	50	30	22		
Desna	26 Apr.	18	14	16	3.3	9.1
Dnepr (Obolon')	3 May	35	19	19		9

[a] 1 Ci = 3.7×10^{10} Bq.

Table I shows the maximum radionuclide concentrations in various rivers for the monitoring period starting from 1 May 1986. The pattern of change in concentration shows that there was a sharp drop when the period of aerosol contamination came to an end (about 6 May). Table II contains data showing changes in radionuclide concentrations in the River Pripyat at Chernobyl.

The total beta activity in the Pripyat fell from 10^{-7} Ci/L in the first days after the accident to $(4-6) \times 10^{-9}$ Ci/L by the end of May.[1] During the period when there was no rainfall (from 6 to 14 June), total beta activity fluctuated within the range $(1-2) \times 10^{-9}$ Ci/L. On 15 June, a short period of rainfall (5.2 mm) recorded by the meteorological station in Chernobyl between 17:50 and 18:25 resulted in a threefold increase in the level of total beta activity for a period of five hours. A second increase in the concentration caused by the arrival of contaminated water from higher lying sections of the catchment area was recorded from 17 to 19 June (up to 5.5×10^{-9} Ci/L). Prolonged rainfall (35–75 mm) over virtually the whole of the River Pripyat catchment area from 20 to 24 June resulted in regular increases in the total beta activity of the water in the Pripyat. On 24 June, the concentration fluctuated by more than an order of magnitude, reaching $(6-7) \times 10^{-8}$ Ci/L. By 28 June, it had decreased to 1.6×10^{-9} Ci/L, by 29–30 June to 1×10^{-9} Ci/L, and in September to $(1-3) \times 10^{-10}$ Ci/L.

---

[1] 1 Ci = 3.7×10^{10} Bq.

By the end of 1986, the concentration of ^{95}Zr, ^{95}Nb, ^{141}Ce, ^{144}Ce, ^{103}Ru and ^{106}Ru in the water of the Pripyat and other rivers was not higher than 1×10^{-11} Ci/L, or three orders of magnitude below the PC$_S$ (the maximum permissible concentration under the Soviet radiation safety standards [1]). The maximum concentration of ^{239}Pu recorded in the Pripyat in the first days after the accident (1×10^{-11} Ci/L) had dropped to 2×10^{-13} Ci/L, or four orders of magnitude lower than the PC$_S$ of 2.2×10^{-9} Ci/L, by August 1986. Consequently, attention subsequently focused mainly on obtaining information on long lived isotopes such as ^{90}Sr and ^{137}Cs.

Determination of the ^{90}Sr content of various waters began on 1 May. The maximum levels were observed in the first days of May in the Pripyat and these reached the PC$_S$ (4×10^{-10} Ci/L). From the end of May and throughout June, the ^{90}Sr content of the river remained at a level of $(3-5) \times 10^{-11}$ Ci/L. Data obtained showed that the ^{90}Sr/^{89}Sr ratio was stable and was in the range 7–14 in the period 14–20 May.

During the flood period in 1987, practically no increase was observed in the ^{137}Cs concentration in the River Dnepr, while in the Pripyat the concentration rose briefly in March–April by a factor of 2–3 by comparison with the low water period. The maximum concentrations of ^{137}Cs in the Pripyat during this period (1.8×10^{-10} Ci/L) were two orders lower than the PC$_S$. The maximum concentration of ^{90}Sr in the Pripyat in March 1987 (1×10^{-10} Ci/L) was four times lower than the PC$_S$. The maximum concentration of ^{137}Cs at the time of the 1987 flood in the river coincided with the maximum snow melt in small rivers within the 30 km zone and was caused both by the washing down of this radionuclide from contaminated catchment areas and, possibly, by the melting of contaminated snow. The total amounts of ^{137}Cs and ^{90}Sr in the snow in the 30 km zone were 450 Ci and 480 Ci,

TABLE II. CHANGES IN RADIONUCLIDE CONCENTRATIONS (10^{-9} Ci/L)^a IN THE RIVER PRIPYAT AT CHERNOBYL

Date (1986)	Ce-144	Ce-141	I-131	Ru-103	Cs-137	Zr-95	Ba-140
1 May	10	11	57	15	6.7	11	38
6 May	—	2.4	22	4.6	4.3	4.5	4.8
3 Jun.	—	—	0.9	0.7	0.6	0.3	—
16 Jul.	1.0	0.4	—	0.4	0.2	1.0	—

^{a} 1 Ci $= 3.7 \times 10^{10}$ Bq.

TABLE III. AVERAGE ANNUAL CONCENTRATIONS OF ^{90}Sr AND ^{137}Cs AND AMOUNTS REMOVED BY THE PRIPYAT AND DNEPR

	River Pripyat, Chernobyl				River Dnepr, Teremtsy			
	Cs-137		Sr-90		Cs-137		Sr-90	
	Concn $(10^{-12}$ Ci/L)[a]	Amount removed (Ci)[a]	Concn $(10^{-12}$ Ci/L)[a]	Amount removed (Ci)[a]	Concn $(10^{-12}$ Ci/L)[a]	Amount removed (Ci)[a]	Concn $(10^{-12}$ Ci/L)[a]	Amount removed (Ci)[a]
1986 (Jun.–Dec.)	600	590	50	100	53	415	13	100
1987 (Jan.–Dec.)	49	410	40	290	19	340	12	250
1988 (Jan.–Dec.)	18	250	40	500	15	260	7	140
1989 (Jan.–May)	11	70	22	150	11	130	5	70
Total (Jun. 1986 – May 1989)		1320		1040		1145		560

[a] 1 Ci = 3.7 × 10^{10} Bq.

respectively. In contrast to 1987, the 1988 flood showed virtually no increase over the low water period in the concentrations of ^{137}Cs and ^{90}Sr in the water of the Pripyat in Chernobyl during the period of maximum water content of small rivers.

In July–August 1988 in the Pripyat there was a historic summer flood as regards water content. This was caused by rain runoff in the Byelorussian alluvial plain (poles'e), which increased the concentration of radionuclides in the river water as a result of surface wash-off both from contaminated catchment areas situated outside the 30 km zone and from partly submerged flood plains located in the area closer to the Chernobyl nuclear power plant. The spring flood of 1989 was the earliest (January–March) since the Chernobyl accident, and there was virtually no surface runoff because of the snowless winter. Table III gives data on the average annual concentrations of ^{90}Sr and ^{137}Cs in the Pripyat and Dnepr from 1986 to 1989 as well as data on the removal of the two isotopes into the Kiev Reservoir by these rivers. The total amounts of ^{90}Sr and ^{137}Cs removed were 1600 Ci and 2465 Ci, respectively. The ^{90}Sr in the river water occurred mostly in solution, the fraction in suspension being less than 5%. The proportion of ^{137}Cs in suspension was significantly higher than for ^{90}Sr, varying between 15% and 40%.

The ratio of the concentration of ^{137}Cs to that of ^{90}Sr fluctuated widely. Before the accident the ratio was less than 0.1. After the accident it increased by two orders of magnitude. In July 1986, the ratio reached 40 in the Pripyat as a result of the washing out of ^{137}Cs with soil particles during rain showers. By 1988–1989, the ^{137}Cs/^{90}Sr ratio had dropped to 0.5 because of two factors: ^{137}Cs adhered more

TABLE IV. COEFFICIENTS OF ^{137}Cs WASHOUT FROM CATCHMENT AREAS OF UKRAINIAN AND BYELORUSSIAN RIVERS IN 1987

River	Catchment area (km^2)	Runoff depth (mm)	Cs-137 in catchment area (Ci)[a]	Washout coefficient	
				For all of 1987	Normalized to 1 mm runoff depth
Pripyat	140 000	53	193 000	2.1×10^{-3}	4.0×10^{-5}
Dnepr	115 000	165	275 000	1.2×10^{-3}	0.8×10^{-5}
Sozh	40 940	150	234 900	1.2×10^{-3}	0.8×10^{-5}
Iput'	10 600	160	68 800	1.9×10^{-3}	1.2×10^{-5}
Besed'	5 580	150	62 100	0.3×10^{-3}	0.2×10^{-5}

[a] 1 Ci = 3.7×10^{10} Bq.

TABLE V.　TRITIUM CONCENTRATIONS (10^{-10} Ci/L)[a], MAY–JUNE 1986

River	May	June
Desna	4.4 ± 0.2	3.5 ± 0.5
Dnepr	4.0 ± 0.4	2.5 ± 0.2
Braginska	—	4.7 ± 0.2
Pripyat	5.1 ± 0.3	4.6 ± 0.2
Uzh	8.8 ± 0.5	4.8 ± 0.6
Teterev	5.5 ± 0.4	4.0 ± 0.8
Irpen'	6.8 ± 1.1	3.2 ± 0.4
Average	5.8 ± 0.7	3.9 ± 0.3
PC_S	3.2×10^{-6} Ci/L	

[a] 1 Ci = 3.7×10^{10} Bq.

firmly to the soil of the contaminated catchment areas than ^{90}Sr, which resulted in a lower (by an order of magnitude) washout factor for ^{137}Cs; and there was an additional intake of ^{90}Sr into the Pripyat from partially submerged flood plains in the area close to the Chernobyl plant.

In monitoring the level of radioactive contamination of rivers in the Russian Soviet Federated Socialist Republic (RSFSR) and the Byelorussian Soviet Socialist Republic, attention was concentrated mainly on those rivers whose catchment areas were located in areas contaminated with ^{137}Cs, i.e. the Rivers Sozh, Iput', Besed', Oka, Zhizdra, Plava, Resseta and Upa. In 1987–1988, the maximum concentrations in these rivers were recorded during the spring flood of 1987 (7–12 April). During this period, the highest concentrations, $(1–1.5) \times 10^{-10}$ Ci/L, were recorded in the Iput' and Besed', while in the other rivers the maximum concentrations did not exceed 6×10^{-11} Ci/L. These figures are 100–300 times lower than the PC_S for ^{137}Cs.

A similar situation occurred at the spring flood of 1988, when the maximum concentrations of ^{137}Cs in rivers in the RSFSR and the Byelorussian Soviet Socialist Republic were lower than in 1987. The maximum level recorded was in the Iput' and was 6×10^{-11} Ci/L.

The data obtained in 1987 on the removal of ^{137}Cs by rivers in the Ukrainian and Byelorussian Soviet Socialist Republics make it possible to determine the coefficients for the washing down of radionuclides from catchment areas. The results of these calculations are presented in Table IV.

As can be seen from this table, the washout coefficients are fractions of a per cent, and washing out cannot therefore lead to a significant change in the radiation situation in caesium contaminated areas. From the results of tests on the River Sozh in 1988, the coefficient of ^{137}Cs washout from the Sozh catchment area during the 1988 flood (March–April) was also calculated: the figure was 2.5×10^{-4} or, normalized to a 1 mm runoff depth, 0.4×10^{-5}.

Measurements were made at different times in 1986–1988 of ^{90}Sr concentration in Russian and Byelorussian rivers contaminated by the Chernobyl accident, and these showed that the concentrations were generally two orders of magnitude below the PC$_S$.

The tritium concentration in rivers flowing into the Kiev Reservoir and in small Russian and Byelorussian rivers was measured in May–June 1986. The results are shown in Table V, from which it can be seen that the tritium concentrations found in these rivers at that time were two or three orders lower than the PC$_S$ and presented no radiological danger.

2. RADIOACTIVE CONTAMINATION OF RESERVOIRS IN THE DNEPR CASCADE

The highest levels of contamination of reservoir water were recorded in the initial period at the time when there was heavy fallout of radioactive aerosols. By way of illustration, Table VI shows the isotopic composition of the radioactive contamination of a sample of water taken on 1 May 1986 from the Kiev Reservoir. It can be seen from this table that the radioactivity of the water during this period was mainly due to particles suspended in the water. Thus for the water sample in Table VI the gamma activity of the filtrate was only about 10% of the total gamma activity. The total beta activity of the water during the aerosol fallout period reached

TABLE VI. ISOTOPIC COMPOSITION $(10^{-9}$ Ci/L)^a OF THE RADIOACTIVE CONTAMINATION OF A WATER SAMPLE TAKEN ON 1 MAY 1986 FROM THE KIEV RESERVOIR (LYUTEZH)

	Te-132	I-131	Ru-103	Ba-140	Cs-137	Cs-134	Zr-95	Nb-95	Ce-144	Ce-141
Suspension	12.2	4.4	17.2	13.3	0.95	—	14.2	13.0	11.4	13.9
Solution	1.6	3.8	0.4	2.5	0.25	0.1	0.2	0.17	—	—

^{a} 1 Ci = 3.7×10^{10} Bq.

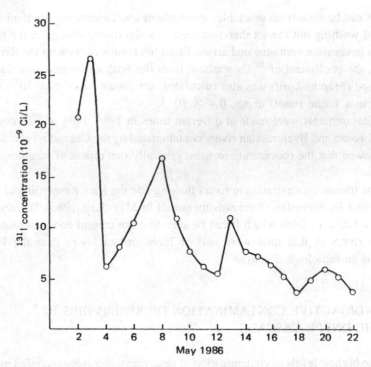

FIG. 1. Concentration of ^{131}I in the water of the Kiev Reservoir in the Dnepr catchment area (1 Ci = 3.7 × 10^{10} Bq).

10^{-7} Ci/L. Thereafter, as a result of a sharp reduction in the rate of release from the source, the contribution of the aerosol component to the contamination of the reservoir also fell sharply, the suspended particles settled fairly quickly on the bottom of the reservoir and by 7 May 1986 the total beta activity of the water in the region of the village of Lyutezh was about 10^{-9} Ci/L. From this time, as data from the gamma spectrometric analysis of water samples have shown, ^{131}I dissolved in the water began to make the greatest contribution to the activity of the water. The highest level of contamination of the Kiev Reservoir by ^{131}I in the Dnepr catchment area resulting from radioactive aerosol was 26 × 10^{-9} Ci/L and was recorded on 3 May 1986 (Fig. 1).

To obtain a rapid assessment of the degree of contamination of the reservoirs of the Dnepr cascade, regular aerial gamma surveys of the surfaces of the reservoirs were conducted from 10 May 1986 onwards. Figure 2 is a sketch map of the reservoirs and indicates the dose rates near their surfaces (on 11–12 May 1986) based on aerial gamma survey results. These made it possible to calculate effectively the degree of contamination of reservoir water.

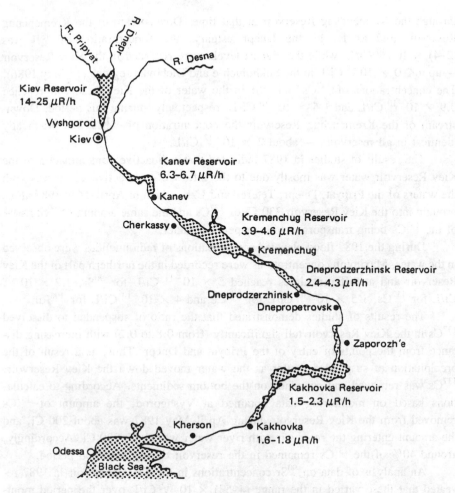

FIG. 2. Distribution of radiation levels above the reservoirs of the Dnepr cascade according to aerial gamma survey data of 11–12 May 1986 (1 R = 2.58 × 10⁻⁴ C/kg).

On the basis of the assumption that no significant fractionation of isotopes took place as the water moved from the Kiev Reservoir down the Dnepr cascade, the degree of contamination of the water in different reservoirs was assessed: the total beta activity of the water was between 10^{-10} and 10^{-9} Ci/L. The aerial gamma survey data for this period were confirmed by the results of gamma spectrometric analysis of water samples taken from the reservoirs.

A survey of the radioactive contamination of all the Dnepr cascade reservoirs was carried out from 14 to 20 May 1986 by boat. This indicated that the radioactivity of the water during this period was due principally to ^{131}I dissolved in water. A 'wave' of radioactive iodine was observed, the lower limit of which was passing through the Kremenchug Reservoir at that time. Downstream of the Kremenchug Reservoir and as far as the Dnepr estuary, the concentration of ^{131}I was $(2-4) \times 10^{-10}$ Ci/L, while the highest levels were recorded in the Kiev Reservoir — up to 2.0×10^{-8} Ci/L in the Sukholuch'e and Glebovka regions (14 May 1986). The concentrations of ^{137}Cs and ^{90}Sr in the water of the Kiev Reservoir reached 0.9×10^{-10} Ci/L and 1.5×10^{-10} Ci/L, respectively, during this period. Downstream of the Kremenchug Reservoir the concentration of ^{137}Cs was practically identical in all reservoirs — about 2×10^{-12} Ci/L.

The results of studies in 1987 indicated that radioactive contamination of the Kiev Reservoir water was mostly due to the inflow of radioactive substances with the waters of the Pripyat, Dnepr, Teterev and Uzh, which in April–May 1987 alone brought into the Kiev Reservoir 330 Ci of ^{137}Cs and the same amount of ^{90}Sr, 40% of the ^{137}Cs being transported with suspended matter.

During the 1987 flood elevated concentrations of radionuclides were observed in the water. Maximum concentrations were recorded in the northern part of the Kiev Reservoir and at different times reached 5×10^{-11} Ci/L for ^{90}Sr, 1.2×10^{-10} Ci/L for ^{137}Cs, 1.5×10^{-10} Ci/L for ^{144}Ce and 4×10^{-10} Ci/L for ^{106}Ru.

The results of studies demonstrated that the ratio of suspended to dissolved ^{137}Cs in the Kiev Reservoir fell significantly (from 0.8 to 0.2) with increasing distance from the points of entry of the Pripyat and Dnepr. Thus, as a result of the precipitation of suspended matter as the water moved down the Kiev Reservoir, ^{137}Cs was removed and deposited on the bottom sediments. According to calculations based on monitoring data obtained at Vyshgorod, the amount of ^{137}Cs removed from the Kiev Reservoir during April–May 1987 was about 200 Ci, and the amount entering the reservoir with river water was about 330 Ci. Accordingly, around 40% of the ^{137}Cs remained in the reservoir during the flood period.

An analysis of data on ^{90}Sr concentrations in the Kiev Reservoir in 1987 revealed that these varied in the range $(4-53) \times 10^{-12}$ Ci/L over the period monitored. The distribution of ^{137}Cs concentration in the water of the Dnepr cascade reservoirs in 1987 was similar to that in 1986, and the ^{137}Cs contamination of reservoir water remained practically at the 1986 level. A significant reduction in the concentration of this radionuclide was observed as the water moved down towards the Dnepr estuary — from $(4-6) \times 10^{-12}$ Ci/L in the Kanev Reservoir to 0.3×10^{-12} Ci/L in the Kakhovka Reservoir. It was established that a substantial proportion (in some cases up to 50%) of the ^{137}Cs in the Dnepr cascade reservoirs was transported with suspended matter.

Whereas ^{137}Cs was removed from the water as it moved down the cascade, the same was not observed for ^{90}Sr. Dilution caused the concentration of ^{90}Sr in the reservoir water to decrease steadily from 30×10^{-12} Ci/L (September–October 1986) in the Kanev Reservoir to 10×10^{-12} Ci/L in the Kakhovka Reservoir. As early as September–October 1987, the ^{90}Sr concentration along the entire cascade was measured at about 10×10^{-12} Ci/L.

The average concentration at the mouth of the Dnepr (from May 1986 to September 1987) was $0.8 \pm 0.15 \times 10^{-12}$ Ci/L for ^{137}Cs and $8.5 \pm 4.0 \times 10^{-12}$ Ci/L for ^{90}Sr. In a water discharge volume of 60 km^3 the amount of ^{137}Cs removed was about 50 Ci and of ^{90}Sr about 500 Ci. In May–June 1986, the tritium concentration in the reservoir water was also determined. A comparison of the distribution of the tritium concentration along the Dnepr cascade with that for other nuclides in the period 13–20 May 1986 showed that, whereas a reduction in the concentrations of ^{131}I, ^{137}Cs and ^{90}Sr was observed as these nuclides moved from the Kiev to the Kremenchug Reservoir, the concentrations of tritium in the Kiev, Kanev and Kremenchug Reservoirs were roughly the same (about 560×10^{-12} Ci) and further downstream the concentrations doubled. The following conclusions can be drawn from this: the main release of tritium into the environment occurred at the start of the accident, after which the tritium was dispersed relatively quickly over large distances (which is confirmed by the high value relative to 1986 of the tritium concentration in May precipitation over practically the whole of the Union of Soviet Socialist Republics) and subsequently entered the reservoirs with precipitation and as a result of molecular exchange with the water surface. This was also the reason why higher levels of tritium were recorded in the lower reservoirs of the cascade, which the contaminated waters of the Kiev Reservoir had not yet reached by 14–20 May.

Data from the gamma spectrometric analysis of bottom sediment samples collected in 1986 enabled three zones of radioactive contamination to be identified:

— The most contaminated section of the Kiev Reservoir, namely that adjoining the mouth of the Pripyat (tens and hundreds of curies per square kilometre);

— The southern parts of the Kiev and Kanev Reservoirs (of the order of 1 Ci/km^2);

— The series of reservoirs south of the Kanev Reservoir as far as the mouth of the Dnepr (< 1 Ci/km^2).

It was established that the main contribution to contamination in the initial period was from radionuclides with moderate half-lives: ^{95}Zr, ^{95}Nb, ^{144}Ce, ^{141}Ce and ^{103}Ru. The contribution of ^{137}Cs to the total gamma activity was about 1%.

Results of the analysis of bottom sediment samples taken between 1 July and 2 August 1986 showed that during this period the Kiev Reservoir exhibited a typical highly patchy distribution for the gamma background of the bottom, this being due to flow rates, the range of depths and the nature of the layer below the bottom. Generally speaking, the highest gamma background levels were observed in deep sections with weak flow, and the lowest in shallow sections with a sandy bottom and on steep bottom slopes. It should be noted that high levels of bottom radioactivity caused by the removal of radionuclides with suspended matter were observed on some days below the dam of the Kiev hydroelectric power plant.

TABLE VII. AVERAGE ^{137}Cs CONTENTS OF RESERVOIR BOTTOMS DURING AUGUST–OCTOBER 1987

Reservoir/location	Average Cs-137 content (Ci/km^2)[a]	Quantity (Ci)[a]	Cs-137 concn in water when turbid (10^{-12} Ci/L)[a]
Kiev	2.5 ± 0.6	2600 ± 620	620
Kanev	0.7 ± 0.2	490 ± 140	190
Kremenchug	0.25 ± 0.10	540 ± 220	40
Dneprodzerzhinsk	0.4 ± 0.2	240 ± 120	100
Zaporozh'e	0.1 ± 0.06	45 ± 30	15
Kakhovka	0.03 ± 0.005	65 ± 10	4
Total		3980 ± 1140	

[a] 1 Ci = 3.7×10^{10} Bq.

A comparison of the results of the gamma spectrometric analysis of several bottom samples collected in May and July 1986 from the Kiev Reservoir indicated that the ^{137}Cs content of the bottom of the reservoir in July was higher than in May by approximately an order of magnitude (average increase from 1 to 15 Ci/km^2). However, in percentage terms the contribution of ^{137}Cs to the total gamma activity was, as before, small (about 2%).

The distribution of nuclides in bottom sediments in 1987 (August) as well as in 1986 was characterized by marked patchiness. The area adjoining the mouth of the River Pripyat again stood out as having the highest levels of contamination. Here the maximum levels of contamination in 1987 were 50 Ci/km^2 for ^{137}Cs and 12 Ci/km^2 for ^{90}Sr.

The calculated average ^{137}Cs contents of reservoir bottoms in August–October 1987 are given in Table VII. From the table it can be seen that when the bottom sediments are stirred up the level of contamination of water in the other reservoirs is considerably lower than in the Kiev Reservoir and is even further below the PC$_S$ since the concentration of ^{137}Cs in the Kiev Reservoir water is only 6% of the PC$_S$ when the bottom sediments are completely turbid.

Contamination of the water by ^{90}Sr will be the most significant when the bottom sediments become turbid. Thus the total amount of ^{90}Sr in September–October 1986 in the bottom sediments of the Kiev Reservoir was 1100 Ci, and under conditions of complete turbidity the concentration of this radionuclide in the water is about 3.2×10^{-10} Ci/L, which is about 80% of the PC_S. The average ^{90}Sr content in the northern part of the Kanev Reservoir for seven samples was 0.31 ± 0.39 Ci/km^2, excluding one very high value of 10.5 Ci/km^2.

An analysis of data on the $^{137}Cs/^{134}Cs$ ratio in reservoir bottoms indicated ratios exceeding 3 downstream of Zaporozh'e. In other words, the level of ^{137}Cs contamination due to Chernobyl is so low in this region that it is comparable with pre-accident contamination from all sources.

3. RADIOACTIVE CONTAMINATION OF SEAS

3.1. Black Sea and Dnepr estuary

The distribution of ^{137}Cs contamination of surface waters of the Black Sea and Dnepr estuary in the first months after the accident was highly uneven because of the non-uniformity of fallout. The eastern part of the sea became more contaminated because of the high fallout levels in this area. Before the accident, the ^{137}Cs content of surface waters was uniform over the water area, the average concentration being about 0.5×10^{-12} Ci/L. The maximum level of contamination recorded in June–July 1986 (14×10^{-12} Ci/L) was almost 30 times higher than the pre-accident level, but in absolute terms this was three orders of magnitude below the PC_S. In other words, the ^{137}Cs levels in the sea water that resulted from the Chernobyl accident did not pose any danger.

By October 1986, because of horizontal and vertical mixing of water masses and a reduction in atmospheric fallout of accident products, the level of surface water contamination by ^{137}Cs had dropped (maximum of 6.7×10^{-12} Ci/L) and the spatial distribution of its concentration in surface waters had become uniform. Winter mixing in 1986–1987 and in 1987–1988 brought about a significant decrease in the concentration of ^{137}Cs in surface waters. In addition, the distribution became practically uniform, with the concentration varying in the range of $(1.1–1.8) \times 10^{-12}$ Ci/L over most of the water area both in 1987 and in 1988.

The situation regarding ^{90}Sr was similar. The maximum concentration in surface waters in June–July 1986 (3.3×10^{-12} Ci/L) was 120 times lower than the PC_S. The effect of ^{90}Sr accident products during this period was perceptible only in the eastern part of the sea, while in the western part the concentration of this radionuclide was practically the same as the pre-accident level of 0.6×10^{-12} Ci/L. By October 1986, the distribution of ^{90}Sr in surface waters had evened out, the average value being about 1×10^{-12} Ci/L. By September 1988 it had dropped to 0.6×10^{-12} Ci/L.

During the studies, detailed vertical profiles of the concentrations of ^{137}Cs (^{134}Cs) and ^{90}Sr were plotted. These demonstrated that both the ^{137}Cs and the ^{90}Sr

originating from the Chernobyl accident were concentrated above the thermocline. The ^{137}Cs and ^{90}Sr depth distributions were used to establish the total quantities of these radionuclides in the water mass of the Black Sea that resulted from the accident, the values being about 45 000 Ci for ^{137}Cs and 7500 Ci for ^{90}Sr. The amounts of ^{137}Cs and ^{90}Sr in the water mass before the accident were about 40 000 Ci each. The contribution of Chernobyl accident products thus resulted in a doubling of the amount of ^{137}Cs in the sea and an increase of 20% in the amount of ^{90}Sr.

The tritium content of the sea water (in the 0–50 m layer) was also determined in June–July 1986. This indicated that the intake of tritium into the Black Sea as a result of the accident had practically no effect on the tritium concentration in the waters of the sea, the average concentration remaining at the pre-accident level of 34 TU.

During the entire period monitored, the ^{137}Cs concentration in the Dnepr estuary was 3–5 times lower than in the adjacent part of the Black Sea, in other words the inflow of the relatively clean Dnepr waters helped to reduce the concentration of ^{137}Cs in the northwestern part of the Black Sea.

The calculations show that from June 1986 to October 1987 about 50 Ci of ^{137}Cs and about 500 Ci of ^{90}Sr were carried into the Black Sea by the Dnepr. Accordingly, removal by the Dnepr represents a total of about 0.1% of the amount of ^{137}Cs and about 7% of the amount of ^{90}Sr from the Chernobyl accident in the sea, and the main source of contamination during the period studied was the atmospheric fallout that took place in the first days after the accident.

3.2. Sea of Azov

The spatial distribution of ^{137}Cs and ^{134}Cs in the water and the bottom deposits of the Sea of Azov was studied in June and November 1987. The range of variation in the concentrations of ^{137}Cs in both periods was not great: $(0.5-1.0) \times 10^{-12}$ Ci/L in June and $(0.6-1.2) \times 10^{-12}$ Ci/L in November. The amount of ^{137}Cs in the sea water was about 250 Ci, of which about 200 Ci (85%) originated from the Chernobyl accident.

It was established that from the mouth of the Don towards the Kerch Strait the ^{137}Cs and ^{134}Cs content of the bottom increased. This was probably due to the sorption of radionuclides by bottom deposits from Black Sea water in near-bottom layers. The total amount of ^{137}Cs in the sea bottom was about 540 Ci, of which 400 Ci were due to the Chernobyl accident.

In the pre-accident period of April 1986, the ^{90}Sr concentration was the same as the preceding year (about 1×10^{-12} Ci/L). In the subsequent period, non-uniformity in the distribution of ^{90}Sr in the surface waters of the sea was observed, the concentration rising to as high as $(1.5-2.0) \times 10^{-12}$ Ci/L in some parts. The total quantity of ^{90}Sr in the water in 1987 was about 420 Ci (before the accident the figure was about 300 Ci).

Another factor, in addition to atmospheric fallout, that may have contributed to the increased concentration of ^{90}Sr and ^{137}Cs after the accident was the inflow of contaminated Black Sea waters from eastern parts of the sea. This process was evidently most significant in May–August 1986, when there was a water mass contaminated by atmospheric fallout in the region of the Black Sea around the Kerch Strait.

On the whole the radioecological situation in the Sea of Azov after the accident was good, the levels of ^{137}Cs and ^{90}Sr in the water being four and two orders of magnitude below the respective PC$_S$ values.

3.3. Baltic Sea

As a result of atmospheric fallout following the accident, a clearly identifiable zone of increased caesium contamination formed in the Baltic Sea, more specifically in the central and eastern parts of the Gulf of Finland and in the Gulf of Bothnia. The maximum concentrations of ^{137}Cs in June 1986 occurred in the Gulf of Finland and were about 25×10^{-12} Ci/L (600 times lower than the PC$_S$). The pre-accident level of ^{137}Cs in the surface waters was about 0.3×10^{-12} Ci/L. The rise in the ^{90}Sr content of the water as a result of fallout was not large. Thus, the average concentrations of this radionuclide in the surface waters of the Gulfs of Finland and Riga in May 1986 were 20% higher than the previous year.

When the radioactive fallout ceased, the level of caesium contamination of the water began to drop slowly as a result of dispersion processes. Our tests in the Kopor Bay of the Gulf of Finland showed that in June 1987 the ^{137}Cs concentration was three times lower than in May 1986 and in June–August 1988 four times lower. Since the water in the surface layers of the Baltic Sea moves from the northern part of the sea towards the Danish Straits, it is to be expected that with the passage of time the area of high ^{137}Cs contamination will shift in this direction. Checks carried out in June–July 1987 indicated that, in qualitative terms, the map of caesium contamination of the sea had changed little by the summer of 1987 in comparison with November 1986 and, as before, the highest levels of contamination (about 6×10^{-12} Ci/L) were observed in the northern part of the Baltic Sea and in the Gulf of Finland. The conclusion is that the removal of activity from the Baltic Sea will be a lengthy process.

The amount of ^{137}Cs in the water mass of various parts of the Baltic Sea in November 1986 was calculated by us, using the data of experts from the Federal Republic of Germany [2], to be the following:

— About 50 000 Ci in the Gulf of Bothnia
— About 30 000 Ci in the open Baltic Sea.

Calculations based on our data indicate that in May–June 1987 the amount of ^{137}Cs in the water mass of the open Baltic Sea was about 25 000 Ci, and in the Gulf of Finland about 3500 Ci. Thus, the amount of ^{137}Cs in the open Baltic remained practically unchanged from November 1986 to June 1987.

The total amount of ^{137}Cs in the sea water after the Chernobyl accident was around 80 000 Ci, of which over 60% was concentrated in the Gulf of Bothnia. According to our [3] and foreign evaluations, the ^{137}Cs content of the water mass of the sea before the accident was about 12 000 Ci. The amount of ^{137}Cs thus rose approximately sevenfold after the accident and the amount of ^{137}Cs due to the accident now accounts for some 85% of the total quantity of this radionuclide in the water.

It should be pointed out that the contribution of ^{137}Cs and ^{90}Sr from Chernobyl to the radioactive contamination of sea water is not large in comparison with the natural radioactivity. For ocean water, beta activity due to natural substances is 9.1×10^{-12} Ci/L per 1‰ salinity [4]. For example, the beta activity of the water in the Black Sea due to naturally occurring ^{40}K is $(170-200) \times 10^{-12}$ Ci/L [4]. Consequently, even the maximum levels of contamination of the Black Sea by ^{137}Cs from Chernobyl in June 1986 were more than an order of magnitude lower than the natural background level.

REFERENCES

[1]   Radiation Safety Standards NRB-76/87, Ehnergoatomizdat, Moscow (1988) (in Russian).

[2]   The Radioactive Contamination of the Baltic Sea by the Chernobyl Fallout, Report submitted by the Federal Republic of Germany at 14th Mtg of Scientific–Technical Committee, Schleswig, 1987.

[3]   VAKULOVSKIJ, S.M., KATRICH, I.Yu., KRASNOPEVTSEV, Yu.V., NIKITIN, A.I., CHUMICHEV, V.B., Radioactive contamination of the Baltic Sea in 1980, Meteorol. Gidrol. No. 9 (1983) 72–78 (in Russian).

[4]   PERTSOV, L.A., Biological Aspects of Radioactive Contamination of the Sea, Atomizdat, Moscow (1978) 160 (in Russian).

VARIABILITY OF RADIOCAESIUM CONCENTRATIONS IN FRESHWATER FISH CAUGHT IN THE UNITED KINGDOM FOLLOWING THE CHERNOBYL ACCIDENT

An assessment of potential doses to critical group consumers

D.R.P. LEONARD, W.C. CAMPLIN, J.R. TIPPLE
Directorate of Fisheries Research,
Ministry of Agriculture, Fisheries and Food,
Lowestoft, Suffolk,
United Kingdom

Abstract

VARIABILITY OF RADIOCAESIUM CONCENTRATIONS IN FRESHWATER FISH CAUGHT IN THE UNITED KINGDOM FOLLOWING THE CHERNOBYL ACCIDENT: AN ASSESSMENT OF POTENTIAL DOSES TO CRITICAL GROUP CONSUMERS.

Radiocaesium deposited from the Chernobyl reactor accident contaminated the aquatic environment in several parts of the United Kingdom. The Directorate of Fisheries Research of the Ministry of Agriculture, Fisheries and Food studies have shown freshwater fish, particularly indigenous brown trout, to be the most important pathway, radiologically. The paper highlights the temporal and spatial distribution of radiocaesium in brown trout between 1986 and 1989. The difference in concentrations for other species is discussed. By comparing theoretical doses assessed on the basis of pessimistically high consumption rates with information provided from specific fisheries, it is concluded that effective dose equivalents are likely to have been less than 1 mSv/a.

1. INTRODUCTION

The monitoring and assessment of the impact of radioactive discharges into the aquatic environment of the United Kingdom are carried out by the Directorate of Fisheries Research (DFR) of the Ministry of Agriculture, Fisheries and Food (MAFF). The deposition of radionuclides from the Chernobyl reactor accident necessitated a considerable expansion of the MAFF(DFR) monitoring programme to keep the radiological consequences under review. The consumption of freshwater fish containing radiocaesium was potentially the most important aquatic pathway [1]. In this paper the variability of radiocaesium concentrations in these fish is considered and potential doses to consumers are reassessed and extended by using new information.

MAFF(DFR) regularly takes samples near the major nuclear sites in England on behalf of the Minister of Agriculture, Fisheries and Food; in Wales on behalf of the Environmental Protection Unit of the Welsh Office; in Scotland on behalf of Her Majesty's Industrial Pollution Inspectorate for Scotland; and in Northern Ireland on behalf of the Department of the Environment for Northern Ireland. The results are published annually [2]. Nuclear sites that are authorized to discharge low level liquid radioactive waste into the freshwater environment are located in North Wales at Trawsfynydd and at locations in the catchment of the River Thames. Radiocaesium discharges into Llyn Trawsfynydd are radiologically more important than those discharged into the Thames owing primarily to the operation of the Magnox power station. This work has demonstrated the importance of radiocaesium uptake via the food chain into the indigenous brown trout and perch. Fallout from nuclear weapons testing has also been studied. The distribution of ^{137}Cs in brown trout flesh was studied between 1961 and 1966 and shown to have a negative correlation with potassium water concentrations [3]. Lake sediment studies have shown that ^{137}Cs deposition from weapons tests fallout reached a peak in 1963 [4]. By the early 1980s, concentrations of ^{134}Cs and ^{137}Cs in freshwater fish due to fallout were generally below 10 Bq/kg (wet). Similar results have been reported for ^{137}Cs in Sweden [5].

2. POST-CHERNOBYL MONITORING

MAFF(DFR) sampling of the freshwater environment was directed towards areas where a large deposition of radiocaesium had occurred and from which indigenous fish, particularly brown trout, might be regularly consumed. Other species, such as pike and perch, which were likely to accumulate similar quantities of radiocaesium but which are radiologically less important because of lower consumption rates, were also sampled. Rainbow trout were sampled because they are eaten regularly but since their mean lake residence time after stocking is short, their radiocaesium uptake is small. Our objective was to catch at least six fish (ideally ten) of each species per sampling period. In practice, this proved difficult to achieve because of resource limitations; often a smaller sample had to suffice. The frequency of sampling varied from fortnightly to annually, reflecting the radiological importance of the species. As concentrations declined, the frequency and number of sampling locations were reduced. Between 1986 and 1988 fish from over 150 UK locations were sampled; this activity involved the analysis of over 3500 samples. Fish were usually counted individually as wet edible flesh in a standard geometry for 3600 s by gamma spectrometry [2, 6]. Table I illustrates the annual mean concentrations of radiocaesium in brown trout from a few of the locations monitored.

Figure 1 shows the sampling locations. The following general observations from the data can be made:

(a)   Mean concentrations were usually very similar in 1986 and 1987, but were usually less in 1988. This may be due to the decreasing availability of caesium due to physical, chemical and biological mechanisms such as a continuing but falling input through the water column, lack of binding onto clay poor soils, hydraulic exchange, uptake through the food chain, and seasonal feeding patterns.

(b)   Some increases in mean concentrations were observed but these are considered to be due to the limited sample size and to the mixing of indigenous with recently stocked fish.

(c)   Total radiocaesium concentrations ranged from a few to over 1000 Bq/kg (wet). Low concentrations were found in fish from low deposition areas, in recently stocked fish and in fish from fish farms. The highest concentrations were found in indigenous fish.

Ennerdale Water in Cumbria, England, was selected for further study since it contains indigenous brown trout that are consumed regularly. No stocking of the lake is necessary as sufficient fish occur there naturally. Because the lake is a source of drinking water, its waters are regularly sampled by the Department of the Environment as part of their responsibilities in the UK [7] and by the North West Water Authority who have responsibilities in Cumbria [8]. That sampling assisted our own water sampling programme as a means of deciding whether fish sampling were necessary, but it was recognized that clay and other caesium binding agents might affect the availability of caesium uptake into the fish, making the fish/water ratio variable. We started catching brown trout in August 1986, a time when concentrations were expected to have reached a plateau. Figure 2 shows a plot of mean total radiocaesium concentrations in brown trout, char and water against time. While radiocaesium water concentrations declined rapidly in 1986, this decrease was not seen in the fish until the latter half of 1987. Brown trout had higher mean radiocaesium concentrations than char, presumably owing to their feeding behaviour and physiological differences. Concentration ratios (fish/water) of ^{137}Cs for brown trout were 4500 in August 1986, 19 000, 12 months later and 11 000, 24 months later; but these are unlikely to be true equilibrium concentration factors. Individual analyses of brown trout have been plotted against time in Fig. 3. These numbers show that concentrations varied by a factor of three about the mean, presumably reflecting differences in individual feeding patterns. Statistical analysis of the data showed that the differences in mean concentration observed between August 1986 and May 1987 were not significant, owing to the limited number of fish analysed. However, after this period a significant reduction in brown trout radiocaesium concentrations was observed. The tabulation of all the Cumbrian fish analysed between May 1986 and April 1988 has been reported separately [6].

TABLE I. ANNUAL MEAN CONCENTRATIONS OF RADIOCAESIUM IN BROWN TROUT FROM ENGLAND, WALES, SCOTLAND, NORTHERN IRELAND AND THE ISLE OF MAN BETWEEN 1986 AND 1988

Location	1986			1987			1988		
	Number of samples	Mean concentration (Bq/kg (wet)) Cs-134	Cs-137	Number of samples	Mean concentration (Bq/kg (wet)) Cs-134	Cs-137	Number of samples	Mean concentration (Bq/kg (wet)) Cs-134	Cs-137
ENGLAND									
Ennerdale Water	28	230	500	124	200	520	68	51	190
Devoke Water	15	390	860	95	250	660	92	80	300
Loweswater	20	360	770	116	220	550	45	39	140
Cogra Moss	2	190	360	1	140	460	3	38	170
Crummock Water	5	110	240	15	110	320	11	33	110
WALES									
Cwmystradllyn Reservoir	11	39	86	6	34	76	1	11	28
Llyn Conwy	22	210	490	44	130	350	35	55	230
Alwen Reservoir	18	69	150	23	66	160	2	20	69
Llyn Trawsfynydd	68	44	220	86	53	310	119	21	150
Llyn Goddionduon	0	—		23	390	1100	39	170	600

SCOTLAND

Loch Dee	7	590	1300	64	440	1200	35	150	550
Loch Doon	14	180	400	45	130	350	11	47	180
Loch Garry, Tayside Region	9	160	350	5	160	430	9	77	300
Ruisdale Water	1	190	490	3	92	250	2	74	324
Sandy Loch Reservoir	1	230	520	1	30	130	6	12	35

NORTHERN IRELAND

River Bush	2	93	250	16	41	140	2	17	84
Altnahinch Dam	12	80	220	19	33	110	2	14	77
Lough Ash	10	120	270	7	25	66	5	2.3	9.7
Movanagher Fish Farm	ND	ND	3.0	6	ND	4.1	1	ND	1.8

ISLE OF MAN

River Druidale	5	160	360	1	200	450	2	23	88
Cornaa River	4	110	220	10	64	160	2	ND	22

ND — not detected.

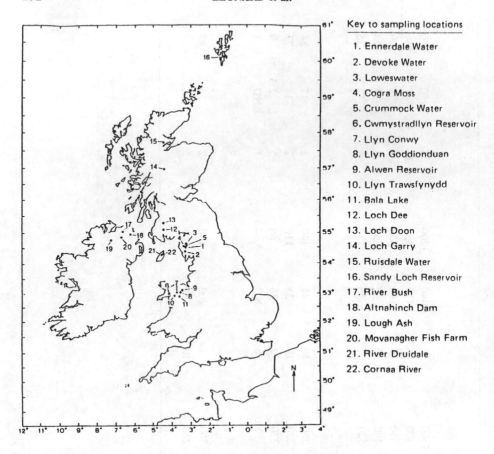

Key to sampling locations

1. Ennerdale Water
2. Devoke Water
3. Loweswater
4. Cogra Moss
5. Crummock Water
6. Cwmystradllyn Reservoir
7. Llyn Conwy
8. Llyn Goddionduan
9. Alwen Reservoir
10. Llyn Trawsfynydd
11. Bala Lake
12. Loch Dee
13. Loch Doon
14. Loch Garry
15. Ruisdale Water
16. Sandy Loch Reservoir
17. River Bush
18. Altnahinch Dam
19. Lough Ash
20. Movanagher Fish Farm
21. River Druidale
22. Cornaa River

FIG. 1. Sampling locations.

At Trawsfynydd there were two main sources of radiocaesium in the lake during 1986. Annual discharges from the power station decreased from 0.063 TBq ^{137}Cs in 1985 to 0.032 TBq in 1986, but this was compensated by Chernobyl radiocaesium. Both direct deposition and runoff from the surrounding mountains resulted in an increase of up to 50% in the ^{137}Cs water concentrations. However, because of the time required for radiocaesium to be transported through the food chain, only approximately 10% of ^{137}Cs in brown trout flesh is considered to have originated from Chernobyl.

A large portion of Scotland received some Chernobyl derived radiocaesium [9]. The highest concentrations recorded during our sampling occurred at Loch Dee, where 3000 Bq/kg (wet) ^{137}Cs was found in one unusually large brown trout in March 1987. In comparison, 18 700 Bq/kg (wet) have been reported as a maximum in Sweden [10] and 55 000 Bq/kg (wet) in Norway [11]. Brown trout from

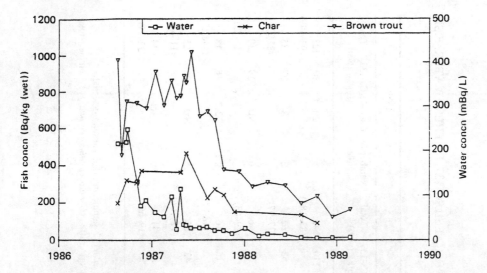

FIG. 2.   Temporal comparison of total radiocaesium concentrations in brown trout, char and water from Ennerdale Water.

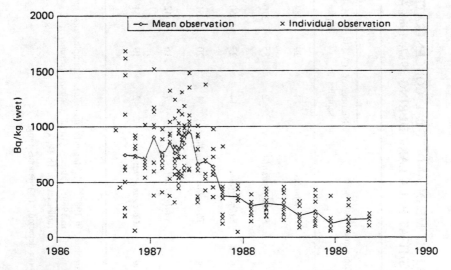

FIG. 3.   Temporal distribution of total radiocaesium in the flesh of individual brown trout from Ennerdale Water.

TABLE II. MEAN AND TOTAL CONCENTRATIONS OF CAESIUM-134 AND CAESIUM-137 IN FRESHWATER FISH CAUGHT AT BALA LAKE DURING 1987, 1988 AND 1989

Common name	Latin name	1987				1988				1989			
		Number of samples	Cs-134	Cs-137 (Bq/kg (wet))	Total (Bq/kg (wet))	Number of samples	Cs-134	Cs-137 (Bq/kg (wet))	Total (Bq/kg (wet))	Number of samples	Cs-134	Cs-137 (Bq/kg (wet))	Total (Bq/kg (wet))
Pike	*Esox lucius*	14	29	84	113 (87–147)	7	32	110	142 (83–181)	6	20	103	123 (92–149)
Perch	*Perca fluviatilis*	2	85	210	295 (50&540)	4	10	52	62 (37–97)	2	8	27	35 (32&37)
Brown trout	*Salmo trutta*	14	23	60	83 (ND–331)	6	10	36	46 (29–63)	6	ND	19	19 (15–25)
Grayling	*Thymallus thymallus*	1	20	51	71	2	ND	16	16 (14&17)	2	2.2	16	18 (14&17)
Roach	*Rutilus rutilus*	7	16	42	58 (47–83)	10	13	42	55 (17–106)	6	ND	25	25 (11–44)
Gwyniad	*Coregonus lavaretus*	2	19	51	70 (63&77)	3	5.1	36	41 (19–66)	6	1.1	13	14 (22–21)

Notes:

Also sampled: in 1987: eel (*Anguilla anguilla*), 1 sample, total radiocaesium 15 Bq/kg; salmon (*Salmo salar*), 1 sample, no radiocaesium detected. in 1989: rainbow trout (*Salmo gairdneri*), 2 samples, total radiocaesium 6.5 Bq/kg; ruffe (*Gymnocephalus cernua*), 1 sample, total radiocaesium 18 Bq/kg.

ND — not detected; ( ) denotes range.

Northern Ireland and the Isle of Man had mean concentrations of ^{137}Cs below 500 Bq/kg (wet).

We have also sampled other species of fish in the UK. For example at Bala Lake in North Wales, at least ten different species are found that may be consumed. Annual sampling has been used since 1987 to compare radiocaesium levels. The total and mean concentrations are shown in Table II. The highest concentrations were observed in pike and perch, i.e. fish occupying the higher trophic levels. Migratory fish such as salmon and eel contained the lowest concentrations of radiocaesium. Similar observations have been made in Finland [12] and Italy [13].

3. DOSE TO MAN

A previous assessment of the individual doses in the UK from consumption of freshwater fish after Chernobyl has suggested that the maximum effective dose equivalent would have been 1.1 mSv/a in 1986 [14]. This estimate was based on a pessimistic consumption rate of 54.75 kg/a for brown trout as well as a selection of radiocaesium measurements which were representative of the highest found in these fish for each region of the UK where sustained consumption could occur.

To improve that assessment, further information has been obtained. At Ennerdale Water and Bala Lake, potential critical group consumers are estimated to consume approximately 18 kg/a of brown trout from Ennerdale Water and 26 kg/a of brown trout, pike and grayling in equal amounts from Bala Lake. Doses are thus estimated at 0.18 mSv/a in 1986 at Ennerdale Water and 0.031 mSv/a at Bala Lake for 1987. Doses would have been similar at Ennerdale Water in 1987 and lower at Ennerdale Water and Bala Lake in subsequent years. Doses from all sources from fish at Llyn Trawsfynydd are 0.17, 0.25 and 0.05 mSv/a for 1986, 1987 and 1988, respectively. A survey at Loch Dee in 1987 showed that critical group consumption rates were much lower than originally assumed at 6.67 kg/a and thus the dose to these people has been reassessed as 0.17, 0.15 and 0.061 mSv/a in 1986, 1987 and 1988, respectively.

4. CONCLUSIONS

Radiocaesium from the Chernobyl nuclear reactor accident has been found in freshwater fish in several areas of the UK, principally associated with areas of generally high deposition. Concentrations covered a wide range, exceptionally up to a few thousand Bq/kg of wet flesh. Despite this, the impact on man from the consumption of freshwater fish has been of only minor radiological significance.

REFERENCES

[1] CAMPLIN, W.C., MITCHELL, N.T., LEONARD, D.R.P., JEFFERIES, D.F.,
     Radioactivity in surface and coastal waters of the British Isles: monitoring of fallout
     from the Chernobyl reactor accident, Aquatic Environment Monitoring Report of the
     Directorate of Fisheries Research of the Ministry of Agriculture, Fisheries and Food,
     Lowestoft 15 (1986) 1–49.

[2] HUNT, G.J., Radioactivity in surface and coastal waters of the British Isles 1987,
     Aquatic Environment Monitoring Report of the Directorate of Fisheries Research of the
     Ministry of Agriculture, Fisheries and Food, Lowestoft 19 (1988) 1–67.

[3] PRESTON, A., JEFFERIES, D.F., DUTTON, J.W.R., The concentrations of
     caesium-137 and strontium-90 in the flesh of brown trout taken from rivers and lakes
     in the British Isles between 1961 and 1966: the variables determining the concentration
     and their use in radiological assessments, Water Res. 1 (1967) 475–496.

[4] PENNINGTON, W., Records of a lake's life in time: the sediments, Hydrobiologia 79
     (1981) 197–219.

[5] EVANS, S., Accumulation of Chernobyl-related Cs-137 by fish populations in the
     biotest basin, northern Baltic Sea, Rep. STUDSVIK-NP-88/113, Studsvik Energi-
     teknik, Nyköping, Sweden (1988) 1–30.

[6] CAMPLIN, W.C., LEONARD, D.R.P., TIPPLE, J.R., DUCKETT, L., Radio-
     activity in freshwater systems in Cumbria (UK) following the Chernobyl accident, Fish
     Research Data Report of the Directorate of Fisheries Research of the Ministry of
     Agriculture, Fisheries and Food, Lowestoft 18 (1989) 1–90.

[7] Levels of Radioactivity in the UK from the Accident at Chernobyl, USSR, on
     26 April 1986, Her Majesty's Stationery Office, London (1986) 1–176.

[8] JONES, F., CASTLE, R.G., Radioactivity monitoring of the water cycle following the
     Chernobyl accident, J. Inst. Water Environ. Manage. 1 (1987) 205–217.

[9] SCOTTISH DEVELOPMENT DEPARTMENT, Chernobyl Accident, Monitoring for
     Radioactivity in Scotland, Scottish Development Department 1(E), Edinburgh (1988)
     1–15.

[10] PETERSEN, R.C., LANDNER, L., BLANCK, H., Assessment of the impact of the
     Chernobyl reactor accident on the biota of Swedish streams and lakes, Ambio 15 6
     (1986) 327–331.

[11] STRAND, T., STRAND, P., BAARLI, J., Radioactivity in foodstuffs and doses to the
     Norwegian population from the Chernobyl fall-out, Radiat. Prot. Dosim. 20 4
     (1987) 211–220.

[12] SAXEN, R., RANTAVAARA, A., Radioactivity of freshwater fish in Finland after the
     Chernobyl accident in 1986, Rep. STUK-A61, Supplement to the Annual Report,
     STUK-A55, Finnish Centre for Radiation and Nuclear Safety, Helsinki (1987) 1–45.

[13] QUEIRAZZA, G., MARTINOTTI, W., Radioattività in matrici ambientali del delta
     del Po, Acqua-Aria 7 (1987) 857–862.

[14] MITCHELL, N.T., CAMPLIN, W.C., LEONARD, D.R.P., The Chernobyl reactor
     accident and the aquatic environment of the UK: a fisheries viewpoint, J. Soc. Radiol.
     Prot. 6 4 (1986) 167–172.

POSTER PRESENTATIONS

IAEA-SM-306/108P

TIME SERIES ANALYSIS
OF TRITIUM PRECIPITATION
IN OSAKA, JAPAN

H. MORISHIMA, T. KOGA, T. NIWA,
H. KAWAI, Y. NISHIWAKI*
Atomic Energy Research Institute,
Kinki University,
Osaka, Japan

1. BACKGROUND

Soon after the Bikini Atoll event in 1954, tritium precipitation began to be measured at the Biophysics Laboratory, Faculty of Medicine of Osaka City University in Japan. At the Atomic Energy Research Institute of Kinki University in Osaka, Japan, monthly tritium precipitation has been measured since 1965.

At first, the concentration of tritium was increased by electrolytic enrichment of rain water prior to measurement. However, the tritium of rain water has been measured recently with a low background liquid scintillation counter directly after distillation. In May 1986, various fission products due to the Chernobyl accident were detected in different parts of Japan. However, no significant tritium increase has been reported. According to the routine measurement of tritium at Kinki University, a peak of tritium precipitation of 244 Bq/m^2 was observed in May 1986. Although the apparent peak was observed, we have conducted a time series analysis of the data and examined the seasonal component, the trends component and the irregular component. If a significant peak which is not regular is mixed, it may appear in the irregular component. After careful analysis it was concluded that the peak of the irregular component in May 1986 may not be significantly higher when compared with the data of the seasonal spring peak of previous years, and that the tritium precipitation due to the Chernobyl accident would be below the detection limit, if there were any precipitation at all. Figure 1 shows the original data of tritium precipitation, the trends component and the irregular component.

---

* Present address: Division of Nuclear Safety, International Atomic Energy Agency, Wagramerstrasse 5, P.O. Box 100, A-1400 Vienna.

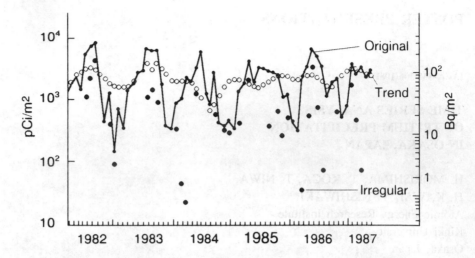

FIG. 1. Tritium precipitation in Osaka, Japan (measured in Ci and Bq; 1 Ci =
3.7 × 10¹⁰ Bq).

2. METHOD OF ANALYSIS

The time series analysis of monthly tritium precipitation in Osaka, Japan, was
carried out on the HITAC computer system at the University of Tokyo Computer
Centre, by the program of additive form of the Census Method IIX_{11}, developed by
the United States National Bureau of Economic Research. The outline of the princi-
ple for the calculating procedure is as follows:

The original time series O is assumed to consist of three components: the trend
factor, T, the seasonal factor, S, and the irregular factor, I:

$$O = T + S + I \tag{1}$$

The first approximate trend factor $(T_1)_i$ for the i-th month is obtained by calculating
the centred 12 term moving average:

$$(T_1)_i = \frac{1}{2} \sum_{k=0}^{1} \left(\frac{1}{12} \sum_{j=-6}^{5} (O)_{i+j+k} \right) \tag{2}$$

The seasonal plus irregular factor $(SI)_i$ is the deviation of $(O)_i$ from $(T_1)_i$:

$$(SI)_i = (O)_i - (T_1)_i \tag{3}$$

For smoothing $(SI)_i$ by adjusting the extreme value, the three term moving average of the three term moving averages for the month is calculated as follows:

$$(S')_i = \frac{1}{3} \sum_{k=-1}^{1} \left(\frac{1}{3} \sum_{j=k-1}^{k+1} (SI)_{i+12j} \right) \tag{4}$$

The $(CS)_i$ is obtained by computing the centred 12 term moving average of $(S')_i$:

$$(CS)_i = \frac{1}{2} \sum_{k=0}^{1} \left(\frac{1}{12} \sum_{j=-6}^{5} (S')_{i+j+k} \right) \tag{5}$$

The first estimate of the seasonal factor $(S_1)_i$ is given by the deviation of $(S')_i$ from $(CS)_i$:

$$(S_1)_i = (S')_i - (CS)_i \tag{6}$$

Accordingly, the first estimate of the irregular factor $(I_1)_i$ is obtained by subtracting $(S_1)_i$ from $(SI)_i$:

$$(I_1)_i = (SI)_i - (S_1)_i \tag{7}$$

All these procedures are repeated several times with empirically determined weighting factors; the final refined values of trend, seasonal and irregular factors are then obtained.

Roughly speaking, the trend factor is represented by the moving average of original data; the deviation from this is the seasonal plus irregular factor, the periodic component of which is the seasonal factor, the rest being the irregular factor.

IAEA-SM-306/63P

MODEL FOR PREDICTING WATER CONTAMINATION IN FALLOUT SITUATIONS

E. ETTENHUBER, M. JURK, M. KÜMMEL, H.-U. SIEBERT
National Board for Atomic Safety
   and Radiation Protection,
Berlin

The contamination of waters in the German Democratic Republic has been monitored for many years. Although the radiation dose due to the contamination of surface waters was negligible after the Chernobyl accident, monitoring of contamination was performed intensively for parameter estimation.

A model for the description of surface water contamination by long lived cationic radionuclides must include the following components which contribute to contamination:

(1)   Input of radioactivity by:
   — Direct deposition of radionuclides on the water surface;
   — Direct transfer of radionuclides deposited on the drainage area to the water (runoff);
   — Long time transfer of radionuclides accumulated in the soil of the drainage area to the water.
(2)   Removal of radioactivity by:
   — Outflow,
   — Sedimentation,
   — Radioactive decay.

The general equation describing the radioactivity balance of waters is given by:

$$\frac{dA}{dt} = E - pE - (\kappa + k)A \qquad (1)$$

where

A   is the radioactivity in water,
E   is the rate of radioactivity input,
p   is the fraction of radioactivity input removed by sedimentation and biological uptake,
κ is the radioactive decay constant,
k   is the flowing off constant, e.g. quotient from flowing off rate and water volume.

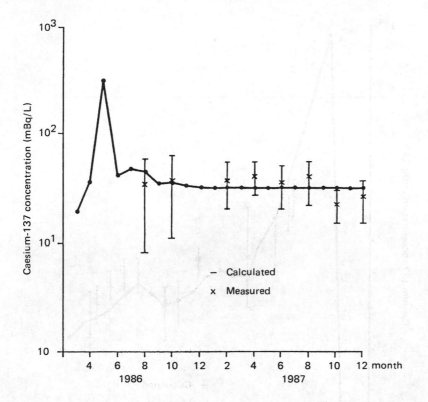

FIG. 1. *Caesium-137 concentrations in the water of the Havel River at Hennigsdorf, German Democratic Republic.*

The rate of radioactivity input includes three components:

$$E = E_1 + E_2 + E_3 \tag{2}$$

E_1 is the deposition of the radionuclide on the water surface:

$$E_1 = F_G\, a_N \tag{3}$$

where

F_G is the area of the water and
a_N is the rate of fallout on the water surface.

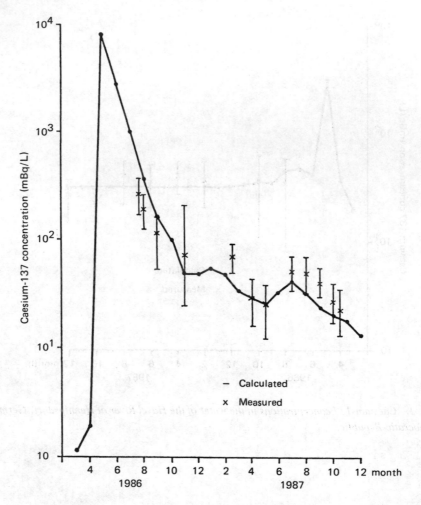

FIG. 2.   Caesium-137 concentrations in the water of Lake Müggelsee.

The runoff (E_2) is given by:

$$E_2 = NF_Ea_{NE} \qquad\qquad (4)$$

where

N   is the rate constant describing the fraction of deposited radioactivity (fallout) removed by runoff,

F_E is the drainage area,

a_{NE} is the rate of fallout in the drainage area.

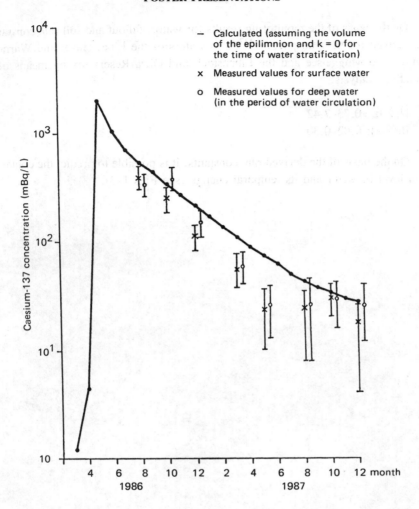

FIG. 3. Caesium-137 concentrations in the surface water of the Ohra Reservoir.

E_3 is the fraction of radioactivity accumulated in the soil of the drainage area which is transported to the water:

$$E_3 = BF_E a_B \tag{5}$$

where

B   is the rate constant describing the fraction of radionuclides accumulated in the soil and transported to the water,

a_B is the total storage of radionuclides in the soil of the drainage area.

On the basis of the monitoring results for water, fallout and soil rate constants were derived for the transport of ^{137}Cs to waters of the Elbe, Havel and Warnow Rivers, Lake Müggelsee and the Eibenstock and Ohra Reservoirs by means of a correlation analysis:

N(%):   0.73–7.42
B(%/a): 0.02–0.30

On the basis of the derived rate constants, it is possible to predict the contamination level of water and its temporal change (see Figs 1–3).

Part II
MONITORING OF RADIOACTIVITY

(a) General

Part III

MONITORING OF RADIOACTIVITY

(a) General

Invited Paper

ASSESSING ENVIRONMENTAL RADIOACTIVITY

O. PAAKKOLA
Finnish Centre for Radiation
   and Nuclear Safety,
Helsinki, Finland

Abstract

ASSESSING ENVIRONMENTAL RADIOACTIVITY.

A great deal of progress has been made in methods for determining radionuclides during the thirty years since the first atmospheric tests of nuclear weapons. With time our understanding of environmental contamination itself has changed. Development of instrumentation has provided new opportunities for detecting radionuclides rapidly and without radiochemical separation. On the other hand, there are also a number of methods which have remained unchanged throughout this thirty year period. The location of the laboratories where measurements will be made is of great importance. A well equipped central laboratory is capable of analysing all known environmental radionuclides, whereas remote local control laboratories are only able to maintain the expertise needed for limited measurement methods. The activity level in the environmental samples determines the methods which can be used. In an emergency situation simple monitoring can be performed on the basis of simple total activity measurements. The estimation of intervention levels in cases of food contamination requires more detailed assessment methods. Reliable methods of sample collection and analysis are necessary in such cases. Direct gamma spectrometric determination is usually sufficient for monitoring purposes. The levels to be determined in radioecological studies, however, are often so low that the most sophisticated radiochemical separation and measurement methods are needed.

1. METHODS OF ASSESSMENT, PAST AND PRESENT

Over the past thirty years the requirements and methods for assessing radioactive substances in the environment have altered. The main goal has been and continues to be the need to evaluate the hazards to man from radioactive fallout. The different sources of radionuclides in the environment have been extensively studied. In 1958 the United Nations Scientific Committee on the Effects of Atomic Radiation (UNSCEAR) published its first comprehensive report [1] to the United Nations General Assembly on the sources, levels and effects of radionuclides. Altogether ten such reports have been published, the latest in 1988 [2]. This latest report includes a summary of estimates of effective dose equivalents (Table I). All sources which are important on a global scale are presented in the summary.

TABLE I. SUMMARY OF ESTIMATES OF EFFECTIVE DOSE EQUIVA-
LENTS [2]

Source of practice	Present annual individual doses (mSv)		Collective dose commitments	
	Per caput (world population)	Typical (exposed individuals)	10^6 man·Sv	Equivalent years of background
Annual			*Per year of practice*	
Natural background	2.4	1–5	11	1
Medical exposures (diagnostic)	0.4–1	0.1–10	2–5	0.2–0.5
Occupational exposure	0.002	0.5–5	0.01	0.001
Nuclear power production	0.0002	0.001–0.1	0.001 (0.03)[a]	0.0001 (0.004)[a]
Single			*Per total practice*	
All test explosions together	0.01	0.01	5 (26)[a]	0.5 (2.4)[a]
Nuclear accidents			0.6	

[a] The additional long term collective dose commitments from Rn and C-14 for nuclear power production and C-14 for test explosions are given in parentheses.

Although several hundred radionuclides are produced by nuclear explosions, or are present in irradiated reactor fuel, only a limited number contribute significantly to human exposure from artificial radionuclides.

The contributions from the different sources to the total amount of radionuclides released are presented in Table II [2–6]. The contamination from atmospheric nuclear weapons tests was distributed much more evenly than contamination from accidents. The fallout from the Windscale and Three Mile Island accidents mostly was contained in the country where the accident occurred. In contrast, the fallout from Chernobyl was distributed over the whole of Europe and the distribution was very uneven.

TABLE II. RELEASES OF RADIONUCLIDES IN REACTOR ACCIDENTS AT WINDSCALE, THREE MILE ISLAND AND CHERNOBYL COMPARED WITH RELEASES FROM NUCLEAR WEAPONS TESTS (10^{15} Bq)

Radionuclide	Half-life	Windscale (1957)	Three Mile Island (1979)	Chernobyl (1986)	Nuclear weapons tests (1945–1980)
Sr-89	50.5 d	0.005	—[a]	80	90 000
Sr-90	28.6 a	0.000 22	—	8	600
I-131	8.05 d	0.60	0.000 55	260	700 000
Te-132	78.2 h	0.60	—	48	70 000
Xe-133	5.25 d	12	370	1 700	1 900 000
Cs-134	2.06 a	0.001 2	—	19	—
Cs-137	30.2 a	0.046	—	38	960
α emitters		0.008	—	6	180

[a] Not detected.

1.1. Atmospheric nuclear weapons tests

Most atmospheric test programmes took place during 1957–1958 and 1961–1962. The total number of atmospheric nuclear weapons tests was 423. A total yield of 545 Mt in equivalent amounts of TNT is estimated to have been exploded, corresponding to a fission yield of 217 Mt [2].

Owing to the high temperature of the detonation the radionuclides from the nuclear explosions were lifted into the upper atmosphere. Thus, very short lived radionuclides had no time to be deposited on the Earth's surface before decaying to an insignificant level. Radionuclides with a half-life ranging from a few days to a few decades were the most significant in terms of the total dose to man.

However, some of the gamma emitters with a relatively short half-life were significant contributors to the external dose. The most important long lived radionuclide from this point of view is ^{137}Cs. The most significant radionuclides contributing to the internal dose are ^{89}Sr, ^{90}Sr, ^{131}I and ^{137}Cs. Theoretically the contribution of ^{14}C to the total dose should be of great importance, but it is distributed so evenly in the environment, and its half-life is so long, that its contribution is difficult to determine.

In the 1950s and 1960s radiochemical separations were also needed when analysing gamma emitters. In the 1950s total beta or gamma measurements were still frequently used. Potassium chloride was commonly used in converting counting data to disintegration rates.

In the 1958 UNSCEAR report [1] one annex dealt with methods of measurement. A brief summary of the determination of the fallout nuclides ^{89}Sr, ^{90}Sr, ^{131}I and ^{137}Cs was given, as well as methods for total beta and gamma emitters. The following instructions were given.

The determination of strontium activity in the various materials described in the annex involved preparation of the sample, separation of strontium and measurement of the activity.

The method for radiochemical determination of ^{90}Sr and also ^{89}Sr was almost the same as that used today. It involved separation of the strontium from interfering radionuclides by several purification steps, measurement of the total strontium activity, determination of the ^{90}Sr activity through its daughter, ^{90}Y, and calculation of ^{89}Sr and ^{90}Sr. Anticoincidence beta counters with a background of about 1 count/min were available. Also, good reference samples were available through the UNSCEAR secretariat for intercalibration.

In some research centres the ^{137}Cs burden of humans was measured in vivo with a whole body spectrometer. Gamma spectrometers were available but they were expensive, and an experienced physicist was needed to operate them. Also, the sensitivity of the detectors was still low.

Radiochemical methods for measuring ^{137}Cs were commonly used. These caesium determination problems were described in the 1958 UNSCEAR report:

> "Radiochemical separation techniques have been described which allow measurement of the caesium beta or gamma activity without energy discrimination. Adequate standards have not been available until recently. An accuracy of ±25 percent may be obtained by comparison of the beta activities of the ^{137}Cs with a ^{90}Sr standard. An intercomparison programme for development of ^{137}Cs standards is desirable."

For ^{131}I both gamma spectrometric and radiochemical methods were used. Adequate standards were commercially available.

In the fallout from atmospheric nuclear weapons tests in the 1950s and 1960s the radionuclides ^{90}Sr and ^{137}Cs were the most important as internal hazards. In fresh fallout, ^{131}I was important, in spite of its short half-life, because of its selective concentration in the thyroid gland.

1.2. Increased nuclear power production and waste disposal

The twenty year period from 1965 to 1985 was a time of increasing nuclear power plant development. From the environmental contamination point of view this period was a time of decreasing contamination of the environment by radionuclides. Radionuclides from earlier atmospheric nuclear weapons tests were decaying, being fixed into the soil and being transported away.

On the other hand new types of radionuclides from reactors were being released into the environment. Corrosion and activation products, such as ^{58}Co, ^{60}Co, ^{51}Cr, ^{54}Mn, ^{124}Sb and ^{110}Ag^{m}, were found in minor amounts in terrestrial and marine environments.

During the latter half of this period greater attention was paid to nuclear waste disposal. The release and transport of some radionuclides with long half-lives have been studied extensively.

Increasingly sensitive methods and detectors were developed to determine these low levels of radionuclides in the environment.

Low background beta measurements are still made by anticoincidence Geiger–Müller counters or proportional counters. The background has decreased over the past thirty years from about 1 count/min to 0.2 count/min for 30 mm detectors. The radiochemical methods for determination of ^{89}Sr and ^{90}Sr are almost identical to the methods developed in the mid-1950s.

For gamma measurements, NaI(Tl) detectors have been largely replaced by semiconductor detectors and multichannel analysers. At the beginning of the 1970s small planar Ge(Li) detectors and 400–512 channel spectrometers were available. At the end of the 1970s computer analysis of gamma spectra was already relatively common. New coaxial Ge(Li) detectors with 4000–8000 channel analysers and a relative efficiency of 15–30% (compared with the efficiency of the 7.5 cm NaI detector) were in use. New, high purity end cap germanium detectors (HPGe detectors) with 40–50% efficiency and a resolution of 2 keV were commonly available in the mid-1980s. The number of channels ranges from 8000 to 16 000. By the end of the 1970s gamma spectrometers were often equipped with computers, but the latest developments have produced gamma spectrometers with only buffer memory directly connected to high powered computers.

Transuranic alpha emitters have been studied extensively over the past thirty years, but it is the problems with high level nuclear wastes that have commonly drawn attention to their behaviour in the environment.

At the end of the 1950s detectors for transuranic alpha emitters were mainly pulse ionization chambers with pulse height analysers. In the 1960s new detectors, at first CsI(Tl) detectors and then new semiconductor detectors, such as silicon surface barrier (SSB) detectors, and in the 1980s passivated implanted planar silicon (PIPS) detectors were developed. Alpha emitting radionuclides can be measured by alpha spectrometric methods after radiochemical separation.

1.3. Chernobyl accident

The explosion at the Chernobyl nuclear power plant on 26 April 1986 was the first accident in which large amounts of radionuclides were spread over several countries.

The radionuclides, identified in aerosol samples, were predominantly volatile elements. The radionuclides detected by gamma spectrometers included ^{99}Mo, ^{99}Tc^{m}, ^{103}Ru, ^{127}Sb, ^{129}Te, ^{132}Te, ^{131}I, ^{132}I, ^{133}I, ^{134}Cs, ^{136}Cs, ^{137}Cs, ^{140}Ba and ^{140}La. The radionuclides ^{131}I, ^{134}Cs and ^{137}Cs made the greatest contribution to the total dose.

In the fallout from the Chernobyl accident strontium isotopes were present in only minor amounts. The ^{137}Cs/^{90}Sr ratio was approximately 100:1, or even less. Transuranic elements were also detected in this fallout, but their concentrations were of the order of some hundreds of microbecquerels per cubic metre. Owing to the development of alpha and gamma spectrometric methods and to readiness to perform radiochemical analysis, various radionuclides were determined relatively rapidly after deposition of the fallout.

In summer 1986 it was evident that the most important radionuclides were ^{131}I, ^{134}Cs and ^{137}Cs. The analytical problems with the Chernobyl fallout were thus mainly gamma spectrometric.

2. LABORATORIES FOR RADIOACTIVITY MEASUREMENT

The radionuclides released into the environment may penetrate our food and drinking water. Detection of these radionuclides depends mainly on radioactivity concentrations. If the radioactivity level is so high that immediate protective measures are needed, it can be monitored with relatively simple instruments and over a short measuring time. The estimation of intervention levels, as proposed by the World Health Organization and the Food and Agriculture Organization of the United Nations, requires more detailed assessment methods. The levels to be determined in radioecological studies are often so low that a well equipped laboratory using sophisticated radiochemical separation and low level spectrometric methods is needed. In any case, for food radioactivity determinations, a very careful sample collection scheme is always necessary. These problems have been extensively discussed in a new guidebook published by the International Atomic Energy Agency (IAEA), entitled Measurement of Radionuclides in Food and the Environment [7].

The programme for monitoring environmental radioactivity requires laboratories of different levels. They may be as follows:

Central laboratory. There should be one or a few in each country. The laboratory should be fully equipped for analysing all kinds of environmental samples of alpha, beta and gamma emitting radionuclides. Its programme should also include radioecological studies on details of mechanisms of radionuclide transfer.

Regional laboratories. According to the size of the country and to the extent of the environmental programme, some well equipped laboratories should be available in addition to the central laboratory. These laboratories should be able to perform low background gamma spectroscopy, and possibly ^{90}Sr determinations.

Local laboratories. These laboratories could be normal food laboratories or customs laboratories, which in an emergency situation should be able to monitor different foods. The same are also needed in large dairies and the food industry. The equipment may be relatively simple gamma spectrometers or integrating gross gamma counters. The main difficulty is in producing consistently correct measurements.

2.1. Central environmental laboratory

According to the IAEA's guidebook [7] a central laboratory should have the facilities, equipment and personnel needed for performing detailed analyses on all types of environmental material and food for the relevant radionuclides associated with any type of nuclear operation which could accidentally release radioactivity to the environment. This means that the laboratory would be capable of performing the required chemical separations and nuclear measurements for determination of:

(a)   Gamma emitters by non-destructive gamma ray spectrometry,
(b) ^{90}Sr and/or ^{89}Sr by radiochemical methods,
(c)   Tritium by liquid scintillation counting,
(d)   Transuranic elements by chemical separations and alpha spectrometry.

The detection limits for different samples depend on measuring time and sample size. Typical values for low level environmental samples are around 0.1 Bq per kilogram or litre of sample.

The central laboratory could also serve as a national reference laboratory for low level radioactivity measurements.

The size of the facilities depends on the number of local nuclear installations (e.g. reactors, reprocessing plants), on food import controls and on the radiation protection programmes of the country in question.

Obviously, organizations that are just beginning to set up monitoring facilities may not be able to establish a full scale operation as outlined here. However, good judgement in selecting the appropriate facilities and equipment will facilitate the founding of a smaller scale yet adequate monitoring laboratory.

2.2. Local laboratories

The laboratory system in Finland is taken as an example of local laboratories (Fig. 1 [8]).

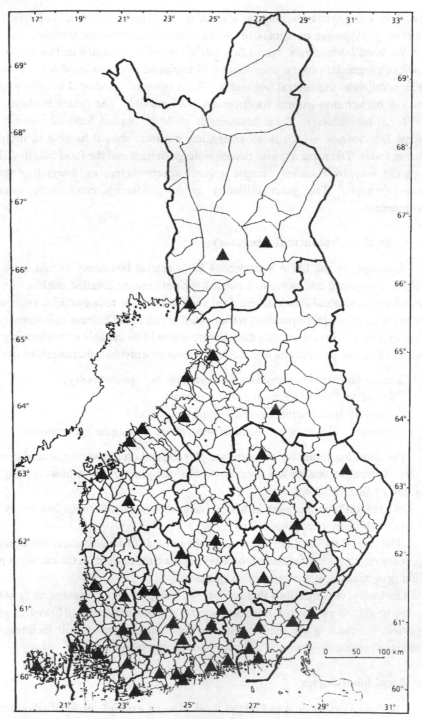

FIG. 1.   The network of local laboratories in Finland for monitoring radioactivity in food. The
laboratories are normal food laboratories equipped with an integrating gamma counter.

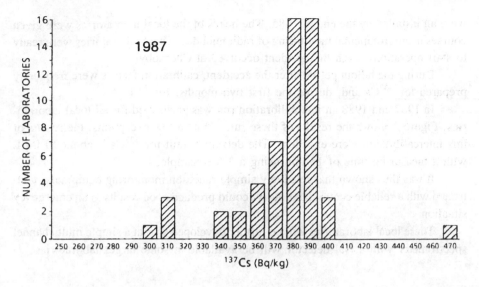

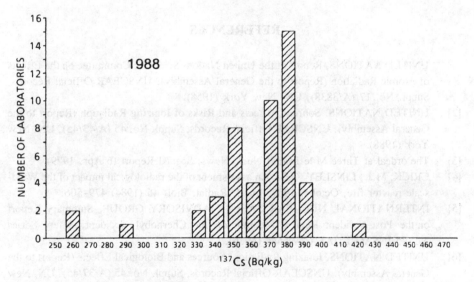

FIG. 2. Intercalibration results of local laboratory gamma monitors. All laboratories measured milk powder with the same activity concentration.

The local laboratories were originally equipped with a Geiger counter during the early 1960s. These counters were later replaced by modern, simple integrating gamma counters. All local laboratories were supplied with identical gamma counters with a small NaI detector. Also, all the detectors were equipped with the same type of background shielding. It was thus possible to calibrate these counters with sufficient accuracy using a gamma spectrometer. These new local laboratory counters were all installed by the end of 1985. The heads of the local laboratories were given courses in environmental monitoring of radionuclides. These laboratories were ready to start operations when the accident occurred at Chernobyl.

During the fallout period after the accident, calibration factors were frequently prepared for ^{137}Cs and, during the first two months, for ^{131}I.

In 1987 and 1988 an intercalibration run was performed for all local laboratories. Figure 2 shows the results of these runs. With a few exceptions, the results of this intercalibration were excellent. The detection limit for ^{137}Cs is about 10 Bq/L with a measuring time of 15 min using a 2.5 L sample.

It was thus shown that relatively simple radiation monitoring equipment, connected with a reliable central laboratory, could produce good results in an emergency situation.

These local laboratories will be further developed so that a simple multichannel spectrometer with a NaI detector will be available in some larger laboratories.

REFERENCES

[1] UNITED NATIONS, Report of the United Nations Scientific Committee on the Effects of Atomic Radiation (Report to the General Assembly), UNSCEAR Official Records, Suppl. No. 17 (A/3838), UN, New York (1958).

[2] UNITED NATIONS, Sources, Effects and Risks of Ionizing Radiation (Report to the General Assembly), UNSCEAR Official Records, Suppl. No. 45 (A/43/45), UN, New York (1988).

[3] The ordeal at Three Mile Island, Nucl. News, Special Report (6 Apr. 1979).

[4] CRICK, M.J., LINSLEY, G.S., An assessment of the radiological impact of the Windscale reactor fire, October 1957, Int. J. Radiat. Biol. **46** (1984) 479–506.

[5] INTERNATIONAL NUCLEAR SAFETY ADVISORY GROUP, Summary Report on the Post-Accident Review Meeting on the Chernobyl Accident, Safety Series No. 75-INSAG-1, IAEA, Vienna (1986).

[6] UNITED NATIONS, Ionizing Radiation: Sources and Biological Effects (Report to the General Assembly), UNSCEAR Official Records, Suppl. No. 45 (A/37/45), UN, New York (1982).

[7] INTERNATIONAL ATOMIC ENERGY AGENCY, Measurement of Radionuclides in Food and the Environment: A Guidebook, Technical Reports Series No. 295, IAEA, Vienna (1989).

[8] PAAKKOLA, O., "Local laboratory network in Finland", paper presented at Mtg on Rapid Instrumental and Separation Methods for Monitoring Radionuclides in Food and Environmental Samples, Vienna, 1988.

RADIOLOGICAL MONITORING AND DECISION MAKING IN HUNGARY IN THE EARLY PHASE OF THE POST-ACCIDENT PERIOD

L.B. SZTANYIK, B. KANYÁR, I. NIKL, D. STÚR
Frédéric Joliot-Curie National Research Institute
    for Radiobiology and Radiohygiene,
Budapest, Hungary

Abstract

RADIOLOGICAL MONITORING AND DECISION MAKING IN HUNGARY IN THE EARLY PHASE OF THE POST-ACCIDENT PERIOD.

Regular monitoring of radiation and radioactive substances in the environment was started in Hungary in the late 1950s in order to detect the appearance and deposition of radioactive fission products released by atmospheric nuclear weapons tests. A more detailed and organized environmental surveillance was initiated around a nuclear power plant in the mid-1970s, but its scope was limited to an area of about 30 km radius. By 1986 the experience which had accumulated during more than a quarter of a century provided the capability of implementing a rather comprehensive environmental monitoring programme practically within hours after the information on the reactor accident at Chernobyl in 1986 was received. Decisions taken for the protection of the public and the information given on the situation and its changes were based on the results of this monitoring.

1. INTRODUCTION

The accident at the Chernobyl nuclear power station in the early morning of 26 April 1986 was by far the worst accident that had ever occurred to any reactor; it was the only power reactor accident to result in extensive contamination of the environs. Although the consequences of the accident induced a worldwide concern among the public, it also provided ample opportunity for countries to test their emergency preparedness, monitoring systems and methods as well as their capabilities to respond adequately. The lessons learned from this event shall be utilized properly to improve the safety of NPPs.

2. THE SYSTEM OF ENVIRONMENTAL MONITORING

Studies of the contamination of the environment with man-made radioactive substances were initiated in Hungary by a research institute in the early 1950s. These

early studies focused on the detection of radioactive fission products of nuclear
weapons tests performed in the atmosphere and on the deposition of these products
on the ground surface [1]. As fallout from testing weapons continued to appear and
resulted in worldwide concern, radiological monitoring networks were set up in the
late 1950s and early 1960s by Hungarian national authorities responsible for different
sectors of the environment. These networks were made up of stations or laboratories
dispersed throughout the country with the aim of measuring the gross beta and gross
gamma activities in aerosol, fallout, water, soil, grass and food samples and of
providing information for civil defence purposes. Nuclide specific analyses could
only be performed subsequently by the head institutes of the networks [2].

After an agreement was signed by the principal nuclear powers in 1963 to ban
open air testing, environmental levels of radioactivity from that source began to
diminish and interest in environmental monitoring decreased accordingly.

Interest in the subject was renewed again in the mid-1970s, when the introduction of nuclear power to Hungary was decided by the Government. An extensive pre-
operational monitoring programme was designed and launched to establish background levels of the environmental radiation and radioactivity in the area, and to
assess the exposure of the population to these sources, soon after the construction
work of the NPP had started.

Responsibility for environmental surveillance around the site of the NPP was
assigned to the operating organization and to the competent authorities, including
those who had already participated in the environmental monitoring for civil defence
purposes. Among the latter, a network for monitoring airborne radioactivity was
operated by the National Meteorological Service. Surface and underground waters
were monitored by laboratories of the National Water Authority. (The two authorities have been incorporated recently into the newly established Ministry of Environmental Protection and Water Economy.) Radioactive substances in soil, agricultural
products and foodstuffs are controlled by institutes of the Ministry of Agriculture and
Food (MOAF). Environmental radiation levels, both indoors and outdoors, and
activity concentrations of radionuclides in drinking water, food products and other
substances having direct relevance to the health of people are subject to control by
the Ministry of Health (MOH) through its Radiological Monitoring and Data Acquisition Network. This network consists of regional radiation hygiene laboratories in
every third public health and epidemiology station under the general guidance of our
institute. The Environmental Radiological Surveillance System of the Authorities is
operated jointly by the authorities involved and its computer assisted Data Evaluation
Centre is operated by our institute for collection, evaluation and interpretation of the
monitoring results.

In addition to the regulatory authority operated networks, the Central Research
Institute for Physics and the Isotope Institute in Budapest, the Institute of Nuclear
Research in Debrecen, the NPP in Paks, and the Mecsek Ore Mining Company in
Pécs have their own environmental monitoring systems.

At the end of April 1986, the monitoring networks of the authorities and the environmental control systems of the research centres were put on alert and requested to switch over to continuous operation. A few days later, when demand in the numbers of measurements exceeded the overall capacity of these systems, some other research laboratories and university institutes also became involved in the environmental monitoring and foodstuff control programmes. These control systems were complemented and assisted by stationary and mobile units of the Civil Defence Organization.

3.   THE SYSTEM OF DECISION MAKING

It is required by the Atomic Energy Act of 1980 and its enacting clause that for the prevention of any dangerous situation that may arise from the operation of nuclear facilities in their surroundings, an emergency plan has to be prepared in co-operation with the appropriate regional and national authorities. This plan shall contain the measures necessary for prevention or alleviation of the danger and must include the names of persons responsible for taking these measures.

Accordingly, the authorities' emergency plan covers all activities that shall be carried out by the authorities or that are under their direct responsibility in a serious emergency situation. The protective measures that can be implemented in case of a nuclear accident are designed to reduce the risk of possible radiation exposure to members of the public. Therefore, decisions to implement any of the protective measures, such as sheltering, iodine prophylaxis, evacuation, etc., are to be made on the basis of projected doses for the potentially exposed individuals.

Responsibility for co-ordination of the national efforts in the event of a major nuclear accident has been assigned to a governmental committee headed by one of the deputies of the Prime Minister and constituted by representatives of the competent national authorities. The committee is assisted by a standing advisory board made up of experts in nuclear safety, radiation hygiene, environmental protection, meteorology, hydrology, medical services, equipment, and facilities of existing research and university institutions, as well as organizations that have permanent emergency functions, such as civil defence and ambulance services. The advisory board is available for the decision makers in regard to protective measures in all phases of an emergency.

Following the Chernobyl accident, assessment of the situation and evaluation of its development were made regularly by the advisory board on the basis of the measurement data provided by the environmental monitoring networks and research centres. As a result of these evaluations, recommendations were also suggested for the authorities and advice given to the population on the measures to be taken to reduce the projected levels of exposure. It was mainly attributable to these measures and advice that no overreaction was experienced on the part of the authorities and no serious panic developed among the population.

4.   MONITORING PERFORMED AND DECISIONS TAKEN IN THE EARLY POST-ACCIDENT PERIOD

It was evident from the very beginning that appropriate assessment of the situation developing in Hungary and adequate decision making on protective measures to mitigate its consequences, if needed, could only be based on careful monitoring of radiation levels and radioactive substances in the environment, in drinking water and in foodstuff samples.

4.1.   Monitoring of the atmosphere

Gross beta activities of aerosol samples taken and measured by stations of the radiometeorological and radiohygienic networks, the environmental surveillance sytems of the NPP at Paks, the Central Research Institute for Physics and our institute in Budapest were the first to detect that radioactively contaminated air masses appeared over the northern and northwestern regions of the country during the night of 29 to 30 April 1986. A somewhat less pronounced second contamination occurred from 3 to 4 May and a more significant third one from 6 to 7 May from the direction of the south and southeast. After an intense rainfall during the night from 8 to 9 May, radioactivity of the aerosols decreased remarkably [3].

These relatively simple and inexpensive measurements of airborne radioactive substances provided valuable information on geographical distribution and temporal variations of the radioactive contamination. It could be concluded that the level of contamination in the area of Budapest represented reasonably well the average level of contamination over the most affected regions of Hungary. This was certainly a great advantage, since the research centres most prepared for precise environmental monitoring were located in Budapest and could assist in evaluation of the situation and could provide expert advice. The disadvantage of the method was that it was lengthy.

Radionuclides identified in aerosol samples by gamma spectrometry indicated the prevalence of fission products such as ^{103}Ru, ^{131}I, I/^{132}Te, ^{134}Cs and ^{137}Cs (see Fig. 1). The particulate fraction of ^{131}I was found to be approximately 30–40% of the total ^{131}I activity, while a much higher proportion of I/^{132}Te was associated with airborne particles. The maximum concentration of ^{131}I in the area of Budapest on 1 May was between 10 and 15 Bq/m^3 which decreased quickly during the subsequent days.

Since the radioiodine concentration in air and the variation of average breathing rate with age were known, the age dependent intake of radioactive iodine via inhalation could be assessed. Because the maximum uptake of ^{131}I was found to be not more than a few hundred becquerels, introduction of iodine prophylaxis by means of the administration of stable iodine preparations was considered to be unwarranted.

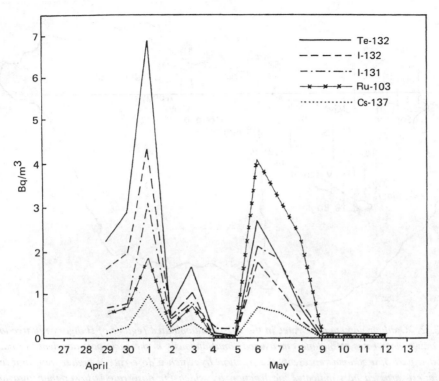

FIG. 1. Daily average activity concentration of the most significant fission products in the atmosphere in Budapest between 30 April and 12 May 1986.

4.2. Outdoor gamma dose rates

Gross beta activities found in aerosol samples also served to select those regions where outdoor gamma dose rate measurements were considered to be important. For these measurements an RSS-111 environmental monitor containing a high pressure ionization chamber was used.

In the area of Budapest, the dose rate of environmental radiation started to increase quickly from its normal value of 95 ± 20 nGy/h on 30 April and reached a maximum level of 430 nGy/h during the night of 1–2 May. Subsequently it decreased gradually with minor recurrences. Dose rates higher than those in Budapest by a factor of less than 2 were found only in a few regions of the country, where the gross beta activity of aerosol samples had also been higher before (see Fig. 2).

The maximum dose rate of environmental radiation measured and its tendency to decrease quickly led us to conclude that any advice to the public to stay indoors and avoid outdoor activities, or to restrict children from playing outdoors, would not be justifiable.

282   SZTANYIK et al.

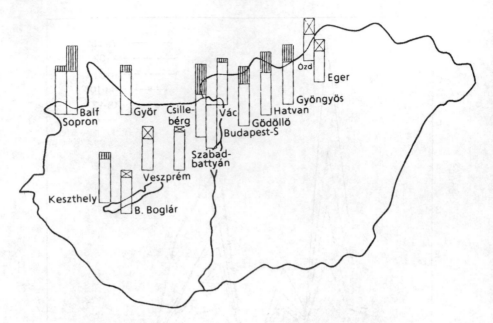

FIG. 2. *Outdoor gamma dose rates in the most contaminated regions of Hungary relative to the actual dose rates at Budapest measured simultaneously in the early post-accident period. Shaded part of a column represents the fraction by which a dose rate is higher than that in Budapest; crossed part indicates the fraction by which the dose rate is lower than that in Budapest.*

4.3. Radionuclide deposition

Deposition of radioactive substances from the atmosphere onto the ground surface was also monitored by the same networks and institutional surveillance systems as the radioactivity of aerosols. The bulk of airborne radioactive particulates deposited on the ground owing to the rainfall on 29 and 30 April 1986, as well as on 8 and 9 May. Gross beta activity of the fallout collected in Budapest during the first two days was above 20 kBq/m^2 and about one half of that was found in the subsequent weeks. For many years before the accident, the monthly average of gross beta activity in fallout had been less than 50 Bq/m^2.

Even the sharply increased deposition densities found were underestimations. The sample preparation method developed in the early 1950s for determination of ^{90}Sr and ^{137}Cs activities in the fallout of nuclear weapons tests turned out to be unsuitable for determining deposition of radioactive substances released by a reactor accident, since a large proportion of volatile radionuclides, such as iodine, was lost in the course of evaporation and subsequent incineration of the residuum.

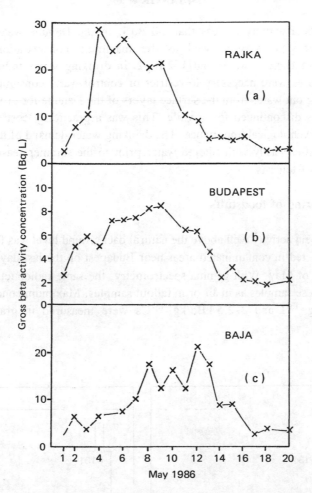

FIG. 3. Gross beta activity in the River Danube water measured in the early post-accident period at the points (a) where the River Danube enters Hungary, (b) flows through Budapest and (c) leaves the country.

4.4. Monitoring of surface waters

All important surface waters, rivers and lakes and sources of drinking water were monitored twice a day mainly by the National Water Authority in the early post-accident period (see Fig. 3). It is worth noting that almost 30% of drinking water consumption in Hungary is provided from the River Danube, partly as filtered water from the wells of the riverside and partly as processed water taken out directly from the surface layers of the river.

284   SZTANYIK et al.

Gross beta activity of less than 30 Bq/L in the Danube water during the first two weeks of May as well as the maximum concentration of ^{131}I of 10–15 Bq/L in Danube water and 1–2 Bq/L in drinking water indicated that no special measures were necessary to restrict or control water consumption. Nevertheless, taking out water from the surface layers of the Danube for processed drinking water was discontinued for a while. This was a practically costless protective measure of psychological importance. The drinking water demand of the capital city could be satisfied easily with filtered water prior to the summer season and to the mass arrival of tourists.

4.5. Monitoring of foodstuffs

Gross beta activity well above the natural background level was found in grass samples collected in contaminated areas near Budapest on the last day of April and the first days of May. With gamma spectrometry, the same radionuclides could be detected in these samples as in air or in fallout samples. Maximum concentrations of 10–12 kBq/kg ^{131}I and 2–2.5 kBq/kg ^{137}Cs were measured in grass on 1 and

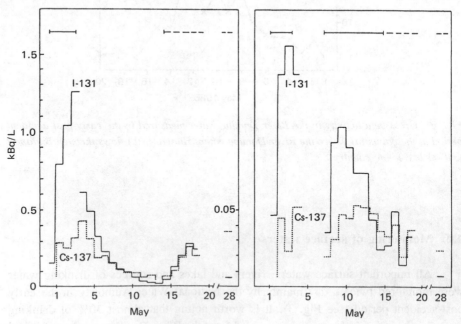

FIG. 4. Radioactivity of milk from cows grazing (——) for a few days, then kept on stored fodder, and thereafter grazing again or fed with fresh grass (---). The concentration of ^{131}I is given on the left hand ordinate, that of ^{137}Cs on the right hand ordinate, both in kBq/L.

2 May. Grazing of farm animals during these days resulted in a rapid increase in the concentration of ^{131}I and ^{137}Cs in fresh milk. A maximum concentration of 1.3–1.5 kBq/L of ^{131}I was found in milk in the first days of May and about 50 Bq/L of ^{137}Cs between 5 and 15 May (see Fig. 4).

On the basis of the results of monitoring fresh milk, State farms and farming co-operatives were requested to interrupt the grazing of cattle. The population was advised to consume only milk and milk products collected, controlled and put on the market by the State dairy industry. Control of milk and milk products marketed by small individual farms would have been an unrealistic task. Radioactive ^{131}I and ^{137}Cs concentrations in blended milk produced by three plants of the dairy industry in Budapest remained significantly lower than those found in some fresh farm milk. The maximum concentration of ^{131}I never exceeded 200 Bq/L and that of ^{137}Cs was less than 50 Bq/L even at the end of the month of May [4].

In all dairies of the country, regular monitoring of incoming milk was performed by the laboratories of the MOAF, and consumer milk and milk products were controlled by the radiohygiene laboratories of the MOH throughout the critical post-accident period. Milk with a high contamination level was withdrawn and either converted into milk powder or cheese, or used for animal feeding.

Significant radioactive contamination was found in leafy vegetable (spinach, sorrel and lettuce) samples, and in fruit (strawberries, cherries and currants) samples taken in the first 2–3 weeks of May. This contamination, however, was superficial and about 25–50% of ^{131}I and 60–90% of ^{137}Cs could be removed from the surface by running water.

Therefore, the attention of the public was called to the importance of careful and repeated rinsing of vegetables and fruit before consumption, in particular those which are usually consumed raw, without any culinary preparation such as cooking or baking.

5.   PREDICTED DOSE

Finally, the population was informed of the level of dose that was likely to be received from environmental sources and from the consumption of contaminated food as a consequence of the Chernobyl accident. According to our preliminary estimate, this was less than one half of the effective dose equivalent that every member of the public receives annually from natural sources, i.e. sufficiently low enough to warrant no further protective measures, beside those introduced by the authorities in the early period. Pregnant women were particularly advised not to seek artificial termination of pregnancy, as recommended by some incompetent general practitioners, because it was unjustified.

Later whole body measurements of incorporated radionuclides [5] confirmed that the majority of the population had even refrained from consuming vegetables and fruit rather than followed the expert advice given. Therefore, intake of radio-nuclides via ingestion was two–three times lower than calculated from the contamination level of foodstuffs and the consumption rate of food assumed conservatively to be the same as the statistical average in the preceding years.

6. CONCLUSIONS

(1) Monitoring of radiation and radioactive substances in the environment, drinking water and foodstuff samples is prerequisite to assessing the developing situation and to taking reasonable decisions following an accident at a nuclear installation.

(2) The risk to the public under such circumstances can only be assessed on the basis of nuclide specific measurements. However, preliminary information on the geographical distribution of the radioactive contamination and on the tendency of its changes can also be gathered by much simpler methods such as gross beta activity and/or gross gamma activity measurements.

(3) For early warning of radioactive contamination of the environment, a system of stations equipped with ionization chambers or Geiger–Müller tube instruments which indicate significant increases in dose rates of environmental radiation is advisable.

(4) In the event that responsibility for different sectors of the environment, e.g. atmosphere, water, soil, agricultural products, etc., is placed on different authorities, close co-operation is indispensable for covering all the necessary activities and for avoiding undue overlapping.

(5) Ad hoc involvement of properly prepared research laboratories and university institutes in an emergency situation seems to be a better way for enlarging the monitoring capacity of a country, rather than establishing beforehand too grandiose and expensive a monitoring system which cannot be used economically in the absence of such accidents.

REFERENCES

[1] SZALAY, A., BERÉNYI, D., Sr., Unusual radioactivity observed in the atmospheric precipitation in Debrecen (Hungary) between Apr. 22 – Dec. 31, 1952, Acta Phys. Hung. **5** (1955) 301–314 (in Hungarian).

[2] SZTANYIK, L.B., et al., "Experience of the Hungarian public health authority in management of the consequences of the Chernobyl accident", Radiation Protection in Nuclear Energy (Proc. Int. Conf. Sydney, 1988), Vol. 2, IAEA, Vienna (1988) 391–399.

[3]   BÍRÓ, T., FEHÉR, I., SZTANYIK, L.B., Radiation consequences in Hungary of the
       Chernobyl accident, Int. Agrophys. **2** (1986) 291–314.
[4]   SZTANYIK, L.B., "Regulation and control of radionuclides in food in a European
       socialist country — Hungary", Radionuclides in the Food Chain (Proc. Int. Conf.
       Laxenburg, 1987), International Life Science Institute Monographs (CARTER, W.W.,
       et al., Eds), Springer-Verlag, Berlin (1988) 421–435.
[5]   KEREKES, A., et al., IAEA-SM-306/76P, these Proceedings, Vol. 2.

EMERGENCY METHODS FOR MONITORING RADIOACTIVITY IN FOOD AND FEEDINGSTUFFS

J.A. BYROM
Food Safety (Radiation) Unit,
Ministry of Agriculture, Fisheries and Food,
London,
United Kingdom

Abstract

EMERGENCY METHODS FOR MONITORING RADIOACTIVITY IN FOOD AND FEEDINGSTUFFS.

Since the Chernobyl accident hand-held sodium iodide detectors have been used to monitor radiocaesium in contaminated sheep and also in feedingstuffs (silage) on the farm. The caesium activity in individual live sheep may be reliably determined by taking a measurement with a sodium iodide detector held against the fleshy part of the buttock of the animal. Since September 1986 over 190 000 sheep have been monitored in this way as part of the Mark and Release scheme of the Ministry of Agriculture, Fisheries and Food (MAFF), in which animals are screened for suitability for entry into the food chain. In vivo monitoring of sheep has also been employed in surveys of fell grazing animals to investigate time and spatial trends in radiocaesium in contaminated animals. Results show that between spring 1987 and spring 1989 average levels on the farms within the area of northern England most affected by the Chernobyl accident went down to 60% of their original value. In order to provide additional public reassurance the sheep monitoring programme has been extended to slaughterhouses. In the first five months after 14 February 1989, 4690 randomly selected carcasses from slaughterhouses in Cumbria and Lancashire were monitored. Results show a mean caesium activity of 60 $Bq \cdot kg^{-1}$. Concern about contamination prompted development of a quick method for monitoring bulk hay and silage with a hand-held detector. This facilitated identification of any portions with an elevated caesium content, from which samples were taken and analysed in the laboratory. In addition to monitoring of individual foodstuffs, MAFF also undertakes duplicate diet surveys in order to estimate more directly doses to particular groups of the population due to food intake. Studies have been carried out at Sellafield in Cumbria, in areas of high and low deposition from the Chernobyl accident (Cumbria, southwest England and London) and at Hinkley Point in Somerset.

1. INTRODUCTION

Following a large scale release of radioactivity affecting the United Kingdom, the initial monitoring response of the Ministry of Agriculture, Fisheries and Food (MAFF) is to supplement routine sampling programmes and obtain milk and vegetation samples from all parts of the country to determine as quickly as possible the scale and nature of any deposition and its implications for the national food supply. Such monitoring will supplement information from dispersion models and any other static monitoring equipment deployed. Monitoring will quickly be concentrated on the areas where contamination is greatest and samples will be obtained daily. Once the situation in respect of these early pathways is established, attention would be given to a wider range of foodstuffs, including animal products. As was demonstrated after the Chernobyl accident, the nature of the deposition and the manner in which it affects food chains in specific situations may require adaptations of monitoring techniques for the rapid screening of contaminated produce.

Duplicate diet studies have been used as a complementary surveillance technique for validating assessments of intake based on conventional environmental monitoring and models.

2.   IN VIVO MONITORING OF SHEEP

The caesium activity in individual live sheep may be reliably determined by taking a measurement with a sodium iodide detector held against the fleshy part of the buttock of the animal [1]. This method can be used routinely to monitor sheep at the rate of one animal per minute. The activity detected by this method is correlated to activity concentration in sheep muscle by calibration against laboratory gamma spectrometry measurements (Fig. 1).

In June 1986 restrictions were imposed on the movement and slaughter of sheep in the areas of the United Kingdom most affected by the Chernobyl accident. Since September 1986 over 190 000 sheep have been monitored as part of MAFF's Mark and Release scheme, in which animals are screened for suitability for entry into the food chain.

2.1. Survey of flocks within and around Cumbrian restricted area

In vivo monitoring of sheep has also been employed in surveys of fell grazing animals to investigate time and spatial trends in radiocaesium in contaminated animals.

Since 1987 ewes from 64 flocks from within the area in northern England under restrictions and from 28 flocks immediately outside it have been periodically monitored using the in vivo technique described. The sheep are monitored when they are brought down from their usual grazing for lambing at spring time and for clipping in the summer. In 1988 the ewes were additionally monitored in the autumn. The programme is continuing through 1989. Measurements are made within four days of the sheep's being moved in order to minimize reduction in radiocaesium activity whilst sheep are feeding on less contaminated pasture; levels of radiocaesium in young sheep moved from fells to lowland sites have been shown to decrease to one

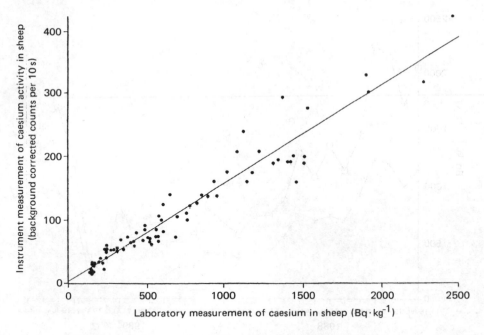

FIG. 1. Calibration graph for hand-held sodium iodide detector.

TABLE I. SURVEY OF FLOCKS WITHIN CUMBRIAN RESTRICTED AREA

	1987		1988			1989
	Spring	Summer	Spring	Summer	Autumn	Spring
Number of sheep	1280	388	900	585	584	609
Mean activity concn (Bq·kg⁻¹)	484	734	374	589	377	289
Max. activity concn (Bq·kg⁻¹)	2449	2237	1637	1848	1749	1966

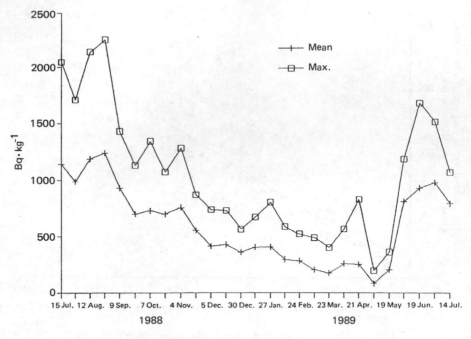

FIG. 2.   *Variation of caesium levels in a single flock of sheep with time.*

half in approximately ten days [2]. Results indicate an overall decrease in caesium
activity: between spring 1987 and spring 1989 average levels on the farms within
the restricted area went down to 60% of their original value (Table I). Overall levels
outside of the restricted area have gone down to about 70% of their spring 1986
value. These reductions are 'true' and are not subject to management influences.
Summer levels are higher than in the preceding spring for 1987 and 1988: this
increase is likely to be due to increased consumption of vegetation (sheep are usually
given feed supplements throughout the winter and spring) and also to increased
caesium activity in the vegetation in the summer [3].

In two separate projects, fortnightly readings have been taken from the
individual animals belonging to flocks grazing contaminated upland pastures. In the
first survey, the ewes were monitored from October 1986 for a period of 300 days.
Radiocaesium levels in these sheep peaked at 1300 Bq·kg^{-1} in November and then
decreased to 240 Bq·kg^{-1} in May 1987 before rising again. The second survey
shows similar results. It began in July 1988 and results for the first year are shown
in Fig. 2: the fall of mean activity in the sheep in September corresponds to the time
that the ewes were brought down from the fells for the gathering of lambs. From
this time the ewes were grazed mainly off the fells until May 1989, when a sharp
rise in activity occurred. Levels in July 1989 were approximately 70% of levels in
July 1988.

3.   SHEEP CARCASS MONITORING

In order to provide a longstop control and check on the effectiveness of the other existing controls, and to provide additional public reassurance, the sheep monitoring programme has been extended to slaughterhouses. Several alternative approaches were considered: these included live monitoring of the animals at the lairage and laboratory analysis of meat samples from the carcasses. The method chosen was an adaptation of the in vivo method using a portable sodium iodide detector. This method has several advantages: it provides a rapid assessment of activity; the monitoring equipment was already available and known to be reliable; and the method causes little disruption to the work of the slaughterhouse.

In December 1988 an initial feasibility study was carried out using 40 sheep from the Cumbrian area. Measurements on the carcasses were made initially with the detector against the top of the leg in a position corresponding to that used for live sheep. However, owing to the vertical orientation of the suspended carcasses and the weight of the detector (3.5 kg), in the method finally adopted the detector was placed against the shoulder muscle. Instrument measurements on the carcasses, together with results of radiochemical analysis of caesium in muscle, were used to provide a calibration.

On 14 February 1989 monitoring of carcasses from six slaughterhouses known to serve the restricted area began; in the first five months, 4690 randomly selected carcasses from slaughterhouses in Cumbria and Lancashire were monitored. The activities range from 0 to 456 $Bq \cdot kg^{-1}$ with a mean of 60 $Bq \cdot kg^{-1}$. The data so far do not indicate any trend in activity levels.

4.   MONITORING OF FEEDINGSTUFFS

Concern about contamination after the Chernobyl accident prompted development of a quick method for monitoring bulk hay and silage with a hand-held detector. In 1986 the first cutting of grass took place in late May and the second around the end of June. As expected, laboratory analysis of silage and hay samples indicated raised levels of radiocaesium. Silage and hay would normally be given as feed supplements to cattle from October and to sheep from December over the winter period until March.

Following initial investigative measurements, a comprehensive programme of silage monitoring using sodium iodide detectors was set up in the Cumbrian restricted area in September 1986. Field measurements were made by holding the detector face against the polythene sheet covering the clamp. For each clamp a mean of three 50 s readings taken from different parts of the clamp was recorded. Samples from 14 clamps received laboratory analysis and thus provided a calibration:

$$x \; Bq \cdot kg^{-1} \; (lab.) \; = \; -12 \; + \; 0.21c$$

where c is the number of counts in 50 s (R^2 = 72%, where R is the correlation coefficient).

The instrumental method is thus useful in providing a guide to activity and in screening for samples where more accurate results would be required. Since milk would be the critical commodity through which caesium in silage would reach man (followed by beef and sheep meat respectively), transfer factors for radiocaesium from silage to milk, together with the calibration, were used to calculate an action limit for silage monitoring. For clamps giving a mean reading of greater than 1200 counts in 50 s a core sample was taken for laboratory analysis and confirmation of caesium concentration.

Between 8 September 1986 and 25 March 1987, 346 samples of silage were monitored. Four samples were found to contain radiocaesium activities above the action limit of 1200 counts/50 s. The instrument readings from this survey showed that activity in the first cut silage was generally higher than in the second cut, which was higher than in the third cut. For the first cut the radioactive deposit would have received less weathering and less dilution by fresh growth of uncontaminated grass than subsequent cuts.

5.   DUPLICATE DIET SURVEYS

In addition to monitoring of individual foodstuffs, MAFF also undertakes duplicate diet surveys in order to estimate more directly doses to particular groups of the population due to food intake. Studies have been carried out at Sellafield in Cumbria, northwest England, in areas of high and low deposition from the Chernobyl accident, and at Hinkley Point in Somerset, southwest England. In each study between 50 and 70 participants, comprising adults and children aged three to five years, were selected by house to house visits on the basis of the amount of fresh locally produced food consumed by the participants. Each participant weighed and set aside a complete duplicate diet of every item of food eaten over a seven day period and kept a diary itemizing the content of each meal, noting all foods known to be of local origin. The diets were analysed for a range of radionuclides and doses to participants calculated. Table II gives a summary of average doses to adult participants in the three surveys, calculated assuming the same rate of intake of contamination over a whole year.

The first study was carried out around the Sellafield nuclear power plant site in two phases (February and August 1986) with the intention of investigating the seasonal variation with the availability of local produce. Participants included individuals living near the Sellafield site, with 'controls' in northwest Lancashire.

TABLE II. DUPLICATE DIET STUDIES: AVERAGE ADULT DOSES (μSv/a)

Sellafield	Phase 1	Study	10
	(February 1986)	Control	4
	Phase 2	Study	40
	(August 1986)	Control	13
Low and high Chernobyl	June 1986	SW England	5
deposition areas		London	8
		Cumbria	35
Hinkley Point	Phase 1	Study	4
	(August 1987)	Control	5

The activity deposited from the Chernobyl accident made a large contribution to the total activity detected in phase 2; average total doses to the Cumbrian participants increased from 10 to 40 μSv for adults, the difference being predominantly due to the caesium component. The doses attributed to americium and strontium in the diet also increased (from <0.3 to <1.1 μSv for americium in adult participants' diets and from 1.9 to 2.1 μSv for ^{90}Sr), owing probably to seasonal availability of local products.

A second study was carried out in June 1986 to examine radionuclide contamination of whole diets as a result of the Chernobyl accident [4]. This study was conducted in two rural areas of the United Kingdom representing high and low levels of Chernobyl fallout (Cumbria and southwest England respectively), and an urban area (London) where the food supply was likely to be derived from a more diverse range of sources. The diets were analysed for caesium, strontium and plutonium. Doses from caesium were 4 to 10 times higher in the Cumbrian group than in the other groups. Little plutonium and no strontium were detected.

The third diet study was conducted within a 9.6 km radius around the Hinkley Point nuclear power plant site in Somerset, with a group of 20 'controls' in south Devon. The average annual adult dose was 4 μSv for the study group and 5 μSv for the control group. Results from the second phase are not yet available.

REFERENCES

[1]   MEREDITH, R.C.K., MONDON, K.J., SHERLOCK, J.C., A rapid method for the in vivo monitoring of radiocaesium in sheep, J. Environ. Radioact. 7 (1988) 209–214.

[2]   HOWARD, B.J., BERESFORD, N.A., Assessment of Methods to Reduce Cs Activity of Sheep in the Field, Progress Report, Inst. of Terrestrial Ecology, Grange-over-Sands, Cumbria (1986).

[3]   COUGHTREY, P.J., KIRTON, J.A., MITCHELL, N.G., Associated Nuclear Services, Epsom, Surrey, Investigations on the Behaviour of Caesium in Agricultural Food Chains — Comparison of Data for September 1986 and September 1987, private communication, 1987.

[4]   MONDON, K.J., WALTERS, C.B., Measurement of radiocaesium, radiostrontium and plutonium in whole diets, following deposition of radioactivity in the UK originating from the Chernobyl nuclear power plant accident, Food Additives and Contaminants (1989) (in press).

POSTER PRESENTATIONS

IAEA-SM-306/48P

MONITORING FOR RADIOACTIVITY IN SCOTLAND
AFTER THE CHERNOBYL ACCIDENT

I.R. HALL, P.R. McGILL
Scottish Development Department,
Edinburgh,
United Kingdom

Monitoring has been carried out in Scotland by or on behalf of the Scottish Office, and by other bodies following the Chernobyl accident. Exposure to radiation from the accident via the following routes was considered possible:

(1) Externally from gamma radiation as the cloud crossed the country
(2) Internally from the inhalation of air
(3) Externally from deposited activity
(4) Internally from directly contaminated drinking water
(5) Internally from directly contaminated foodstuffs
(6) Internally from indirectly contaminated foodstuffs.

A summary of the assessed doses is presented in Table I.

In anticipation of the expected arrival of the radioactive cloud on 2 May 1986, operators of nuclear installations and the National Radiological Protection Board (NRPB) were asked to intensify their routine programmes of environmental monitoring, including the measurement of radioactivity concentrations in air and rain. The results indicated that deposition was highest in those areas of the country, mainly on the west coast, with heaviest rainfall as the cloud passed. Early estimates of the dose to the population indicated that countermeasures with respect to the radioactivity in air and the amount deposited on the ground were unnecessary.

Public water supplies were first sampled on 4 May. Levels of radioactivity were found to be very low and well within the limits at which action would have been required. However, it was considered prudent to advise, for a few days only, against consumption of rain water collected directly from roofs or similar structures.

From the outset it was known that a primary effect of surface deposition was the contamination of vegetables. Monitoring in early May centred on those areas where spring vegetables were produced commercially and subsequent check monitoring was carried out on a range of fruit and vegetables as they became available for market. All levels were found to be below the action levels.

In the case of Chernobyl, early monitoring indicated that ^{131}I, ^{134}Cs and ^{137}Cs were of most importance. Decisions about the need for restrictions on foodstuffs were less urgent than those for other countermeasures. For some foodstuffs, such as milk products and meat, there may be a delay of several weeks between radioactivity being released to the environment and its appearance in food, and it is possible to base decisions on direct monitoring measurements over relatively long periods. Other foodstuffs need immediate consideration although even for milk, which may be expected to show a peak ^{131}I concentration within one or two days after pasture has been contaminated, it will normally be possible to obtain some measurements before deciding whether to intervene on public supplies.

TABLE I. SUMMARY OF ASSESSED DOSE (mSv) BY MAIN EXPOSURE ROUTE[a]

Exposure route		Southwest Scotland		Rest of Scotland	
		Critical group	Average group	Critical group	Average group
(1) Cloud gamma radiation	— all	<0.01	<0.01	<0.01	<0.01
(2) Inhalation	— infant	0.01		<0.01	
	— child	0.02		<0.01	
	— adult	0.01		<0.01	
(3) Deposited activity	— adult	0.10	0.03	0.07	0.02
(4) Water	— all	<0.01	<0.01	<0.01	<0.01
(5) Directly contaminated foodstuffs	— infant	0.02	—	<0.01	—
	— child	0.02	—	<0.01	—
	— adult	0.01	—	<0.01	—
(6) Indirectly contaminated foodstuffs[b]					
(i) Cow milk	— infant	0.87	0.57	0.34	0.23
	— child	0.40	0.19	0.16	0.08
	— adult	0.20	0.10	0.08	0.04
(ii) Cow milk products	— infant	0.04	0.01	0.02	<0.01
	— child	0.03	<0.01	0.01	<0.01
	— adult	0.02	<0.01	<0.01	<0.01

TABLE I. (cont.)

Exposure route			Southwest Scotland Critical group[c]	Rest of Scotland Average group[c]
(iii)	Sheep meat	— infant	0.02	<0.01
		— child	0.06	0.02
		— adult	0.05	0.01
(iv)	Beef	— infant	0.03	<0.01
		— child	0.06	0.02
		— adult	0.05	0.01
(v)	Venison	— infant	0.26	<0.01
		— child	0.60	<0.01
		— adult	0.46	<0.01
(vi)	Wild trout	— adult	0.30	<0.01

[a] Assessments are based on intakes during the first 12 months after the accident.

[b] Not all the foodstuffs monitored are included in this table. Excluded are those foodstuffs for which the assessed dose to the population was insignificant.

[c] Separate assessments for southwest Scotland and the rest of Scotland were not made since these foodstuffs are generally consumed over the country as a whole and not just where reared, slaughtered or caught.

It was considered that the most significant dose to the public in the short term would be internal exposure from indirectly contaminated foodstuffs, particularly among young children drinking above average quantities of milk from areas of highest deposition, and an extensive programme of milk monitoring began on 2 May. For most foodstuffs, not contaminated directly, levels of activity build up reasonably slowly. This is not the case for milk, where levels of ^{131}I increase and then decrease rapidly, and decisions on countermeasures have to be made quickly before peak levels have been reached. To assist in the decision making, the NRPB has produced [1] a model of radioactivity uptake by cows from pasture and the resultant levels in milk for a number of radionuclides deposited at different rates; using the results of this work it was possible to predict likely levels of ^{131}I and caesium in cow milk from the earliest monitoring results. These indicated that action would not be

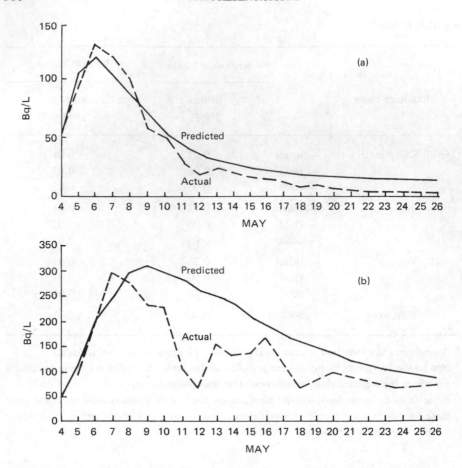

FIG. 1. Levels of (a) ^{131}I, (b) ^{134}Cs and ^{137}Cs in milk in Stranraer from 4 to 26 May 1986.

necessary. Figure 1 shows, for monitoring at Stranraer in the county of Dumfries and Galloway, comparisons between the predictions made from early milk monitoring results and the actual activity levels recorded. In general the differences were small, although actual cumulative levels were less than predicted.

The highest activity levels were found in the southwest of Scotland. For the critical group the assessed doses were high compared with exposure from other routes; for critical group infants, the assessed effective dose equivalent for the full year after the accident was about 17% of the 5 mSv/a control level adopted.

As well as cow milk, early monitoring of cow milk products and goat and ewe milk and milk products was undertaken.

Monitoring levels of activity in other foodstuffs began only about three or four weeks after deposition. On 23 May the Commission of the European Communities approved a limit, applicable to all major items of diet, of 1000 Bq/kg of combined ^{134}Cs and ^{137}Cs on contamination in foodstuffs affected by the Chernobyl fallout.

Sheep meat was first monitored on 28 May, and extensive sampling of lambs prior to slaughter took place from June onwards. It became clear that activity would, in some cases, exceed the 1000 Bq/kg limit, and action was taken on 24 June to restrict the movement and slaughter of sheep within certain areas of the country. Subsequently the areas affected were progressively reduced where monitoring of live lambs showed that activity had fallen to satisfactory levels. Since most of the lambs moved from the affected areas were moved to lowland ground for fattening, levels of caesium fell quickly after the lambs had left the designated areas. In total some 1.5 million sheep on 2900 farms were affected by restrictions. Levels in lambs have continued to decline although movement restrictions are still in place in certain areas.

The range of activity levels in other foodstuffs monitored in general gave no cause for concern.

REFERENCE

[1]   LINSLEY, G.S., et al., Derived Emergency Reference Levels for the Introduction of Countermeasures in the Early to Intermediate Phases of Emergencies Involving the Release of Radioactive Materials to the Atmosphere, Rep. NRPB-DL10, Natl Radiological Protection Board, Chilton, UK (1986).

IAEA-SM-306/68P

ENVIRONMENTAL MONITORING IN ITALY
FOLLOWING THE CHERNOBYL ACCIDENT

M. LOTFI*, S. MANCIOPPI, S. PIERMATTEI,
L. TOMMASINO, D. AZIMI-GARAKANI**
Directorate of Nuclear Safety and Radiation Protection,
National Commission for Nuclear and
   Alternative Energy Sources (ENEA),
Rome, Italy

The surveillance of environmental radioactivity in Italy is carried out by several laboratories under the co-ordination of the Directorate of Nuclear Safety and Radiation Protection of the National Commission for Nuclear and Alternative Energy Sources (ENEA-DISP). The major goal of the monitoring programme is to assess dose equivalents to the general population from a number of sources distributed on a regional scale. The programme, which has existed since the mid-1950s, has been strengthened as the result of the Chernobyl accident.

Estimating the effective dose equivalent (EDE) from dietary intakes of radionuclide contaminated foods of the average consumer in different age groups has been the aim of the present food basket study. According to the World Health Organization (WHO), foods consumed regularly are assumed to be those whose annual intakes exceed 20 kg. Considering the pattern of food consumption in Italy, such foods are cereals, vegetables, fruits, meat, milk and milk products. The intake rates of foodstuffs were taken from the food consumption statistics published by the Nuclear Energy Agency of the Organisation for Economic Co-operation and Development in 1987 [1].

Measurements of radioactive caesium in different foodstuffs were mainly carried out by the environmental radioactivity laboratories in three administrative districts. The sampling points were chosen to monitor both foodstuffs produced locally and those imported in large quantities. The samples can therefore be considered to be representative of the average national situation. For quality control of the measurements, intercomparison exercises among the various laboratories have been carried out regularly. The samples are directly analysed by HPGe detectors and the activity is referred to the fresh weight of the sample measured. The contamination

---

    * Present address: ACC/SCN, X.50, World Health Organization, CH-1211 Geneva 27, Switzerland.
    ** Present address: Paul Scherrer Institut, CH-5232 Villigen PSI, Switzerland.

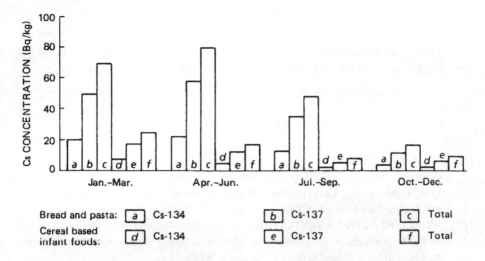

FIG. 1. Radiocaesium contents of bread and pasta (quarterly, 1987).

of wheat is not considered here, since it was the subject of another investigation and will be reported in detail elsewhere [2]. Figure 1 shows the variation of caesium activity in bread and pasta detected in samples of the most common brands and also in samples of cereal based baby foods.

The age specific doses per unit intake factor used here in the calculations of EDE are those recommended in 1988 by the WHO Working Group on guideline values for derived intervention levels for radionuclides in foods [3].

The results showed that the total individual EDE received via food ingestion in 1987 was 175, 113 and 71 μSv for adults, children and infants, respectively. The corresponding values in 1988 were 16, 12 and 20 μSv, respectively.

The most important lesson of the Chernobyl accident was the necessity for environmental monitoring networks in all countries, as most countries demonstrated lack of preparedness for large scale monitoring following a major radiation accident [4]. These networks should operate according to a well defined protocol and standardized procedures for carrying out measurements on selected environmental samples. Obtaining such data and using environmental transfer models would allow estimation of the degree of radioactive contamination of the most important food-stuffs. In this context we wish to recall another important item on which a general consensus should be reached, i.e. the definition of a set of intervention levels to be applied on a worldwide basis.

ACKNOWLEDGEMENT

One of the authors (M.L.) took part in this work with the support of the ICTP Programme for Training and Research in Italian Laboratories, Trieste, Italy.

REFERENCES

[1]   NUCLEAR ENERGY AGENCY OF THE OECD, The Radiological Impact of the Chernobyl Accident in OECD Countries, OECD/NEA, Paris (1987).
[2]   LOTFI, M., et al., Concentrations of radioactive caesium in durum wheat and its products in Italy after the Chernobyl accident, J. Environ. Radioact. (in press).
[3]   WORLD HEALTH ORGANIZATION, Derived Intervention Levels for Radionuclides in Foods. Guidelines for Application after Widespread Radioactive Contamination Resulting from a Major Radiation Accident, WHO, Geneva (1988).
[4]   MIETTINEN, J.K., Lessons of Chernobyl: A Commentary, Environ. Int. **14** (1988) 201–203.

IAEA-SM-306/21P

^{137}Cs WHOLE BODY MONITORING AMONG IAEA STAFF

R. OUVRARD, R. HOCHMANN
Division of Nuclear Safety,
International Atomic Energy Agency,
Vienna

From 1 January 1986 to 15 August 1989 about 850 whole body gamma measurements were performed at the International Atomic Energy Agency's Laboratory at Seibersdorf. These routine measurements, performed mainly on safeguards inspectors, have been of great interest for evaluating the consequences of the Chernobyl accident as far as ^{137}Cs incorporation through the food chain is concerned.

Figure 1 shows the evolution of the ^{137}Cs body burden. Data on the ^{137}Cs committed dose equivalent are shown in Table I. This quantity has been calculated using the average daily intake during equal time intervals. At the end of the whole period, the total intake of ^{137}Cs was about 11 100 Bq. With a conversion factor of 0.0136 $\mu Sv \cdot Bq^{-1}$, the committed dose equivalent for the monitored group would be about 150 μSv.

TABLE I. COMMITTED DOSE EQUIVALENT

Year	Incorporation (Bq)	Committed dose (μSv)
1986	3 450	47
1987	6 520	85
1988	1 120	15
1989	410	5.5
Total (calculated)	11 100	150

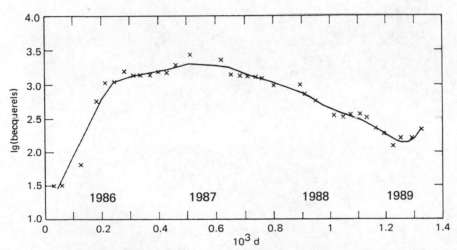

FIG. 1. ^{137}Cs body burden (average monthly values) (day 0 is 1 May 1986).

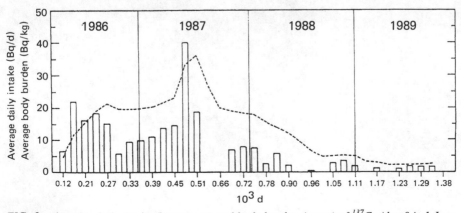

FIG. 2. Average daily intake (histogram) and body burden (curve) of ^{137}Cs (day 0 is 1 January 1986).

Monthly average daily intake and body burden are presented in Fig. 2. Values were calculated using the average Cs whole body content at the beginning and at the end of each monthly period and the value of the International Commission for Radiological Protection for the effective biological half-life (110 d).

Part II

MONITORING OF RADIOACTIVITY

(b) Strategies and Policy

ORGANIZATION OF RADIATION MONITORING OF THE ENVIRONMENT IN THE CHERNOBYL NUCLEAR POWER PLANT ACCIDENT ZONE

E.I. IGNATENKO
Ministry of Atomic Energy,
Moscow

V.I. KOMAROV
World Association of Nuclear Operators,
Moscow

A.G. PROSKURYAKOV
'Kombinat' Industrial Association,
Pripyat

Union of Soviet Socialist Republics

Abstract

ORGANIZATION OF RADIATION MONITORING OF THE ENVIRONMENT IN THE CHERNOBYL NUCLEAR POWER PLANT ACCIDENT ZONE.

The radiation monitoring problems which presented themselves in acute form immediately after the Chernobyl accident are still of vital importance today. A comprehensive radiation monitoring system was set up at the time which is still operating and undergoing further refinement. It has a number of subsystems: ground based operational monitoring, integral monitoring aided by airborne sampling techniques, spectral analysis and data analysis using a unified mathemetical modelling system. ASKRO, an automated system for monitoring of the radiation situation, has a special role to play in this complex of activities. The following tasks were set for the development of ASKRO: development, approval and optimization of regulations governing the monitoring of the radiation situation in natural media during the post-accident period; elaboration of strategies for deploying a widespread monitoring network under conditions in which the main transport arteries and lines of communication are disrupted, there is a total loss of power supply to the 30 km zone, and so on; development of equipment and high speed techniques for the mass analysis of environmental samples; development of automated radiation monitoring systems; generalization of the results obtained by the different organizations concerned with a view to establishing a databank on the environmental situation; development and use in the 30 km zone of the ASKRO model; and practical implementation of mobile radiospectrometry facilities. A hybrid monitoring system was selected which uses information on vent stack release parameters and on the technological processes and fuel core cycle of the power plant's units; and information obtained from: the monitoring sensors covering the plant building and distribution lines; the network of monitoring stations dispersed over the monitoring area; the external dosimetry laboratory carrying out monitoring by means of sampling techniques; and the mobile units with on-board spectrometers. On the basis of experience acquired with the Chernobyl ASKRO, proposals have now been put forward for the creation of a generalized standard ASKRO for all nuclear power plants — proposals which lay down the main principles governing the structure of such a system.

1.   INTRODUCTION

The urgent problems of monitoring the radiation situation which arose immediately after the accident at Unit 4 of the Chernobyl nuclear power plant continue to be of vital significance. They necessitated the establishment of an integrated radiation monitoring system, which is still in use and continues to be refined. This includes the 'Kombinat' Industrial Association's subsystem for ground based operational monitoring, the USSR State Committee for Hydrometeorology's integrated monitoring subsystem using airborne sampling techniques, the spectral analysis subsystem developed by the Nuclear Research Institute of the Ukrainian Academy of Sciences and data analysis using a mathematical modelling system. Also developed and partially implemented during the operations to eliminate the consequences of the accident was an automated system for monitoring of the radiation situation, called ASKRO.

To enable the necessary operations to be carried out a radiation monitoring network was deployed, including stations for the one-time or seasonal sampling of objects in the environment. The monitoring network is used for the following tasks:

— Monitoring of radioactive contamination in the 30 km zone around the power plant and compilation of detailed maps of the radiation situation in the zone to help with the performance of all emergency, restoration and environmental protection work;
— Monitoring of radioactive contamination in soils;
— Monitoring of radioactive contamination in the atmosphere;
— Determination of radioactive contamination levels in groundwater and surface water;
— Determination of radioactive contamination levels in bottom deposits of water bodies.

The strategic tasks include:

— Development, approval and optimization of regulations governing the monitoring of the radiation situation in natural media during the post-accident period;
— Elaboration of strategies for deploying a widespread monitoring network under conditions in which the main transport arteries and lines of communication are disrupted, there is a total loss of power supply to the 30 km zone, and so on;
— Development of equipment and high speed techniques for the mass analysis of environmental samples;
— Development of automated radiation monitoring systems;

— Generalization of the results obtained by the different organizations concerned with a view to establishing a databank on the environmental situation;
— Development and use in the 30 km zone of the ASKRO model;
— Practical implementation of mobile radiospectrometry facilities.

For the purpose of monitoring radioactive contamination, regulations were drawn up to govern the collection of samples of dry and wet atmospheric fallout, soil samples from representative sites and samples of flora and fauna.

Water samples were collected from the drainage and water supply wells in the 30 km zone down to the Quaternary and the Buchansko-Kanevskij and Senoman–Nizhnemelovoj aquifers. The water throughout the Dnepr Basin was kept under strict surveillance. Samples of bed deposits, clinging water plants, animal plankton and fish were regularly analysed.

These samples were taken to radiochemistry laboratories and subjected to dispersion and structural analysis. In air samples a pulmonary fraction of particular danger to human beings was identified. All preparations were tested for the presence of 'hot particles'. The quantity, dispersion, composition and activity of the hot particles were determined. The bulk of the samples were sent to spectroscopic units in order to determine the concentrations of individual radionuclides. The large volume of samples involved necessitated the creation of automated systems capable of determining concentrations, producing maps depicting the element specific contamination of individual land areas by products released from the reactor, and evaluating the radiation situation. In addition to the work already described, program packages were developed for Soviet-made computers of the SM1420 type, including programs capable of producing maps in two and three dimensional co-ordinates covering the entire area under consideration, based on the structure of the databank and the package of service programs developed by the Nuclear Research Institute of the Ukrainian Academy of Sciences. Programs were created to determine the scale of an accident on the basis of measured ground level concentrations of radionuclides at monitoring points situated at various distances from the power plant, including the system of ASKRO stations.

The Chernobyl accident and a number of accidents in other countries have shown the need for a monitoring system capable of determining in the shortest possible time the radiation situation in the area adjacent to a given nuclear power plant and of forecasting its development for the purpose of calculating the radiation burdens in respect of staff and public prior to taking an official decision on the protective measures to be implemented.

2.   ASKRO

In this part of the paper we shall consider questions relating to the establishment of and initial operating experience with the Chernobyl ASKRO as a system for the prompt detection of changes in and monitoring of the radiation situation.

The radiation monitoring methods that were in use before the Chernobyl accident (i.e. manual sample collection and laboratory measurements) proved to be of little effect in an emergency situation requiring monitoring and prediction operations on a realistic time-scale — something which can only be achieved through the use of automated systems.

There are two principal methods of assessing the radiation situation:

— Direct measurement on-site and in the monitoring area by means of a network of measuring instruments dispersed over this area;
— Measurement of parameters relating to radioactivity discharges through the vent stack and other possible channels through which radioactive substances could be released, measurement of meteorological parameters and calculation of the radiation situation in the zone.

Both methods have inherent values and shortcomings. Thus, in order to arrive at a picture of the radioactive contamination in the zone a large number of monitoring points are necessary. Furthermore, with direct measurement, the information is received only after the cloud has reached the sensors.

By measuring vent stack release and meteorological parameters it is possible to estimate the radiation situation in advance, without the need for a widespread network of sensors. But inherent in this method are shortcomings such as low accuracy, particularly where the local relief structure is complex, and the inability to make predictions in the event of a radioactivity leakage resulting from damage to distribution lines and building structures, as occurred at Chernobyl.

For the purpose of assessing the nature of an accident, determining the exact isotopic composition of the radioactive contamination and calculating the isofields for the corresponding forms of irradiation, the radioactive sampling method must be used.

The most promising avenue is therefore hybrid monitoring, which uses information:

— On the vent stack release parameters;
— On the technological processes and core cycle of the plant's units;
— From the monitoring sensors covering the plant building and distribution lines;
— From the network of monitoring stations dispersed over the monitoring area;
— From the external dosimetry laboratory carrying out monitoring by means of sampling techniques;
— From the mobile units with on-board spectrometers.

For collecting data in the area of the Chernobyl plant the Tunets monitoring system was used. This has a hierarchical, three level structure and includes physical information sensors with communication equipment, as well as data acquisition and display consoles.

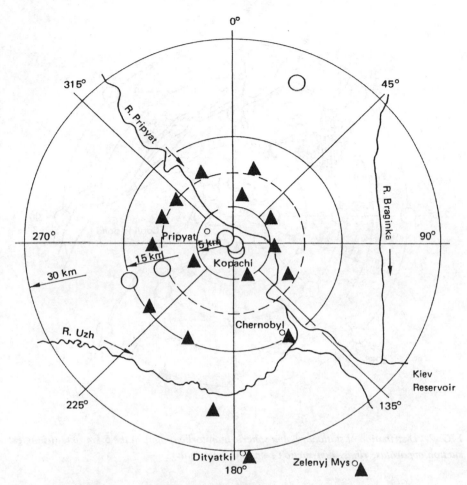

FIG. 1. Distribution of stationary atmospheric monitoring points in the 30 km zone (triangles: suction apparatus; rings: non-suction sampling apparatus).

The computer equipment used in the Chernobyl ASKRO comprises two SM1634 type minicomputers as data switchers and concentrators, and ARM 2.01 and SM1420.22 complexes for processing, databank maintenance, forecasting and generation of contamination maps, diagrams and routes.

ASKRO now has 23 measurement stations (Figs 1, 2), 6 of which are located on the plant site (1 being in the coolant pond), 3 in the town of Pripyat, 3 in the town of Chernobyl and 2 in the monitoring area. Using the ASKRO system it is possible to measure the gamma radiation level in the ranges 100 μR/h to 0.1 R/h and 0.5 R/h to 10 000 R/h, with an error no higher than 60%.[1] The information is transmitted by wire and radio channels.

---

[1] 1 R = 2.58 × 10^{-4} C/kg.

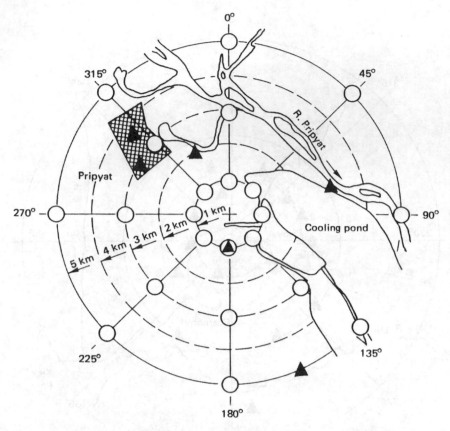

FIG. 2. *Distribution of stationary atmospheric monitoring points in the 5 km zone (triangles: suction apparatus; rings: non-suction sampling apparatus).*

In the near future it is planned to connect up the meteorological station and the radiation monitoring system, which provides information on the parameters of the releases through the vent stack belonging to Unit 3 and the sarcophagus of Unit 4 of the Chernobyl plant.

Initial operating experience with the Chernobyl ASKRO has served to confirm the effectiveness of automated monitoring systems: on more than one occasion, an increase in the gamma dose level has been detected and subsequently confirmed by manual measurements. At the same time, the data received have been characterized by a low level of information content and reliability, this being mainly so in the case of data received by radio link.

When data are transmitted by radio link the decision to transmit is taken only from the measurement station. This means that the interval between contacts is two to three hours, at which time the result of a single measurement is transmitted. Considerable fluctuations, the reason for which is hard to explain, are observed in the readings from measurement stations which communicate by radio link.

Information on the radiation situation is received from the ASKRO stations, which are the local data acquisition centres. They are housed in special buildings with grid and emergency power supplies and heating. The monitoring stations are currently equipped with:

— Detectors for measuring the gamma dose;
— Suction apparatus;
— Anemometers and thermometers;
— Rainwater collection vessels (on the ground beside the stations).

In the near future the stations will also be equipped with:

— Microprocessor controllers for hardware control, for the acquisition and preliminary analysis of data and for communications;
— Automatic spectrometric units attached to the suction assemblies to provide a rough estimate of the concentration of radioactive aerosols (of iodine).

Wire or cable lines and telephone links serve as the principal means of communication between the monitoring stations and the data acquisition centres. Radio serves as a backup link up to a distance of 50 km.

It is planned to consolidate the network of radiometric parameter measurement points through the use of portable measurement stations, each with two or three sensors with overlapping measurement ranges, control apparatus based on a single crystal microcomputer, and data transmission equipment. For communication with monitoring stations or with the data acquisition centres, low power radio stations with a range of 5–8 km, and also wire links or the existing electric communications, may be used. Such measurement stations can also be effective in the period following an accident.

A data acquisition centre performs the following functions:

— Acquisition, preliminary processing and buffering of data from the monitoring stations;
— Analysis of the radiation situation in the event of an accident;
— Control of communication links;
— Control of operating modes of the monitoring stations;
— Diagnosis of the functioning of the measurement apparatus and lines of communication;
— Control of the notification system and presentation of operational data concerning the radiation situation.

The data acquisition centre transmits the accumulated data to the processing centre through a communication link on a periodic basis, and also on request. As a reserve option it is also possible to supply the data on magnetic tape. Expansion into a regional system for monitoring the radiation situation is envisaged for the future.

The output data on the radiation situation are presented in tabular form on alphanumeric terminals, and in the form of cartograms of the activity distribution on graphic printout devices and monitors. Should an accident or other event calling for an immediate response be detected, the alert and protection systems are activated.

The most rational manner in which to present data on the radiation situation to distant users with assigned decision making functions is to link personal computers of the IBM PC class, or the equivalent Pravets ES1841 or SM1810, into the network. It will thus be possible to receive the required data in what, to all intents and purposes, amounts to real time, and also, in normal power plant operation, to put the personal computers to other productive uses.

The Chernobyl ASKRO software for the SM1420 computer is based on a PSX-2 operating system (version 4.2), adapted GSK and GRAFOR interactive graphics packages for the IBM PC, and a package comprising statistical data processing and a large number of service programs.

The mathematical simulation aids currently include two models of vent stack release distribution and a model of wind-borne dust transfer. However, since there is no automatic measurement of either meteorological parameters or the activity concentration in the vent stack releases, the models cannot yet be used for forecasting.

Initially, ADABAS was used as the database control system, but in view of its bulkiness the decision was subsequently taken to use a PCDB type relational database or equivalent.

The databases accumulate and store information on the radiation situation, meteorological conditions, models and their parameters, as well as normative, reference and geodesic information. There are plans to create a database of situation specific decisions, containing previously elaborated alternatives of optimum decisions aimed at ensuring radiation protection of staff and public in the event of various types of accident.

There have been several cases of intentional damage to measuring apparatus by persons unknown, and there is therefore a need to develop the appropriate means of protection against this not insignificant social element.

The setting up and equipping of the monitoring stations for the Chernobyl ASKRO is under way, and the task of connecting 'Gorbach' apparatus to the data acquisition centre is receiving attention.

3.   CONCLUSIONS

Using our initial experience of operation of the Chernobyl ASKRO, we have been able to identify the following top priority problems which must be solved in order to enhance ASKRO's operational effectiveness:

— Preparation of a full list of the parameters measured and of optimum layouts for the arrangement of measuring equipment;
— Use of inexpensive means of communication which permit the transmission of information upon initiative both from below and from above;
— Development and implementation of equipment to monitor the activity concentration through the vent stack, and also the rate of discharge and temperature of the gases thus released;
— Equipping of monitoring stations with microprocessor controllers capable of providing on the spot preliminary data processing and evaluation;
— Deployment and connection to ASKRO of a meteorological station;
— Development of softare and new models; debugging and improvement of the accuracy of existing facilities;
— Switching in to ASKRO of information on the core cycle and other technological processes and on damage to buildings and distribution lines.

On the basis of experience from other countries and that acquired through operation of the Chernobyl ASKRO, the 'Kombinat' Industrial Association has prepared a document entitled Proposals Regarding the Development of a Standard ASKRO for Nuclear Power Plants and the Further Development of the ASKRO for the Chernobyl Nuclear Power Plant. The document specifies the aims, tasks and main construction principles of the system.

The functions of ASKRO are specified as follows:

— Rapid detection of accidents involving the release of radioactive substances;
— Rapid supply of information to the plant services and management concerning the radiation situation within the plant and on the plant site, and provision of information to the public;
— Forecasting of the radiation situation and production of recommendations concerning implementation of the Plan of Measures for the Protection of Staff and Public in the Case of an Accident at a Nuclear Power Plant.

The following are the specific tasks that ASKRO must perform:

— Acquisition, processing, storage and presentation of data on the radiation situation in the monitoring area;
— Identification of places with a worsened radiation situation as a potential source of contamination;
— Determination of the dynamics of change of the radiation situation in the monitoring area;
— Forecasting of the radiation situation at the plant site and in the monitoring area under normal plant operation conditions and in the event of a possible accident;
— Detection of accidents and assessment of their nature and scale;
— Computation of the gamma exposure rate distribution and the concentration of radioactive aerosols in the monitoring area at any given time;

— Assessment of the radiation situation in the control area of the plant and short term forecasting of the dose burdens to staff and public;

— Prompt transmission of data concerning all aspects of the accident situation to the different departments of the plant and to the higher regional and national competent bodies, including recommendations on implementation of the Plan of Measures for the Protection of Staff and Public in the Case of an Accident at a Nuclear Power Plant;

— Computation of optimum post-accident routes to operations areas and computation of the distribution of staff engaged in operations to eliminate the consequences of the accident;

— Computation of the total, monthly and annual gas aerosol release, the balance of water-borne radionuclides and the accumulation of radionuclides in natural media;

— Computation of the transfer of radioactivity from the contaminated area by natural and anthropogenic means.

The automated data acquisition part of ASKRO should be supplemented with the results of manual sampling and laboratory measurements, with a view to achieving an accurate spectral analysis.

NUCLEAR EMERGENCY PLANNING AND MONITORING STRATEGIES IN THE NETHERLANDS

A.H. DAL
Ministry of Housing, Physical Planning
    and the Environment,
Leidschendam

J.F.A. VAN HIENEN
Netherlands Energy Research Foundation,
Petten

D. VAN LITH
National Institute of Public Health
    and Environmental Hygiene,
Bilthoven

Netherlands

Abstract

NUCLEAR EMERGENCY PLANNING AND MONITORING STRATEGIES IN THE NETHERLANDS.
   Following the Chernobyl accident the Dutch Government initiated an extensive project in order to review nuclear emergency response in the Netherlands. The project resulted in the National Plan for Nuclear Emergency Planning and Response (draft), which is described in general terms. The proposed structure and organization of technical information are considered in more detail. The paper focuses on monitoring strategies and discusses topics such as dose contributions, measurement zones, mobile monitoring teams and stationary networks. Since the National Plan still has a draft status, the figures presented in the paper are only preliminary.

1.   HISTORICAL BACKGROUND

   As a result of the Chernobyl accident in April 1986 the Netherlands was confronted with radioactive contamination of large parts of the country. In responding to this accident the Government was forced to improvise since the existing emergency regulations did not take into account the many national and transboundary aspects encountered.

In view of the established shortcomings, and given the fact that within 1000 km from the centre of the Netherlands 120 nuclear power stations are operational and 40 under construction, the Dutch Government initiated an extensive project in 1987, in which not only all tiers of Government participated, but in which also were reflected the interests of the Dutch nuclear industry and, in particular, the institutes specialized in radiological monitoring.

The objective of the project was to describe and specify in operational form the actions to be taken by the Government in the event of accidents involving radioactive substances in the Netherlands and abroad which might affect the Dutch population, and to indicate the provisions, in terms of both equipment and personnel, and the regulations required.

The project resulted in the formulation of the National Contingency Plan for Nuclear Accidents, which was presented to the Dutch Parliament in February 1989. In anticipation of political discussion and approval an implementation programme has already been started. This programme includes a study of monitoring strategies, some preliminary results of which are discussed in this paper.

2. THE NATIONAL PLAN FOR NUCLEAR EMERGENCY PLANNING AND RESPONSE

2.1. Main features

Although the National Contingency Plan devotes particular attention to nuclear power stations in the Netherlands and bordering areas, all types of accidents involving nuclear facilities and activities have been reviewed. Depending on the potential scale of the accidents and the related consequences the co-ordination of the emergency response will be effected at either the national or the local level.

At the centre of the national emergency organization, defined for severe accidents with national or international implications, a Policy Team consisting of ministers and/or high ranking civil servants will take decisions with regard to necessary protective actions, and will co-ordinate execution of the measures. The Policy Team is supported by a (Public) Information Group, responsible for providing information to the public and the press, and by an advisory team of (radiological) experts, the Technical Information Group (TIG), responsible for collection, processing and interpretation of relevant radiological and meteorological data. The TIG has to provide the Policy Team with radiological information and advice on countermeasures.

Response to accidents with lesser impact will be co-ordinated by the local government. If needed, the national emergency organization will render technical assistance in these cases.

Apart from the organization, the National Plan describes the countermeasures which can be taken in the early, intermediate and recovery phases of nuclear accidents, and sets the goals for monitoring strategies by providing intervention levels for direct protective actions and derived intervention levels for the ingestion pathways, e.g. via food and drinking water [1].

2.2. Typology of accidents

As stated above, the consequences of accidents involving all relevant facilities and activities have been examined. On the basis not only of criteria such as the potential scale of the radiological impact, but also of the scale of the measures to be taken, the size of the area potentially affected, public anxiety, the need for central public information, political and economic factors, international obligations, etc., the facilities and activities have been divided into Categories A and B, indicating the governmental level responsible for the co-ordination of the nuclear emergency response.

The Government will co-ordinate the emergency response in the case of accidents concerning facilities and activities in Category A:

— Nuclear installations,
— Ships using nuclear energy,
— Spacecraft using nuclear energy,
— Nuclear defence equipment,
— Facilities and activities in a foreign country.

Response to accidents with supposedly local impact will be co-ordinated by the municipality. The following facilities and activities have been listed in Category B:

— The uranium enrichment installation of UCN (Ultra Centrifuge Netherlands) in Almelo,
— The COVRA (Central Organization for Radioactive Waste) installations for processing and storing radioactive waste in Petten,
— Installations producing radioactive substances and sources,
— Sites (fixed and movable) where radioactive substances and sources are used,
— Civil transport of radioactive material.

2.3. Emergency classification and planning zones

For the nuclear power plants located within the Netherlands the emergency classification system of the International Atomic Energy Agency (IAEA) has been adopted. This provides four different ratings of emergencies, namely emergency stand-by, plant emergency, site emergency and off-site emergency. In this system notification of an emergency class is based on both plant status and emission criteria.

TABLE I. EMERGENCY PLANNING ZONES
(described in terms of distance of zone perimeter from plant)

Nuclear power plant capacity (MW(e))	Organization zone (km)	Direct measure zones		
		Evacuation zone (km)	Iodine prophylaxis zone (km)	Sheltering zone (km)
≤ 100	5	0	4	7
≤ 500	10	5	10	20
≤ 1000	15	5	15	30

TABLE II. CRITERIA FOR IAEA EMERGENCY CLASSIFICATION SYSTEM

Emergency stand-by	Plant emergency	Site emergency	Off-site emergency
No significant release; no measures off-site	Release ≥ 10 times allowed daily release; no measures off-site	Contamination ≥ 5000 Bq/m^2 iodine; indirect measures off-site	Off-site dose ≥ 5 mSv H_{eff} or 50 mSv H_{th}; (in)direct measures off-site

Note: H_{eff}: effective dose; H_{th}: thyroid dose.

With regard to countermeasures for the protection of the public that may be feasible under emergency conditions, two groups can be discerned:

— Direct measures, involving direct protective measures for the public, such as evacuation, iodine prophylaxis and sheltering: execution of these measures may be expected in the case of an off-site emergency.
— Indirect measures with regard to water resources management, drinking water supply, agricultural and food supplies, etc.: these measures are conceivable for both site and off-site emergencies.

To prepare for the complex direct measures, emergency planning zones have been defined, using the PWR-5 source term from the Rasmussen WASH-1400 study (retention time 2 h, release time 4 h) as reference source term for light water reactors [2]. For meteorological conditions the prevailing Pasquill class D2 (neutral, wind speed 4 m/s) was chosen.

The calculations resulted in the definition of three different zones for the above mentioned direct measures, and an organization zone in which a combination of direct measures might be expected (Table I).

The emission criteria for the emergency classes are linked to the measures (Table II), using the lowest intervention level at the site boundary for indirect measures (restrictions on grazing land) as a threshold for the site emergency, and the lowest level for direct measures (sheltering) for the transition between the site and off-site emergency classes.

3. TECHNICAL INFORMATION

3.1. Organizational aspects

The organization of technical information involves services and organizations active in this field, which are supervised by the TIG in the early and intermediate phases of an accident (Fig. 1). The TIG consists of (radiological) experts on the environment, public health, agriculture and foodstuffs, (drinking) water, reactor safety, labour protection and emergency response, and an expert on meteorology. The chairman is provided by the State Health Inspectorate of the Ministry of Housing, Physical Planning and the Environment (VROM).

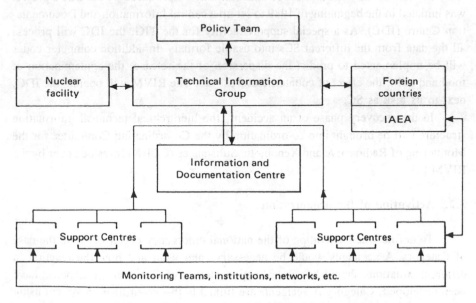

FIG. 1. The organization of technical information.

The following institutes, which under normal conditions also have tasks in co-ordinating radiological monitoring, have been appointed as Support Centres (SCs) and are represented in the TIG:

— National Institute of Public Health and Environmental Protection (RIVM)
— Royal Netherlands Meteorological Institute (KNMI)
— Government Institute for Quality Control of Agricultural Products (RIKILT)
— Institute for Inland Water Management and
   Government Institute for Sewage and Waste Water Treatment (DBW/RIZA)
— National Commodity Inspectorate (RKvW)
— Ministry of Home Affairs (BiZa)
— Ministry of Housing, Physical Planning and the Environment (VROM)
— Ministry for Welfare, Health and Cultural Affairs (WVC)
— Nuclear power stations in the Netherlands.

These centres will, under supervision of the TIG, collect and validate the measurement data in their specific fields, e.g. air, ground and water. These data will be provided to the TIG in a preprocessed and compiled form.

The nuclear plants have been added to the list of SCs to supply data on plant status and the source term. Under accident conditions inspectors of the nuclear inspectorate of the Ministry of Social Affairs and Employment will be present in the plant to provide a direct link with the TIG. The Royal Netherlands Meteorological Institute will furnish the meteorological data needed to make an estimate of the accident consequences.

In view of the large amount of data that may be produced by the SCs a project was initiated in the beginning of 1989 to set up a central Information and Documentation Centre (IDC). As a special support facility for the TIG, the IDC will process all the data from the different SCs into usable formats. In addition computer codes will be implemented to predict the dispersion of the release, the contamination of food and milk, the effect of countermeasures, etc. The RIVM will operate the IDC, next to its task as SC.

In the recovery phase of an accident, the hierarchical technical information structure will be brought into co-ordination by the Co-ordinating Committee for the Monitoring of Radioactive and Xenobiotic Substances (CCRX), presided over by the RIVM [3].

3.2. Activation of the organization

To define when activation of the national emergency organization in the case of Category A accidents would be necessary, and what and how the reaction in different situations should be, co-ordination classes of Government response have been developed. Category A accidents are linked to this co-ordination system using existing emergency plans and accident classification systems and taking into account

TABLE III. RESPONSE MATRIX

	0	Co-ordination classes[a]			
		I	II	III	IV
Borssele Dodewaard Petten	Emergency stand-by	Plant emergency	Site emergency	Off-site emergency	
Doel (5 km)[b]	Unusual event, level 0	Site emergency, level 1	Site emergency, level 2	General emergency, level 2	General emergency, level 3
Tihange (40 km)	Unusual event, level 0; site emergency, level 1	Site emergency, level 2	General emergency, level 2; general emergency, level 3		
Emsland (20 km)[c]		Kata-strophen-voralarm	Kata-strophen-alarm		
Kalkar (15 km)		Kata-strophen-voralarm	Kata-strophen-alarm		
Remaining reactors		All notifications			
Ships		Category 1	Category 2	(BARK plan)	
Satellites		All notifications			
Nuclear defence material				All notifications	
Unknown		All notifications			
Scale-up of Category B	All notifications	Notification scale-up			

[a] Class 0: action by nuclear inspectorates; Class I: separate action by involved departments; activation of an assessment team; minor co-ordination necessary; Class II: interdepartmental co-ordination (accent on radiological aspects); Class III: interdepartmental co-ordination (accent on public order and safety); Class IV: disaster. The complete emergency response organization (including the TIG) will be activated starting from Class II.

[b] According to the draft Belgian National Emergency Plan (1988).

[c] According to the Federal German Gemeinsames Ministerialblatt (February 1989).

TABLE IV. STATIONARY MONITORING NETWORKS IN THE NETHERLANDS

Support Centre	Measurement data on:	Number of stations	Permanently on stand-by
RIVM	Environment	58	Yes
BiZA	Air, deposition	300	Yes (planned)
RIKILT	Food, milk	125	No
RKvW	Consumer goods	15	No
DBW/RIZA	Surface water	15	No
VROM	Drinking water	86	No
WVC	Contaminated persons	7	No

the relation between distance and effect. The resulting preclassification is summed up in a response matrix (Table III). To carry out appraisal of actual accident consequences an Assessment Team will be activated upon verification of an emergency notification.

On the basis of the IAEA accident classification system the national emergency organization, including the TIG, will for example be activated if a site or off-site emergency occurs in a nuclear power plant within the Netherlands.

For Category B accidents technical support for the local authorities is guaranteed by the National Institute of Public Health and Environmental Protection, the Ministry of Housing, Physical Planning and the Environment, and the Ministry of Social Affairs and Employment. Possible scale-up of Category B accidents to the national level has been taken into account.

3.3. Monitoring capacity

An inventory has been made of the present or planned monitoring equipment and specific measurement techniques of the institutes and services participating in the technical information organization. Much effort has been and is still being put into providing workable means, in terms of mobile and of stationary monitoring. Monitoring vans have been improved; a monitoring plane capable of analysing radioactive clouds is on constant stand-by. Detection limits for radioactive sources were established during the Kosmos 1900 operation for different altitudes and ground speeds, using fixed wing aeroplanes (10 mSv/h at 1 m) and marine helicopters (1 mSv/h at 1 m) equipped with 1.5 L NaI crystals.

The networks for the monitoring of food, milk and consumer goods have been automated and provide on-line measurement results upon activation. A fully automated network for environmental monitoring is at present operational (Section 4.4), and will be expanded in the near future (Table IV).

Moreover, the RIVM is working out stand-by contracts with leading universities and institutes in the radiological field to augment the capacity for specific measurements such as alpha spectrometry and the measurement of strontium in the case of an accident.

Important for Category B accidents is the fact that the 51 regional fire brigades have been newly equipped.

Finally, to optimize the acquired information, a project on standardization and harmonization of sample collection and monitoring techniques has been set up to ensure comparable measurement results [4].

4.   MONITORING STRATEGIES

A monitoring strategy describes which information should be collected, depending on the type and the phase of the accident and the meteorological circumstances. It also describes (in general terms) the place and frequency of the measurements needed to be able to decide upon appropriate countermeasures. Defining such a strategy in advance has enormous advantages since it enables preliminary estimates to be made of the requirements with respect to personnel, equipment and training, and can save precious time when an accident actually occurs.

In spite of the fact that monitoring strategies are of such importance in nuclear emergency planning, relatively few publications on this subject are available. Therefore, the Government decided to start a follow-up project on monitoring strategies, focused on the collection of data in the case of accidents involving Category A and Category B facilities and activities. When this paper was being written the project was still under way. Therefore, only some preliminary general results are described here.

4.1. Source terms for planning purposes

The approach which was adopted in the project can be described as follows. In order to acquire insight into potential dose, contamination and distance effects, source terms were applied to the different objects listed in Categories A and B, in some cases only using one predominant nuclide (Table V). As stated earlier, the PWR-5 source term was chosen for the nuclear power stations. For the nuclear defence material critical masses had to be used. In the Netherlands the laboratories

TABLE V. SOURCE TERMS FOR NUCLEAR ACCIDENTS

Object	Activity (Bq)			
	Noble gases	Volatile compounds	Less volatile compounds	Actinides
Nuclear power plant, 500 MW(e) (PWR-5)	1.95×10^{18}	4×10^{17} (I) 3×10^{15} (Cs)	2×10^{15} (Ru)	$\sim 1 \times 10^{12}$
Satellites:				
Kosmos 945	—	—	—	9.2×10^{13}
SNAP 9A	—	—	—	6.3×10^{14} (Pu-238)
Laboratory, Class A, fuel rod, 200 MW·d	2.5×10^{12}	9.3×10^{12} (Cs)	7×10^{12} (Ru) 1.9×10^{13} (Ba)	8×10^{11}
Laboratory, Class B[a]	—	7.4×10^{10} (I-131)	—	1.7×10^{7} (Pu-238)
Nuclear defence material:[b]				
U-235	—	—	—	3.8×10^{7}
Pu-239	—	—	—	2.3×10^{11}
Type A transport package	3.7×10^{13} (Xe-133)	1.48×10^{12} (I-131)	—	—

[a] Twenty per cent of the stored activity is released.
[b] Critical masses for U-235 and Pu-239 have been used; 1% of the activity is released as 1 μm particles.

involved in radiological work are divided into four classes depending on the activity (differing between classes by a factor of 1000) that they are allowed to store and process. Only the two highest classes (A and B) are relevant here.

Upon considering the range of effects for all the objects, a division could be made into anticipated local, regional and national monitoring strategies (Table VI).

4.2. Dose contributions and monitoring

If accidents occur, measurements have to be conducted to check whether intervention levels for countermeasures are or will be exceeded. Analysis of the population dose calculated, for instance, for a PWR-5 source term shows that inhalation of airborne radioactivity contributes largely to the total received dose (Table VII).

The inhalation dose rate depends on the degree of contamination and the radiotoxicity of the nuclides that are released. The presence of alpha and beta emitters, however, is hardly revealed by exposure rate meters, while contamination monitors do not provide adequate information. The difference, for example, between the measured exposure rate due to ground and air contamination and the inhalation dose rate might well be a factor of 20 for 'fresh' fission products from nuclear plants, and even a factor of 10 000 for contamination by fission products from cooled down fuel elements. Since the fire brigades equipped with relatively simple monitors will conduct the first measurements, this observation shows that it would be advisable to

TABLE VI. MONITORING STRATEGY FOR FACILITIES AND ACTIVITIES
(a: intensive monitoring in combination with countermeasures; b: intensive monitoring; c: mainly food and water monitoring)

Object	Monitoring strategy			
	Local		Regional	National
	On-site	Off-site	Off-site	Off-site
Category A				
Nuclear power plant	a	a	a	a
Research reactor	a	a		
Ship	a	a	a	c
Spacecraft		a	b	c
Nuclear defence equipment		a	b	
Object abroad				a
Category B				
Uranium enrichment	a	a	c	
Laboratory, Class A	a	a	c	
Laboratory, Class B	a	b		
Processing and storage	a	b		
Type A transport package		a	(c)	
Type B transport package		a	b	

TABLE VII. DOSE CONTRIBUTIONS (mSv) FOR EFFECTIVE, THYROID
AND RED BONE MARROW DOSE, AFTER 24 h EXPOSURE IN OPEN FIELD
TO PWR-5 SOURCE TERM (500 MW(e)), FOR PASQUILL CLASS D2

	Internal radiation by inhalation		External radiation			
			Cloudshine		Groundshine	
	Distance from source:		Distance from source:		Distance from source:	
	1 km	20 km	1 km	20 km	1 km	20 km
H_{eff} (dry)	640	7	36	1.1	180	1.8
H_{th} (dry)	17 000	180	40	1.2	200	2
H_{rbm} (dry)	25	0.28	32	0.98	160	1.7
H_{rbm} (rain)	25	0.18	32	0.82	280	6

augment the measured exposure rates with a safety factor when deposition is taking
place. It also follows, however, that early performance of nuclide specific measurements is imperative to determine the influence of the nuclide composition on the total
received dose. In addition, these measurements provide information on the source,
e.g. the disclosure of ruthenium and zirconium can point to severe core damage.

4.3. Measurement zones

Mobile monitoring teams have to be safeguarded against high doses during and
after the release. To present an example of an overview relating potential
consequences for all the facilities and activities to dose effects for monitoring teams,
measurement zones can be defined on the basis of (not yet established) permissible
dose:

— *An exceptive zone*, where the dose will exceed 100 mSv in 24 h. This zone
   might be entered, for instance, to collect urgently needed information.
— *A restrictive zone*, where the dose will be between 5 and 100 mSv in 24 h. This
   zone might be entered for quick surveys.
— *A free zone*, where the dose will stay below 5 mSv in 24 h. Access to this zone
   could be more or less unlimited. The zone will extend to the edge of the contaminated area, based on ^{131}I contamination exceeding 5 kBq/m^2 or
   contamination by other nuclides in excess of limiting values.

TABLE VIII. MEASUREMENT ZONES DURING AND AFTER RELEASE, CALCULATED FOR PASQUILL CLASS D2

| Object | Distance from source to zone perimeter (km) | | | | |
| | Exceptive zone | | Restrictive zone | | Free zone |
	During release	After release	During release	After release	
Nuclear power plant, 1000 MW(e) (PWR-5)	7	3	63	16	400
Nuclear power plant, 500 MW(e) (PWR-5)	5	2	35	11	250
Nuclear power plant, 50 MW(e) (PWR-5)	1	0	7	3	80
Transition from site to off-site emergency class	0	0	0.1	0	7
Research reactor	0	0	0	0	0.1
Spacecraft	n.a.	0.02	n.a.	0.02	n.a.
Nuclear defence material	1	0	6	4[a]	20[b]
Laboratory, Class A, fuel rod, 200 MW·d	0	0	3	2[a]	20 (Cs-137) 6 (I-131) 62[b]
Transport of I-131, 1.5 TBq (gas)	0.1[c]	0	0.2[d]	0.04	11

Note: The criteria for the exceptive and restrictive zones are respectively 100 and 5 mSv effective dose during the release and in the first 24 h after the release.

[a] Restriction based on high ground contamination (alpha activity $\geq$ 3.7 kBq/m^2).

[b] Distance based on I-131 contamination $\geq$ 5 kBq/m^2 and/or alpha activity $\geq$ 200 Bq/m^2. The distance for Cs-137 is based on contamination of 3100 Bq/m^2, which might lead to a dose of 5 mSv/year upon continuous use.

[c] Value based on H_{th} = 1000 mSv.

[d] Release of complete contents in 10 min; the value is calculated for H_{th} = 50 mSv.

TABLE IX. DENSITY OF SOME RADIOACTIVITY MONITORING
NETWORKS IN EUROPE

Country	Number of posts	Typical separation (km)	Surface covered per station (km²)
Germany, Fed. Rep.	1 000	16	250
Austria	336	16	250
Finland	300	34	1 120
Switzerland	51	28	800
Sweden	26	130	17 300
Denmark	10	66	4 300
Netherlands:			
installed	58	25	620
planned	300	11	120

The distances between the outer limits of the measurement zones and the source have been calculated for Category A and Category B objects using selected source terms (Table VIII).

To prevent a prolonged stay of monitoring teams in areas of high dose rates during a release, a cloud might be approached from both sides by six teams entering the cloud simultaneously at three different distances from the source. Measurement results can then be extrapolated in the tangential and radial directions. For this purpose it would be best to use predesignated measurement locations for fixed objects to ensure that background levels are known and that there should not be a blackout in radio communications. Locations for this purpose can be chosen on concentric circular routes around the object, with intervals of 400 to 500 m close to the object increasing to 1000 m at 10 km, thus ensuring a factor of 10 difference between the measurement results according to a Gaussian dispersion under Pasquill class D2.

An alternative to mobile teams would be to employ automatic monitoring stations (networks) placed around the object.

4.4. Monitoring networks

Monitoring networks can be used for several purposes apart from replacing mobile teams. First of all, automatic stations can be utilized to indicate increases of airborne radioactivity not otherwise notified, and secondly a network can actually determine exposure levels or beta activity. In setting up automatic networks certain factors have to be taken into consideration, such as the typical distance between posts, the distance to potential sources, the detection limits of the equipment and the type and frequency of measurements.

An optimum has to be found in the relation between the type of equipment employed and the required maintenance (quality), and also between density and uncertainty (costs). The density of automatic measurement stations to detect and measure radioactivity has been given attention in the project, since the variety in the set-ups of foreign networks indicated no answer with regard to an obvious optimal solution (Table IX).

In the Netherlands 58 early warning monitoring posts constantly measuring gamma exposure rates are in operation, a number that will be increased to 300 for civil defence purposes. Of the operational stations, 14 measure airborne beta activity. At the RIVM one station is constantly measuring iodine concentrations and specific gamma activity of airborne radioactivity.

To introduce a cost–benefit approach for the number of monitoring stations that have to be installed in order to measure a stochastic dispersion after a release with an acceptable degree of certainty, computer simulations were carried out at the Netherlands Energy Research Foundation (ECN) [5], based on the Gaussian plume model. By altering the concentration in the plume axis and the width of the plume for differing weather conditions, a relation between grid size and relative error was found (Fig. 2). For the Dutch set-up these calculations show that the relative error of about 35%, using 58 stations, will be diminished to 12% upon increasing the number to 300 stations.

Other computer simulations based on the puff model were carried out at the RIVM [6] in order to examine the detection characteristics of the present network. The results have shown that for a random location in the Netherlands and a source term of 10 TBq the probability of detecting the event is approximately 50%, using

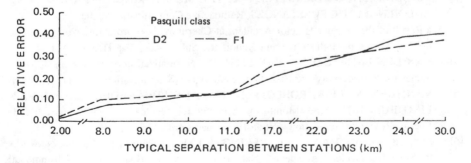

FIG. 2. Variation of relative error with measurement station density.

the 14 installed beta monitoring stations. Using all 58 stations, a source of 10^6 TBq is detected with a probability of 70% for the nuclear power plant at Borssele and 90% for the plant at Dodewaard. The figure for Borssele, which is lower for geographical reasons, can be improved by installing a local ring network. The analysis of the density of such a set-up shows an optimum at 10 stations located approximately 5 km from the source.

5.   GENERAL CONCLUSIONS AND OBSERVATIONS

The Netherlands is on a steady course towards an adequate nuclear response organization and the provisions, in terms of personnel and equipment, that have to go with it. Much attention is focused on creating a sound base for continuity not only by providing training and educational programmes and considering longer term expenses in addition to initial costs, but also by utilizing legislative means.

In the field of technical information a strong tendency can be observed towards using and developing supportive computer programs for planning and actual emergency response. Upon applying these programs in order to make an estimate of the effects of postulated source terms on the environment, and relating the findings to the directives for countermeasures issued by the European Community, the impact of contamination appeared to be predominant compared with dose effects. This observation will affect the monitoring strategies for food and water which will be tackled next.

REFERENCES

[1]   MINISTRY OF HOUSING, PHYSICAL PLANNING AND THE ENVIRONMENT, National Contingency Plan for Nuclear Accidents, Ministry of Housing, Physical Planning and the Environment, The Hague (1989).

[2]   NUCLEAR REGULATORY COMMISSION, Reactor Safety Study: An Assessment of Accident Risks in US Commercial Nuclear Power Plants, WASH-1400 (NUREG-75/014), NRC, Washington, DC (1975) Appx IV.

[3]   CO-ORDINATING COMMITTEE FOR THE MONITORING OF RADIOACTIVE AND XENOBIOTIC SUBSTANCES, Radioactive Contamination in the Netherlands as a Result of the Nuclear Reactor Accident at Chernobyl, Co-ordinating Committee for the Monitoring of Radioactive and Xenobiotic Substances, The Hague (1986).

[4]   VAN DEN ENDE, C.A.M., VAN LITH, D., Normalization of Measurement Techniques for Radioactivity and Radiation, Netherlands Normalization Inst., Delft (1988).

[5]   VAN HIENEN, J.F.A, ROELOFSEN, P.M., DEURWAARDER, C.P., KEVERLING BUISMAN, A.S., Monitoring Strategy for Nuclear Accidents, Netherlands Energy Research Foundation, Petten (in preparation).

[6]   SLAPER, H., DE HAAN, B.J., VELING, E.J.M., Density of Monitoring Networks to Signalize Nuclear Accidents, Natl Inst. of Public Health and Environmental Protection, Bilthoven (1989).

THE UNITED KINGDOM RESPONSE PLAN FOR NUCLEAR ACCIDENTS OVERSEAS: IMPLEMENTATION OF THE NATIONAL RADIOACTIVE INCIDENT MONITORING NETWORK (RIMNET)

M.W. JONES
Her Majesty's Inspectorate of Pollution,
Department of the Environment,
London,
United Kingdom

Abstract

THE UNITED KINGDOM RESPONSE PLAN FOR NUCLEAR ACCIDENTS OVERSEAS: IMPLEMENTATION OF THE NATIONAL RADIOACTIVE INCIDENT MONITORING NETWORK (RIMNET).

The United Kingdom Government has formulated a National Response Plan to deal with the consequences of nuclear accidents overseas. The Department of the Environment, through Her Majesty's Inspectorate of Pollution, has responsibility for carrying through the implementation of the plan. The paper describes the first phase of the plan: the establishment of a national Radioactive Incident Monitoring Network (RIMNET). Within the National Response Plan, the Phase 1 RIMNET system provides a national means for monitoring the radiological effects of any nuclear incident overseas and for informing many organizations and the public of its implications. The system will detect any overseas nuclear incidents affecting the United Kingdom, even if these incidents are not formally reported to the United Kingdom or there are delays in notification. Data from the RIMNET system will be available to the appropriate lead Government department in the event of a domestic nuclear incident.

1. INTRODUCTION

The United Kingdom has long established plans for dealing with the consequences of nuclear accidents within its own shores. However, following a careful review of the Chernobyl accident, the Government decided that a new contingency plan was necessary for dealing with nuclear accidents overseas. This is called the National Response Plan.

The Department of the Environment (DOE) is responsible for co-ordinating the design and implementation of the National Response Plan. A key part of this plan is the establishment of a network of continuously operating radiation monitors capable of detecting independently any radioactivity arriving over the United Kingdom. This network is known as RIMNET, the acronym for Radioactive Incident Monitoring Network.

RIMNET is currently being installed progressively in phases [1]. The first phase of the RIMNET system has now been set in place, providing a means of detecting radiation arriving over the United Kingdom and handling the response to it [2]. This paper gives details of the implementation of the operation of RIMNET and describes the way in which it will be used.

Government departments, local authorities and the nuclear industry in the United Kingdom have all participated in the planning of the RIMNET programme. The work of these organizations has been co-ordinated by a national co-ordinating committee — the Radioactive Incident Monitoring Co-ordinating Committee (RIMCC). The RIMCC is chaired by a senior officer in Her Majesty's Inspectorate of Pollution (HMIP). It will continue to guide the operation of the Phase 1 system, and to co-ordinate the development of later phases. It will also review exercises designed to test the performance of RIMNET.

The current RIMNET system is expected to operate for a period of two to three years. During this period the more extensive and fully automated later phases will be planned, and will progressively replace Phase 1.

International agreements negotiated since the Chernobyl accident should ensure that the United Kingdom Government is notified, through HMIP, of any nuclear accident overseas resulting in a significant release of radioactivity. Operation of the RIMNET system will mean that, even if these notification arrangements fail, any unexpected increases in radiation levels over the United Kingdom of the kind that could result from an overseas accident can be detected immediately.

2.   NATIONAL RESPONSE PLAN ARRANGEMENTS

The structure of the National Response Plan arrangements for handling overseas nuclear incidents is shown in Fig. 1.

The Government's response will be managed by a Technical Co-ordination Centre (TCC) staffed by officials from Government departments including the Meteorological Office and the National Radiological Protection Board (NRPB). An inspector from HMIP will manage the TCC.

Staff in this centre will have access to data stored on the Central Database Facility (CDF) and will be responsible for liaising with the departments and agencies which they represent to obtain radiological assessments. The TCC will provide a fully co-ordinated response. In particular, it will ensure that all information and advice bulletins issued by Government departments and other statutory bodies are consistent.

If necessary additional Government monitoring programmes will commence. The Ministry of Agriculture, Fisheries and Food (MAFF) is responsible for monitoring food, livestock, crops and fish and for taking appropriate action to restrict the sale and distribution of contaminated food. People and goods coming into the country

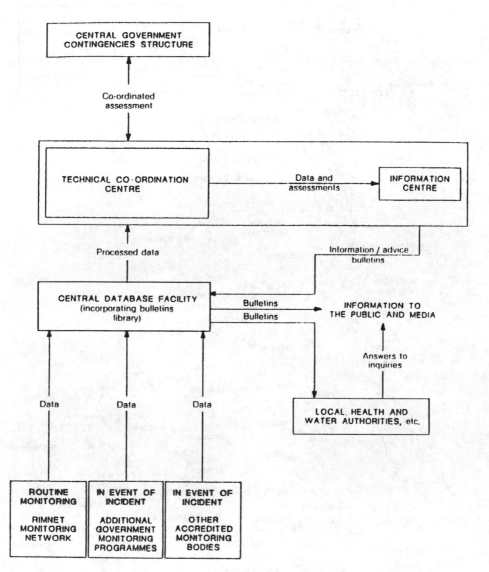

FIG. 1. Form of National Response Plan systems. The Technical Co-ordination Centre may include representatives from the Department of the Environment, the Ministry of Agriculture, Fisheries and Food, the Department of Trade and Industry, the Department of Energy, the Department of Health, the Ministry of Defence, the Meteorological Office, the National Radiological Protection Board and the Scottish, Welsh and Northern Ireland Offices.

FIG. 2. Phase 1 RIMNET monitoring sites and bulletin regions.

from affected areas overseas will be monitored by the relevant Government department. Provision of advice on the contamination of drinking water is a DOE responsibility, and plans for monitoring of supplies in the event of an incident have been drawn up.

Other monitoring data supplied by accredited organizations will also be collated in the CDF to assist the TCC with its work. Further data links with other Government departments to enable the necessary information to be accessed remotely are being planned.

On the basis of the information provided by the TCC, staff in the DOE Information Centre (IC) will prepare regular press releases. National and regional information from the IC will also be broadcast on viewdata and teletext systems (CEEFAX and PRESTEL) to ensure that the public are kept up to date. Regional information will be based on the ten areas shown in Fig. 2. The use of predefined regions will enable advice to be related more specifically to the estimated route of any radioactive cloud.

Data held on the CDF and additional information and advice will be supplied daily by electronic mail to official bodies such as local and water authorities offering to help keep the public informed. This will help these organizations answer inquiries from the public on local radiation matters. Local authorities may wish to act as the main focal point for local public inquiries in their area.

Where an individual Government department has direct statutory responsibility for a specific matter, that department may also supply its own media and public briefings provided that the TCC is kept informed.

The TCC and IC facilities are located in the DOE operations room in London.

3.   IMPLEMENTING RIMNET

The framework of the National Response Plan was announced by the Prime Minister, Mrs. Thatcher, on 30 June 1987. The initial work of developing the details of the plan was put in hand immediately. A team of consultants was put together, which by October 1987 had produced, in consultation with a number of Government departments, a detailed programme for implementing the plan.

It was estimated that complete implementation of the plan would take two years. In view of the length and complexity of this task it was decided to implement the plan in two phases, usually referred to as Phase 1 and Phase 2 RIMNET.

Specifically RIMNET provides a national system for detecting and measuring radiation levels in the air over the United Kingdom. It also includes modern information technology facilities for timely dissemination of information and advice concerning an accident to a wide range of organizations and the general public.

Phase 1 of the RIMNET system is based on gamma ray dose rate monitoring equipment installed at 46 field stations throughout the United Kingdom in June 1988. The locations, which are all meteorological observatories, are shown in Fig. 2.

The radiation detector, monitor and line driver are housed in a weatherproof cubicle located within a meteorological field station enclosure. The detector and monitor assembly is connected by cable to the observer's office nearby, where results are printed on a hard copy printer.

Regular gamma ray dose rate readings are taken by the meteorological observer, usually every hour. The readings are transmitted to the headquarters of the Meteorological Office at Bracknell along with other meteorological data and then transmitted to the CDF, which is located at the DOE in London, and to a backup computer installed at a DOE office in Lancaster.

Data from all RIMNET sites are analysed within the CDF. An alarm will automatically trigger in the event of any unexpected rise in gamma ray dose rate readings satisfying one of three algorithms covering different eventualities. The CDF computer and its associated communications equipment were handed over to the DOE by the installation contractor at the end of 1988 and commissioned early in 1989.

In the event of an overseas nuclear accident being reported or detected, the provisions of the National Response Plan will be implemented to assess the effect of the accident on the United Kingdom. The Secretary of State for the Environment will be the lead Government minister. The Director of HMIP will be responsible for advising ministers about any subsequent actions.

If it is clear that there will be an effect on the United Kingdom, Parliament, relevant official bodies and the public will be informed, and appropriate alert messages will be issued. National Response Plan arrangements described in the earlier sections of this paper will be initiated.

4.   ROLE OF GOVERNMENT DEPARTMENTS AND OTHER ORGANIZATIONS

Government departments with relevant responsibilities will be represented in the TCC. These may include the DOE, MAFF, the Department of Health (DH), the Ministry of Defence (MOD) and the three territorial departments — the Scottish Office, Welsh Office and Northern Ireland Office. The NRPB and the Meteorological Office will also be represented.

Following an incident, Government departments will continue to discharge their normal statutory responsibilities. However, the departmental representatives at the TCC will ensure that decisions and actions are fully co-ordinated. These representatives will also provide the details and assessments necessary for the TCC and IC to prepare information and advice bulletins.

To discharge their wide range of responsibilities in Scotland, Wales and Northern Ireland, and to provide local information and advice, the three territorial departments will set up their own incident control centres in Edinburgh, Cardiff and Belfast. The work of these centres will be co-ordinated with the national response through their TCC representatives.

If departments commission supplementary monitoring programmes for their own purposes, the results, and assessments of them, will be available to the TCC.

4.1. Role of local, health and water authorities

In England, Scotland and Wales the local authorities may wish to act as the main focal point for local public inquiries in the event of an overseas nuclear incident. A number of lead local authorities have been identified with the help of the local authority associations, and will be amongst the DOE contact points for the initial incident alert message. They may disseminate the initial alert information to other authorities. Similar arrangements are being planned with the water authorities. Health authorities will be informed by the appropriate Government departments.

In Northern Ireland, a Government department group, established to advise on peacetime radioactive emergencies, will act as the main focal point for public inquiries. In addition local authorities will be kept informed and will assist by dealing with inquiries at more localized levels.

Information officers in local, health and water authorities will have access to the regular information and advice bulletins issued via viewdata and teletext systems. Lead local authorities will also have access to the more detailed information distributed by electronic mail. This will assist the authorities in dealing with public inquiries at the local level.

Local, health and water authorities may collect radiological monitoring data of their own during an overseas incident. Where such data are available they will be used to obtain a more detailed analysis of the distribution of radioactivity across the United Kingdom.

4.2. Role of the National Radiological Protection Board

In the event of an overseas nuclear accident the NRPB will advise Government departments on the interpretation of radiological monitoring data. The NRPB will be represented at the TCC, and have direct access to the CDF from its headquarters at Chilton. The NRPB will also supply monitoring data which it collects to the CDF.

4.3. Role of the Meteorological Office

The Meteorological Office will use a model to estimate and forecast radionuclide concentrations in air and surface deposition activity. The model will draw on meteorological and radiological data to predict the movement of the radioactive material released from the accident over Europe and the sea around the United Kingdom. The results will be sent to the Meteorological Office representative in the TCC to enable assessments to be made both on the likely movement of any plume of radioactive material across the United Kingdom, and on the pattern of deposition of radioactive material on the ground.

4.4. Role of the nuclear industry

In the event of any future overseas nuclear accident, the Central Electricity Generating Board (CEGB), British Nuclear Fuels plc (BNFL), the United Kingdom Atomic Energy Authority (UKAEA) and the South of Scotland Electricity Board (SSEB) have expressed their willingness to collect supplementary monitoring data at their various sites to assist with the response. They would also assist with the analysis of samples from elsewhere.

5.   ACCREDITATION OF SUPPLEMENTARY DATA SUPPLIERS

An accreditation system will establish a register of supplementary data suppliers able to provide specific monitoring results to the CDF. This will enable additional information on radionuclide concentrations in the environment to be obtained.

6.   INTEGRATION WITH OTHER CONTINGENCY PLANS

The National Response Plan provides for a response to a nuclear accident abroad. Plans exist for dealing with nuclear accidents within the United Kingdom. These are based on the 'lead department' concept. According to the nature of a nuclear accident in the United Kingdom, the Department of Energy (DEn), the Scottish Office, the MOD or the Department of Transport will take direct responsibility as appropriate. Arrangements for handling nuclear accidents in the United Kingdom are well publicized in a number of Government publications. Data from RIMNET will be available to the relevant Government department in the event of a nuclear accident in the United Kingdom.

7.   PROPOSALS FOR LATER PHASES

Phase 2 of the RIMNET programme will increase the number of monitoring stations to about 85. Additional monitoring facilities will be provided at some sites. These will include automatic air sampling and deposition measurement. All monitoring stations will operate automatically and supply readings to the CDF.

The CDF will have additional analytical and display facilities to provide electronic assessment and the automatic preparation of draft viewdata pages and press releases.

REFERENCES

[1]   DEPARTMENT OF THE ENVIRONMENT, Nuclear Accidents Overseas. The National Response Plan and Radioactive Incident Monitoring Network (RIMNET). A Statement of Proposals, Her Majesty's Stationery Office, London (1988).

[2]   DEPARTMENT OF THE ENVIRONMENT, The National Response Plan and Radioactive Incident Monitoring Network (RIMNET). Phase 1, Her Majesty's Stationery Office, London (1989).

METHODOLOGY FOR EMERGENCY ENVIRONMENTAL MONITORING USED BY THE EMERGENCY TECHNICAL SUPPORT CENTER OF THE CHINA INSTITUTE FOR RADIATION PROTECTION

Zunsu HU
China Institute for Radiation Protection,
Taiyuan, Shansi,
China

Abstract

METHODOLOGY FOR EMERGENCY ENVIRONMENTAL MONITORING USED BY THE EMERGENCY TECHNICAL SUPPORT CENTER OF THE CHINA INSTITUTE FOR RADIATION PROTECTION.

An Emergency Technical Support Center has been set up at the China Institute for Radiation Protection, Taiyuan. The activities of the centre are described in general and methodologies with regard to emergency environmental monitoring in particular. Owing to the geographic location of the centre, its emergency monitoring activities are mainly focused on the later phases of an accident. The requirements of a mobile laboratory are outlined, sample measurements and personal monitoring are discussed and some equipment is described in more detail.

1. INTRODUCTION

Despite accidents such as occurred at Three Mile Island and Chernobyl, the safety record of nuclear energy is still the best in comparison with the records for fossil fuel derived energies. Nevertheless, practice shows that accidents are not like a 'bolt from the blue', but are part of normal operation [1]. No matter how small the probability is, accidents cannot be ruled out completely. An almost continuous range of abnormalities can be considered and attempts are made to quantify and control their probabilities.

Thus, emergency planning and preparedness are justified as part of normal operation to reduce the possible consequences of a nuclear accident. In China, from the very beginning of the development of nuclear power, safety has been given top priority. On the one hand, quality control in design, manufacture and construction and in training of operational staff is emphasized to minimize the probability of an accident. On the other hand, emergency planning and preparedness are actively pursued to cope with any possible accidental occurrence.

In the course of our work on emergency planning and preparedness, the International Atomic Energy Agency has extended its assistance in the form of training courses, expert services, fellowships and equipment. Relevant regulations and guidelines have been preliminarily developed on the basis of IAEA recommendations, and comprehensive research work on accident consequence analysis, emergency response, protection measures, etc., is being carried out. Emergency plans in support of the Qinshan Nuclear Power Plant in east China and the Daya Bay Nuclear Power Plant in south China are being prepared. At the China Institute for Radiation Protection (CIRP) in Taiyuan, an Emergency Technical Support Center has been set up and a programme on emergency planning and preparedness, supported by the IAEA, is being implemented. This programme includes research work on emergency techniques and relevant aspects of radiation protection and extension of radiological emergency assistance to areas affected by an accident. The following activities are included:

— Assistance to competent authorities in drafting regulations and guidelines,
— Assistance to power plant licensees for working out emergency plans,
— Training courses for emergency personnel,
— Accident consequence analyses,
— Emergency monitoring,
— Medical aid,
— Protective measures.

This paper presents some general considerations with regard to emergency monitoring and discusses the basics of the CIRP Emergency Technical Support Center.

2.   GENERAL CONSIDERATIONS ON EMERGENCY MONITORING

Because of the vast size of China and the existing transportation conditions, it is likely to be difficult to reach the scene of an accident from Taiyuan in a short time after its occurrence. Therefore, emergency monitoring to be performed by the Emergency Technical Support Center is mainly concentrated on the radiological characteristics of releases in the intermediate and late phases.

An emergency radiological monitoring group is being organized at the centre, members of which are from different units of the CIRP, but mostly from the Environmental Monitoring Station. The group is planning to equip a mobile laboratory with field dosimeters, samplers for gases, aerosols and other kinds of sample, apparatus for sample preparation, equipment for sample assessment, surface contamination monitors, personal dose monitors with presettable alarm thresholds, a simplified whole body counter and so on. The mobile laboratory will also be equipped with an electricity generator and an air cleaning system.

Because of the present lack of a modern highway system, a large vehicle would not be appropriate for the mobile laboratory. This limits the total weight and size of the equipment. For the same reason, other means of transportation may have to be considered for the monitoring equipment, and hence this should be packed in such a way that it can be carried conveniently by train or by aircraft.

In view of the rareness of major nuclear accidents, and considering that it would be improbable for China to have a large number of nuclear power plants before the year 2000, it would be difficult and expensive to keep all the monitoring equipment functioning properly while practically never putting it to use. It would therefore seem an ideal alternative to make the emergency monitoring equipment in such a way that it would also be appropriate for routine applications, though it may not be the best and most economical choice for this purpose. In this case the equipment would be used for routine measurements and be tested and calibrated periodically without incurring any extra burden. The measuring range of such monitors should be wide enough, but the higher part of the range may never be used normally. It is therefore necessary to have simple ways of checking the higher ranges while a monitor is in everyday use. Modern developments in electronics and detector technology make it possible to check all the functions of an instrument automatically without interfering in its normal operation.

The monitoring instruments should be powered by built-in power units as far as possible and have minimal power consumption. Batteries used in all instruments should be of a common type if possible so that they are easily exchangeable.

Complicated sample preparation procedures must be avoided, so gamma spectrometric analysis is preferable. Because of the possible shortage of liquid nitrogen supplies, NaI(Tl) detectors may be the best choice for making emergency measurements of samples. In particular, a Ge(Li) detector should not be used because of damage to the detector should the liquid nitrogen supply be exhausted.

Surface contamination monitors with gas flow counters must also be avoided, owing partly to possible shortages in gas supply and partly to their fragility.

Because the radiation levels are usually not readily controllable in emergency situations, personal dose monitors with presettable alarm thresholds are very useful. Such monitors should be convenient to carry in the breast pocket. The display should be clearly readable without any manual operations, and the audible alarm should be loud enough to be heard in noisy surroundings.

3.   MONITORING OF RADIATION LEVELS

In the later phases of an accident, radiation is mainly due to radionuclide deposition. Beta/gamma survey meters with ionization chambers are used for measurements of radiation levels. A NaI(Tl) type FD-71 scintillation counter was considered for radiological surveillance. Although rather tough, this counter shows so strong an energy dependence that it is not appropriate for dose assessment. An environmental dosimeter has been developed with a pressurized ionization chamber and a computerized electrometer to measure and record environmental radiation continuously. Data recorded in the last 8–24 h are stored in the computer memory, the variance and trends of variations in environmental radiation levels are analysed and results are printed in text, tabular or graphic form by the computer. This system has been applied at the CIRP Environmental Monitoring Station for routine measurements. Improvements are being made to extend its measuring range for the purpose of evaluating dose and dose rate more effectively under emergency situations. A scintillation environmental dosimeter with tissue equivalent scintillator has been developed which has a sufficiently good energy response and whose measuring range is from 0.1 μSv/h to 0.1 Sv/h [2]. A gamma survey meter with a telescopic probe handle has been developed which can be used to search for hot spots.

4.   MEASUREMENTS OF SAMPLES

Samples to be measured include aerosols, fallout, soil, water, milk and other biological samples.

A glass fibre filter is used for aerosol sampling. For collection of iodine in elemental form or in organic compounds, charcoal filter paper is used in addition to the glass fibre filter. The collection efficiency of charcoal filter paper for elemental iodine is about 30%.

A far infrared water vaporizer for rapid water sample preparations and a rapid ashing furnace for biological sample preparations were developed at the CIRP [3, 4]. In the ashing furnace, NO_2 is used as oxidizer. The ashing speed is three to seven times faster than that of a conventional muffle furnace, the quality of the ashed samples is better and the power consumption is two to four times less. This furnace is more appropriate for treatment of biological samples under emergency conditions.

A multiprobe low level alpha counter with very low power consumption has been developed which is assembled in a standard NIM bin and can simultanously measure up to six samples independently, with data being printed out automatically [5]. The built-in NiCd rechargeable batteries can work as long as 500 h before another charge becomes necessary. Being free from interference from the main power network, the instrument can work stably in complex situations and is suitable for use in mobile laboratories. A similar multiprobe assembly for beta counting is under development.

Gamma spectrometry is mainly carried out with NaI(Tl) spectrometers. In addition, an HPGe detector with a 1 L Dewar vessel is equipped both for sample measurements and for in situ assessment. Each detector works in connection with an independent portable multichannel analyser which is modified with a pocket microcomputer. A centralized computer system has been avoided for higher flexibility and reliability.

Samples spiked with ^{152}Eu have been made as standard samples for efficiency calibration of the gamma spectrometers. The different energies of the gamma photons emitted by ^{152}Eu are used to fit an efficiency curve for a spectrometer to a sample with specified geometry and medium. The standard ^{152}Eu source is provided by the China Academy of Metrological Sciences and spiked samples are made at the CIRP.

5. PERSONAL MONITORING

In addition to thermoluminescent dosimeters for routine personal monitoring, several types of directly readable personal dose monitors have been developed. One of them is the type GJ-1 personal dose monitor [6], which consists of a GM counter, a voltage converter with very low power consumption and a counting circuit. Its weight, including two AA size batteries, is 150 g, and the battery life span is three months under continuous operation in natural background conditions. An audible alarm is given when the cumulative dose exceeds a presettable threshold. The dose rate range is from background level to 10 mSv/h with an overflow indication. Another digitized personal dosimeter with an ionization chamber is appropriate for a dose rate up to 1 Sv/h; its built-in batteries can work continuously for several years.

A simplified whole body counter with shadow shielding was set up at the CIRP. Its background count rate is close to that of a fully shielded whole body counter in the energy range above 300 keV and twice as high below this energy, and could be used for personal internal contamination screening in emergency situations. However, it weighs 5 t and is therefore too heavy to be carried by a medium sized vehicle. Actually, for emergency internal contamination measurements, the most relevant parts of the body are the chest and thyroid, so a chair shaped chest counter might be enough.

For monitoring radioiodine in the thyroid, a general purpose NaI(Tl) detector with a collimator is used. In addition, some beta/gamma surface monitors are equipped for the measurement of surface contamination on the human body and on clothes.

6. CONCLUSIONS

A nuclear energy programme is being implemented in China. The Qinshan Nuclear Power Plant is expected to be in operation by the end of 1990, and the Daya Bay Nuclear Power Plant to be put into operation in 1992. As a result of this situation, an Emergency Technical Support Center has been set up at the China Institute for Radiation Protection. This serves to concentrate our existing knowledge and resources and also to help establish a unified emergency management system in China.

The Emergency Technical Support Center is mainly intended to provide guidance and services for emergency planning and preparedness for the nuclear power plants in advance. During an accident, the centre will provide large scale accident consequence assessments through telecommunication systems and provide emergency assistance, including medical aid, in the intermediate and late phases.

Radiation monitoring plays an important role in emergency response. The Emergency Technical Support Center is mainly assigned to carry out those measurements that cannot usually be performed by nuclear power plant personnel or local authorities. For this reason, the setting up of a mobile laboratory is considered beneficial.

The emergency monitoring equipment is expected to work both for routine measurements and for emergency use. For this purpose, a wide measuring range, high reliability, portability and built-in power sources are required. Also, it is important to be able to check automatically during normal operations the higher ranges which will be used only during emergency monitoring. Developments in microelectronics and radiation detection technology have made all this possible. Efforts are being made to meet these requirements.

Further effort is desired to be able to provide assistance with robots and teleoperator controlled devices for emergency treatments, sampling, measurements, etc.

Stations for environmental measurements have been established at the Qinshan Nuclear Power Plant and at the CIRP. Routine environmental monitoring is performed to detect variations in background radiation and to identify possible accidental contamination, either regional or global. Routine monitoring includes the collection and analysis of different kinds of samples and the measurement of radiation levels. In addition, meteorological towers have been set up for local observations. It is planned that regional or global meteorological data will be obtained from the national meteorological authorities to back up a real time accident consequence assessment system. The effectiveness of environmental monitoring stations has been proved by the measurements carried out in the time of the releases from the Chernobyl accident [7], but the emergency technical system still has to be completed and its effectiveness needs to be improved by practice and drill.

REFERENCES

[1]   LINDELL, B., Radiation protection — A look to the future: ICRP perceptions, Health Phys. 55 (1988) 145–147.

[2]   LIU, Y., et al., A gamma scintillation dosemeter, Radiat. Prot. Lett. (Taiyuan) **3** (1983) 26–32 (in Chinese).

[3]   JIN, M., et al., Rapid ashing equipment for biological sample preparation, Radiat. Prot. (Taiyuan) **5** (1985) 366–370 (in Chinese).

[4]   JIN, M., et al., Study of a rapid dry ashing method for biological samples, Radiat. Prot. (Taiyuan) **6** (1986) 357–364 (in Chinese).

[5]   DAI, Z., et al., A 6-channel α-counter with automatic recording, Radiat. Prot. Lett. (Taiyuan) **3** (1986) 19–23 (in Chinese).

[6]   LU, X., et al., An X, γ audible personal dosimeter, Radiat. Prot. Lett. (Taiyuan) **4** (1986) 18–20 (in Chinese).

[7]   HU, Z., "Environmental assessment of the Chernobyl releases in China", Mesures à décider et à mettre en œuvre en cas d'accident nucléaire, Collection «Sessions», Travaux du thème da la session académique de juin 1987, Paris, Académie du Royaume du Maroc (1987) 84–106.

       ....
       ....
       ....

METHODES D'EVALUATION DE L'IMPACT RADIOECOLOGIQUE DE L'ACCIDENT DE TCHERNOBYL SUR LE FLEUVE RHONE

A. LAMBRECHTS, L. FOULQUIER, M. PAILLY
Laboratoire de radioécologie des eaux
  continentales,
Centre d'études nucléaires de Cadarache,
Saint-Paul-lez-Durance, France

Abstract–Résumé

METHODS OF EVALUATING THE RADIOECOLOGICAL IMPACT OF THE CHERNOBYL ACCIDENT ON THE RIVER RHONE.

As a result of radiological monitoring of the Rhône, it has been possible to compare the impact of the Chernobyl accident with that of the nuclear facilities located along its course. Water, sediment, plant and fish samples have been collected and analysed. Chemical analyses by atomic absorption or neutron activation techniques have made it possible to determine the stable elements. The ^{3}H, ^{90}Sr and Pu contents have been determined radiochemically. Other nuclides have been detected by Ge gamma spectrometry. The information collected is being stored in a computerized database.

METHODES D'EVALUATION DE L'IMPACT RADIOECOLOGIQUE DE L'ACCIDENT DE TCHERNOBYL SUR LE FLEUVE RHONE.

Le suivi radioécologique du Rhône permet de comparer l'impact de l'accident de Tchernobyl à celui des installations nucléaires implantées sur son cours. Des échantillons d'eau, de sédiment, de végétaux et de poissons sont prélevés et conditionnés. Des analyses chimiques par absorption atomique ou activation neutronique permettent de doser les éléments stables. Les teneurs en ^{3}H, ^{90}Sr et Pu sont déterminées par radiochimie. Les autres nucléides sont détectés par spectrométrie gamma-Ge. Une base de données informatisée regroupe les informations.

1. INTRODUCTION

Le Rhône est l'un des premiers fleuves au monde par la variété et le nombre de ses installations nucléaires. Depuis quinze ans, un suivi radioécologique permet de recenser et de mesurer les radionucléides issus des effluents liquides des installations nucléaires, des retombées des essais atmosphériques et de Tchernobyl.

2. LE RHONE ET LES ZONES DE PRELEVEMENTS

Le Rhône est équipé de six sites nucléaires regroupant 17 réacteurs de puissance et deux usines du cycle du combustible. Sur deux de ses affluents sont situés deux centres de recherche nucléaire. On compte 70 stations de prélèvements réparties en amont et en aval des différentes installations et permettant d'en mesurer l'impact. On peut définir trois zones de prélèvements (fig. 1).

3. METHODES DE PRELEVEMENTS

Les informations permettant de définir les conditions de collecte des échantillons (date, localisation, poids, volume ou nombre d'individus, etc.) sont soigneusement notées. Les techniques d'échantillonnage doivent être reproductibles [1].

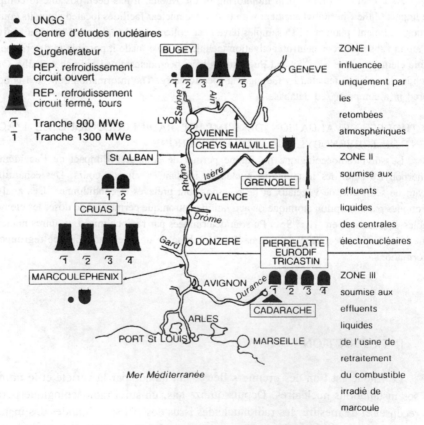

FIG. 1. Le Rhône et ses installations nucléaires.

L'eau est collectée à l'aide d'un appareil autonome [2]. Il assure le prélèvement d'eau brute, sa filtration à 0,8 μm, la rétention des éléments dissous par des résines échangeuses d'ions (cationique et anionique) et celle des matières organiques sur adsorbant. Le sédiment est ramassé manuellement, au cône de Berthois. Les végétaux immergés sont cueillis à partir de la berge ou d'un bateau. Les poissons sont collectés par pêche électrique.

4.   PREPARATION DES ECHANTILLONS

Les échantillons sont rapidement transportés au laboratoire. L'eau, les sédiments et les végétaux sont conservés dans une chambre froide à 4°C. Les poissons sont congelés à −40°C.

La détermination des faibles niveaux de radioactivité dans les échantillons prélevés dans la nature nécessite de concentrer au maximum la plus grande quantité possible de matière.

Le conditionnement permet de réduire le volume par séchage ou incinération, d'homogénéiser le produit obtenu et d'en réaliser, si besoin est, une fraction aliquote.

Les filtres et les résines résultant des prélèvements d'eau sont séchés à l'étuve à 110°C puis incinérés au four à 560°C.

Pour le sédiment, une fraction est séchée à 40°C en vue des analyses chimiques, granulométriques et minéralogiques. Une autre fraction est séchée à 110°C pour les analyses d'activation neutronique et les mesures de radioactivité.

Les végétaux sont soigneusement lavés, pesés frais et séchés à 110°C jusqu'à obtention d'un poids sec constant. Les produits secs sont pesés et broyés. Une partie est conservée pour les mesures du tritium et du ^{99}Tc, le reste est incinéré dans un four dont la montée en température jusqu'à 550°C suit un cycle de 24 heures. Les poissons sont décongelés et pesés frais. Certains peuvent être disséqués, d'autres sont préparés entiers. Leur traitement est identique à celui des végétaux.

5.   METHODES DE MESURES

Les dosages chimiques des métaux sont effectués par absorption atomique ou par activation neutronique. Cette dernière méthode permet le dosage simultané d'une trentaine d'éléments en particulier des éléments traces difficilement dosables par voie chimique. Toutefois, elle ne permet pas l'évaluation des éléments H, C, N, O et Pb, qui ne sont pas activables, ni Al, Ar, Mg, Mn, S, Ti et V, qui ont des périodes radioactives trop courtes. Pour les sédiments, on détermine la granulométrie, les éléments hydrosolubles et échangeables ainsi que la capacité d'échange cationique.

La radioactivité est mesurée par des méthodes éprouvées depuis des années assurant des résultats reproductibles. Les mesures des activités alpha et bêta globales ne donnent qu'une indication grossière des niveaux de contamination des échantillons. Nous leur préférons les analyses par radiochimie où l'échantillon est solubilisé et l'élément recherché est isolé chimiquement. C'est en particulier le cas pour ^{90}Sr et ^{99}Tc mesurés par comptage bêta, ou pour le plutonium et les transplutoniens mesurés par spectrométrie alpha [3].

Pour les mesures du tritium, l'eau est distillée et mélangée à de l'instagel pour sa mesure dans un spectromètre bêta à scintillation liquide. La séparation de l'eau interstitielle et du sédiment proprement dit (par pressage) donne l'activité du tritium de l'eau libre et celle du tritium lié à la matière organique. Dans le cas des végétaux et des poissons, la mesure se fait sur l'eau obtenue par brûlage du produit sec.

Les autres radionucléides sont détectés par spectrométrie gamma dans un laboratoire équipé de sept détecteurs germanium situés en sous-sol dans une chambre blindée, couplés à sept spectrographes de 8000 canaux et interconnectés à un mini-ordinateur «MULTI 20» [4].

Les résultats des mesures des analyses chimiques ou radioactives sont exprimés en g ou $Bq \cdot L^{-1}$ pour l'eau, en g ou $Bq \cdot kg^{-1}$ sec pour les sédiments et les végétaux, en g ou $Bq \cdot kg^{-1}$ frais pour les poissons. Les rapports moyens poids frais/poids sec (14,2 pour les végétaux, 3,9 pour les poissons) permettent de comparer nos résultats avec ceux exprimés différemment par d'autres auteurs.

6.   GESTION DES DONNEES

Les données relatives aux prélèvements, au conditionnement et aux résultats de mesures sont gérées sur un ordinateur IBM 370 du réseau CEANET. La base de données nommée IDOMENEE (Informatisation des données des mesures d'environnement) utilise le modèle SQL/DS (Structure Query Language/Data System). Elle contient au total 225 rubriques adressables concernant les mesures effectuées sur près de 5000 échantillons. Le modèle permet l'évolution de la base, la sélection de données et des calculs arithmétiques ou statistiques simples.

Un microordinateur Compaq 386s sert de terminal actif qui permet d'effectuer non seulement les saisies, les modifications et les interrogations de la base mais également des traitements plus complexes et d'assurer:
— le suivi des niveaux de radioactivité dans les compartiments du milieu aquatique;
— la corrélation des mesures radioactives aux données physiques et chimiques du milieu;
— des comparaisons d'une station à l'autre dans un même cours d'eau, dans des lacs ou des bassins fluviaux différents;
— l'évaluation de l'impact des différentes installations nucléaires en situation de rejets contrôlés comme en situation accidentelle.

7. BILAN DE LA RADIOACTIVITE DU RHONE

A l'aide de ces techniques nous avons pu mettre en évidence l'impact des différents termes-sources [5]. Avant l'accident de Tchernobyl, la zone 1 du Rhône montre, outre la radioactivité naturelle, l'influence des retombées des essais nucléaires atmosphériques caractérisés par la présence de ^{137}Cs et ^{90}Sr. La zone 2

TABLEAU I. RADIOACTIVITE DES COMPARTIMENTS DU RHONE EN FONCTION DES TERMES-SOURCES, AVANT L'ACCIDENT DE TCHERNOBYL

Radioactivité (zones du fleuve)	Sédiment (Bq/kg sec)	Végétaux (Bq/kg sec)	Poissons (Bq/kg frais)
Naturelle (zones 1–3)	2250	1700	110
Artificielle			
— retombées (zone 1)	10	6	2
— centrales (zone 2)	29	280	28
— Marcoule (zone 3)	1058	1636	35

TABLEAU II. RADIOACTIVITE DES PRINCIPAUX RADIONUCLEIDES ARTIFICIELS EN MAI 1986 DANS LA ZONE 1 DU RHONE[a]

Nucléides	Sédiments (Bq/kg sec)	Végétaux (Bq/kg sec)	Poissons (Bq/kg frais)
Cs 134	122 (1)	192 (1)	$4,2 \pm 1,6$ (14/14)
Cs 137	240 (1)	389 (1)	$8,8 \pm 3,2$ (14/14)
Ru 103	160 (1)	594 (1)	$2,1 \pm 1,4$ (6/14)
Ru+Rh 106	120 (1)	385 (1)	
Total	642	1560	$15,1 \pm 6,2$

[a] Les moyennes sont affectées de leur dispersion pour une probabilité de 95% et du nombre de valeurs significatives par rapport au nombre de mesures.

TABLEAU III. RADIOACTIVITE DES PRINCIPAUX RADIONUCLEIDES ARTIFICIELS EN MAI 1986, DANS LA ZONE 2 DU RHONE[a]

Nucléides	Sédiments (Bq/kg sec)	Végétaux (Bq/kg sec)	Poissons (Bq/kg frais)
Cs 134	90 ± 68 (7/7)	122 ± 30 (7/7)	4,2 ± 1,1 (44/44)
Cs 137	178 ±132 (7/7)	326 ± 221 (7/7)	9,2 ± 2,4 (44/44)
Ru 103	257 ±229 (7/7)	1964 ± 671 (7/7)	3,0 ± 1,2 (24/44)
Ru+Rh 106	152 ±133 (7/7)	1154 ± 355 (7/7)	—
Ag 110m	6,1 ±4,2 (7/7)	17 ± 11 (7/7)	1,1 ± 3,4 (3/44)
Total	684 ±565 (7/7)	3618 ± 1000 (7/7)	15,1 ± 3,8 (44/44)

[a] Les moyennes sont affectées de leur dispersion pour une probabilité de 95% et du nombre de valeurs significatives par rapport au nombre de mesures.

est marquée par une douzaine de radionucléides caractéristiques des effluents liquides des centrales nucléaires. Parmi ceux-ci, les $^{134,\ 137}$Cs, $^{57,\ 58,\ 60}$Co sont présents dans tous les compartiments; les ^{110m}Ag, ^{106}Ru+Rh, ^{54}Mn et ^{95}Zr sont moins fréquents. Dans la zone 3, les nucléides caractéristiques de l'usine de retraitement du combustible irradié apparaissent: $^{103,\ 106}$Ru, ^{144}Ce+Pr et des traces de Pu et d'Am. La radioactivité artificielle des divers compartiments du fleuve est inférieure ou de l'ordre de grandeur de leur radioactivité naturelle (Tableau I).

Immédiatement après l'accident de Tchernobyl (mai-juin 1986), on constate une élévation globale du niveau de radioactivité des compartiments du fleuve particulièrement visible dans la zone 1, uniquement influencée par les retombées atmosphériques (Tableau II).

Dans les autres zones du Rhône, la radioactivité due à l'accident s'ajoute à celle des effluents des installations nucléaires masquant même l'effet de plusieurs années de fonctionnement (Tableau III).

Après disparition des nucléides à vie courte et moyenne (iodes, tellures, etc.), on a continué à détecter les $^{134,\ 137}$Cs, les $^{103,\ 106}$Ru et le ^{110m}Ag provenant de Tchernobyl. Un an après l'accident, le ^{103}Ru avait complètement disparu et le ^{106}Ru n'était encore visible, à des concentrations faibles, que dans le sédiment et les végétaux. Fin 1988 et début 1989, seuls les $^{134,\ 137}$Cs sont toujours mesurables sur l'ensemble du cours d'eau. Le suivi in situ des concentrations en radiocésium dans les poissons permet de vérifier que la période biologique évaluée au laboratoire (200 jours environ) concorde avec les observations de terrain.

8. CONCLUSION

L'évaluation de l'impact de l'accident de Tchernobyl sur le Rhône est possible grâce aux 15 années de mesure radioécologiques (près de 5000 spectrométries gamma, quelques centaines d'analyses radiochimiques) qui ont été effectuées précédemment et qui constituent un état de référence. Cela a nécessité la mise au point d'une méthodologie de prélèvement fondée sur une définition des stations, des compartiments à prélever et des conditions de conservation et de traitement des échantillons. La continuité d'un travail d'équipe — entre ceux qui prélèvent, préparent, mesurent et interprètent — a permis d'avoir une méthodologie reproductible et de donner des valeurs directement comparables avant et après l'accident [6].

REMERCIEMENTS

Les prélèvements d'échantillons ont été effectués avec l'aide des groupements de Lyon et d'Aix en Provence du CEMAGREF et du Département de zoologie de l'Université Claude-Bernard à Lyon. Les analyses chimiques sont réalisées par le Laboratoire Solaigue de Nîmes, la radiochimie par le laboratoire LARB du Centre d'études nucléaires de Fontenay-aux-Roses, la spectrométrie gamma et les activations neutroniques par le LMEI d'Orsay. Nous les assurons de notre gratitude.

REFERENCES

[1]   AGENCE INTERNATIONALE DE L'ENERGIE ATOMIQUE, Measurement of radionuclides in food and the environment: A guidebook, Collection des Rapports techniques n° 295, AIEA, Vienne (1989).

[2]   MAUBERT, H., PICAT, P., Appareil de prélèvement des eaux douces en vue de la mesure radioactive, Note CEA-N-2205, CEN-Cadarache, Saint-Paul-lez-Durance, France (1982).

[3]   WILLEMOT, J.M., VERRY, M., JEANMAIRE, L., Détermination analytique du plutonium et des transplutoniens dans l'environnement, Rapport CEA-R-5460, CEN-Saclay, Gif-sur-Yvette, France (1988).

[4]   FOULQUIER, L., PALLY, M., Données sur la teneur en tritium lié des poissons des grands fleuves français, Ann. Assoc. Belg. Radioprot. 7 3–4 (1982).

[5]   FOULQUIER, L., PHILIPPOT, J.C., BAUDIN-JAULENT, Y., Métrologie de l'environnement. Echantillonnage et préparation d'organismes d'eau douce. Mesure des radionucléides émetteurs gamma, Rapport CEA-R-5164, CEN-Saclay, Gif-sur-Yvette, France (1982).

[6]   FOULQUIER, L., DESCAMPS, B., LAMBRECHTS, A., PALLY, M., «Analyse et évolution de l'impact de l'accident de Tchernobyl sur le fleuve Rhône», 24th Congress of the International Association of Theorical and Applied Limnology, Munich (août 1989).

STANDARDIZATION OF MEASUREMENTS AS A POWERFUL TOOL FOR IMPROVING THE EXCHANGE OF INFORMATION FOLLOWING A NUCLEAR ACCIDENT

J. GELEIJNS, A.V. BAERVELDT
Netherlands Standards Institute,
Delft

B.F.M. BOSNJAKOVIC
Ministry of Housing, Physical Planning
    and the Environment,
Leidschendam

D. VAN LITH
National Institute of Public Health
    and Environmental Hygiene,
Bilthoven

Netherlands

Abstract

STANDARDIZATION OF MEASUREMENTS AS A POWERFUL TOOL FOR IMPROV-
ING THE EXCHANGE OF INFORMATION FOLLOWING A NUCLEAR ACCIDENT.
    In 1988 the Netherlands Standards Institute (NNI) started a project, called NORM-
STAR, which aims at the standardization of radioactivity measurements within three years.
The standardization will be of help especially when large amounts of measured values have
to be collected, processed and interpreted by means of nuclear accident information systems.
To be standardized are off-site measurements of radioactivity, including measurements in the
environment and in the food chain. The status, organization, programme and planning of the
project are presented in the paper. The NORMSTAR project is being carried out on a national
scale. However, international standardization is considered highly necessary and therefore
possibilities for achieving this are also discussed.

1. BACKGROUND

During and after the accident at the Chernobyl nuclear reactor it became evident that the availability of reliable and well defined measurement results played a
key role in the exchange of information both on a national and on an international
scale. The lack of a standard format for the measurement results was a serious
handicap. Much precious time was (and still is) invested in interpreting and assessing
the comparability of the various data.

361

2.   NORMSTAR

As a result of the experience gained since the Chernobyl accident, several projects on nuclear emergency planning and response were started in the Netherlands. One of them is NORMSTAR, which was started in 1988 by the Netherlands Standards Institute (NNI). Its objective is the standardization, within three years, of methods of making off-site measurements of radiation and radioactivity for nuclear accident information systems. The project is concerned with sampling, sample preparation and measuring techniques related to the environment and the human food chain. Attention is also paid to the reporting of the results.

3.   ORGANIZATION OF NORMSTAR

The NORMSTAR project receives funding from the Ministry of Housing, Physical Planning and the Environment, the Ministry of Home Affairs, the Ministry of Transport and Public Works and the Ministry of Agriculture and Fisheries. Project management is conducted by an interdepartmental steering committee in which the above mentioned ministries are represented. The project is carried out by the NNI on the basis of the experience of the institute in achieving consensus on measuring methods in general and on measuring methods for radioactivity and radiation in particular. A project group of the NNI writes the standards and works in co-operation with the 'Radioactivity Measurements' technical committee of the NNI. In this committee experts from specialized research institutes, industry and Government have been appointed. An 'Agricultural Products' working group has been set up to assist the technical committee in the realization of the NORMSTAR project.

4.   STANDARDIZATION

Standardization is the establishing of provisions meant for common and repeated use; it is aimed at achieving an optimum degree of order in a given context. Essential in the process of standardization is the consultation of all interested parties. First a basic proposal and a working draft of a standard are written by the project group, in co-operation with the technical committee. Next the draft of the standard is sent to the relevant bodies for so-called external comments. Finally, after evaluation of these comments, the standard is published. Standardization is usually a slow process if it concerns national standardization and even more so if it concerns international standardization. However, the NORMSTAR project aims at producing a large number of national standards within the relatively short period of three years.

The NORMSTAR project will be finished in the middle of 1991. The standards written are pre-standards, i.e. prospective standards for provisional application in technical fields where there is an urgent need for guidance. They have to be reviewed, and if necessary revised, three years after publication. The validity of a pre-standard is limited to a maximum of six years.

Standardization of measurements is a powerful tool for improving the exchange of information after a nuclear accident since, if standardized methods are applied, the following conditions are supposed to be fulfilled:

— Sampling, sample preparation and measuring are clearly and precisely defined.
— The methods are based on the state of the art and on agreement between the different authorities involved.
— The descriptions of the methods (i.e. the standards) are easily available.
— The reporting is systematic and complete, and indicates the inaccuracy due to the statistical nature of radioactive decay and the determination limit.

The measurement results will be suitable for rapid exchange, computerization and rapid interpretation. Only by referring to a standard is information given about the way the measurement has been carried out. This information concerns, for example, the procedure applied, the method of calibration of the counter and the correction applied for radioactive decay.

In addition, standards can be used when an emergency organization is trained or tested by intercomparison runs.

5. DEFINITION STUDY

At the start of the project a definition study was carried out to formulate a proposal for the project programme [1]. This study consisted of an inventory and analysis of the existing situation in the Netherlands and other countries. On the basis of this study it was established which measurements are necessary for assessing the seriousness, extent and evolution of an accident situation.

In the determination of the priorities for the measurements to be standardized several aspects were considered, such as the importance of the measurements from a radiological protection point of view, the role they play in decision making, economic aspects, the number of monitoring authorities concerned with a particular type of measurement, and existing international obligations and agreements (e.g. the Convention on Early Notification of a Nuclear Accident and the Convention on Assistance in the Case of a Nuclear Accident or Radiological Emergency).

For radiological protection purposes measurements of exposure rate and radioactivity in air, deposition, water, milk and fresh vegetables and fruit are particularly relevant. For the assessment of radiation exposure ^{131}I, ^{132}Te/^{132}I, ^{134}Cs, ^{137}Cs, ^{103}Ru, ^{106}Ru, ^{89}Sr and ^{90}Sr are the most important nuclides to be determined.

Economic aspects mainly involve trade in agricultural products. In these products measurements of the following nuclides are essential: [131]I and ^{90}Sr, the alpha emitting nuclides [239]Pu and [241]Am and other nuclides with a half-life greater than 10 d, especially [134]Cs and [137]Cs.

6.   REALIZATION OF PROJECT

At an early stage it was decided to make a systematic distinction between standards for sampling, sample preparation and measuring techniques, so generally two or three separate standards may be required for one measuring result. However, in the long run the standardization process and the application of the standards will become considerably more efficient for various reasons:

— It will be possible for the same measuring technique to be used on samples of different origins in accordance with one standard.
— It will be easier to modify standards in line with changes of opinion or on the basis of experience.
— Subcommittees or working groups involved in the standardization process will be able to achieve results more efficiently.

Table I presents the measurements to be standardized with the highest priority.

TABLE I.   MEASUREMENTS TO BE STANDARDIZED WITH THE HIGHEST PRIORITY

Sampling	Sample preparation	Measuring technique
Water and constituents in water [2]	Water [6]	Semiconductor gamma spectrometry [7]
Fresh leafy vegetables and fruit	Constituents in water	NaI gamma spectrometry
Meat	Fresh leafy vegetables, fruit and grass	Gross alpha counting [8]
Fishery products	Meat	Gross beta counting [9]
Deposition	Fishery products	Alpha spectrometry
Grass [3]	Sr separation method	Measurement of exposure rate
Airborne substances [4]	Pu separation method	Measurement of surface contamination
Airborne iodine [5]	Urine	Measurement of air filters
Milk		
Dried milk		
Urine		

7.   HARMONIZATION WITH WORK OF INTERNATIONAL
      ORGANIZATIONS

The events associated with the Chernobyl accident have clearly shown that the
exchange of information is of vital importance, not only on a national but also on
an international level. The need for international agreement on the measuring
methods to be applied has therefore been mentioned explicitly in the NORMSTAR
project proposal. Furthermore, the use of intervention levels and of radioactivity
limits on an international scale also calls for international agreement on the measuring methods.

The infrastructure for international standardization exists in the form of two
main organizations, the International Organization for Standardization (ISO) and the
International Electrotechnical Commission (IEC). International standardization in
the field of radioactivity measurements already takes place within several Technical
Committees. The most important of these are:

— ISO/TC 85   Nuclear energy
— ISO/TC 147/SC 3   Water quality, radiological methods
— ISO/TC 190/SC 6   Soil quality, radiological methods
— IEC/TC 45B   Nuclear instrumentation, radiation protection
                                instrumentation.

The carrying out of activities within the several ISO TCs differs from the
integral approach in the Dutch project, where one Standards Committee co-ordinates
the standardization within the entire field of radioactivity measurements. In addition,
the modular approach as chosen in the NORMSTAR project is not followed within
these TCs. With several TCs involved in the standardization process inconsistencies
may occur. Liaison between the different ISO TCs may prevent this.

Within the framework of the European Committee for Standardization (CEN),
no standardization has been carried out yet in the field of radioactivity measurements. Thus the concentration of all activities within one technical committee and
the modular approach could be applied here from the beginning.

It is obvious that activities carried out by other international organizations such
as the International Atomic Energy Agency, the World Health Organization and the
Food and Agriculture Organization of the United Nations are of great importance for
a national standardization project. The IAEA recently finished a project called
Fallout Radioactivity Monitoring in Environment and Food. A new co-ordinated
research programme on Rapid Instrumental and Separation Methods for Monitoring
Radionuclides in Food and Environmental Samples has been started. When writing
the standards within the NORMSTAR project our aim is to keep at least in line with
international activities and international standardization.

8.   PROPOSAL

It is proposed that a discussion be started on what is the most effective way
to achieve international standardization in the field of radioactivity measurements.

REFERENCES

[1]   VAN DEN ENDE, C.A.M., VAN LITH, D., Definition Study of the NORMSTAR
       Project, NNI, Delft (1988).
[2]   Radioactivity Measurements — Sampling and Preservation of Water and Particulate
       Matter, NVN 5625, NNI, Delft (1989) (in Dutch).
[3]   Radioactivity Measurements — Sampling of Grass, NVN 5624, NNI, Delft (1989) (in
       Dutch).
[4]   Radioactivity Measurements — Sampling of Airborne Substances, NVN 5621, NNI,
       Delft (1989) (in Dutch).
[5]   Radioactivity Measurements — Determination of the Activity Concentration of Iodine
       in Ambient Air, NVN 5631, NNI, Delft (1989) (in Dutch).
[6]   Radioactivity Measurements — Sample Treatment of Water, NVN 5626, NNI, Delft
       (1989) (in Dutch).
[7]   Radioactivity Measurements — Determination of the Activity of Gamma Ray Emitting
       Nuclides in a Counting Sample by Means of Semiconductor Gamma Spectrometry,
       NVN 5623, NNI, Delft (1989) (in Dutch).
[8]   Radioactivity Measurements — Determination of Specific Gross-Alpha Activity of a
       Solid Counting Sample by Means of the Thick Source Method, NVN 5622, NNI, Delft
       (1989) (in Dutch).
[9]   Radioactivity Measurements — Determination of Specific Gross-Beta Activity and
       Specific Rest-Beta-Activity of a Solid Counting Sample, NVN 5627, NNI, Delft (1989)
       (in Dutch).

POSTER PRESENTATIONS

IAEA-SM-306/5P

ENVIRONMENTAL RADIOACTIVITY
MONITORING SYSTEMS IN ITALY

G. MONACELLI, V. PICCARDO, L. CRESCIMANNO,
M. LUCCHESE, S. SECCHI
Direzione Generale Servizi Igiene Pubblica,
Ministero della Sanità,
Rome, Italy

The spread of radioactive contamination due to the accident at the Chernobyl power plant and the consequent necessity for constant monitoring of the pattern of contamination in Italy to support decision making on initiatives for safeguarding public health have been a severe test of the response capability of the environmental monitoring systems operating in the country.

At the end of the 1950s, the national monitoring systems were organized principally to check for contamination subsequent to nuclear experiments. Such systems have evolved over time and currently comprise [1]:

— A system for monitoring of radioactivity in the air at ground level, measurement of total beta radioactivity and gamma ray spectrometry.
— Research into radionuclides which have major implications for human health, such as ^{90}Sr and ^{137}Cs in fallout associated with rain.
— Radiation monitoring of environmental matrices, such as soils and waters (e.g. sea water, river water, lake water and water destined for irrigation and drinking). As the data relating to such ecosystems are not easily interpretable, fish, molluscs, weeds and sediments are also monitored with the aim of improving the modelling of radioactivity transfer from water to humans.
— Systems for monitoring of activity in the main foodstuffs such as milk, meat, bread and several food products.

The Chernobyl accident indicated the necessity to have an alarm system which could signal in a short space of time the increase of radioactivity in the air, with sampling points located in such a way as to permit the quick identification of the contamination path.

Also under study is the possibility of adding the measurements from the ground stations to measurements carried out with helicopters and aeroplanes [2].

Past experience has also revealed the importance of local factors, such as orography and microclimate, in the spread of contamination. Hence there is a need to implement the monitoring systems at regional level. With this aim the Health Minister in 1987 issued some directives in order to unify and co-ordinate initiatives in this field [3].

In particular, the establishment of 21 laboratories, one for each of the departments into which the Italian territory is administratively divided, is being implemented. Such laboratories will be adequately organized and equipped so that they will also be able to respond to exceptional events, both on a local scale and in a larger context.

REFERENCES

[1] Rapporto Annuale sulla Radioattività Ambientale in Italia, Vol. 1, Reti Nazionali, Anni 1986–87, DISP-ARA-RAM/01/89, ENEA, Rome (1989).

[2] Il rischio ambientale nella produzione di energia: risultati sperimentali, calcoli e riflessioni dopo Chernobyl, Ann. Ist. Super. Sanità 23 2 (1987).

[3] MINISTERO DELLA SANITA, Direttive agli Organi Regionali per l'Esecuzione di Controlli sulla Radioattività Ambientale, Circolare No. 2/87, Gazzetta Ufficiale, Ministero della Sanità, Rome (1987).

IAEA-SM-306/11P

STRATEGIES FOR MONITORING ENVIRONMENTAL CONTAMINATION IN PAKISTAN FOLLOWING A MAJOR NUCLEAR ACCIDENT

S.S. HASAN, M.A. RAHMAN, Z.A. BAIG,
N.P.K. LODHI, S. AHMAD
Directorate of Nuclear Safety
    and Radiation Protection,
Pakistan Atomic Energy Commission,
Islamabad, Pakistan

For the monitoring of environmental radioactivity in Pakistan, a number of high velocity air samplers have been installed at different locations all over the country. Airborne particulates are collected on special types of filters and subjected to gamma spectrometric analysis at the laboratories of the Pakistan Institute of Nuclear

Science and Technology (PINSTECH). The air samplers operate continuously and the normal frequency of filter change is every 15 days. Analysis of the samples yields the normal background of various radionuclides in the local environment. In the case of a major nuclear accident the frequency of filter change would be increased and filters could be changed daily.

Food samples, including meat, milk and vegetables, are analysed monthly at the health physics laboratories of PINSTECH. For this purpose facilities for radiochemical analysis are available at PINSTECH. In the event of a major nuclear accident food samples would be analysed with increased frequency. During and after the Chernobyl accident various environmental samples were analysed to ascertain the activity levels in various food items. These levels were found to be insignificant [1]. Regarding contamination levels in food products, the International Radionuclide Action Levels for Food (IRALF), recommended by the Food and Agriculture Organization of the United Nations, are followed in Pakistan.

REFERENCE

[1] PERVEEN, N., et al., "Post-Chernobyl radioactivity measurement in Pakistan", paper presented PAEC–KfK Sem. on Post-Accident Management, Islamabad, 1987.

IAEA-SM-306/111P

PROYECTO DE LA RED NACIONAL DE VIGILANCIA RADIOLOGICA AMBIENTAL EN ESPAÑA

J.L. MARTIN-MATARRANZ, J.M. MARTIN-CALVARRO,
G. LOPEZ, J.L. BUTRAGUEÑO
Consejo de Seguridad Nuclear,
Madrid, España

Se presenta el Proyecto de la Red Nacional de Vigilancia Radiológica Ambiental, cuyos objetivos son:

1) Disponer de una red de vigilancia radiológica a nivel nacional, que funcione de modo continuo y que permita conocer la distribución y evolución de los radisótopos presentes en el medio ambiente, así como los niveles de radiación ambiental en términos de dosis.

2) Disponer en todo momento de datos experimentales que permitan realizar estimaciones del posible riesgo radiológico a que está sometida la población como consecuencia de una posible contaminación radiactiva del medio ambiente, y servir de apoyo a las decisiones a tomar.

3) Difundir periódicamente los resultados de esta vigilancia radiológica e informar puntualmente a la opinión pública.

La Red de Vigilancia Radiológica Ambiental comprende dos sistemas o redes parciales bien diferenciados:

a) La Red de Estaciones de detección y medida en continuo que, a su vez, consta de:

— medidores en continuo de radiación ambiental en términos de tasa de dosis,
— medidores en continuo de concentración de radioyodos y de aerosoles.

El número medio de estas estaciones se adaptaria a las características específicas de cada emplazamiento (poblaciones en la zona, topografía de la zona, etc.). La localización de estas estaciones coincide con las estaciones de la Red de medida de parámetros meteorológicos de que dispone el Instituto Nacional de Meteorología (INM).

Los datos se transmiten a través de las líneas de comunicación entre las estaciones y el INM, y desde aquí hasta el Consejo de Seguridad Nuclear (CSN) mediante una línea dedicada, establecida entre ambos centros.

b) La Red de Estaciones en las que se llevará a cabo un programa periódico de muestreo y análisis:

En dicha red se realiza el muestreo y medida de los parámetros radiológicos asociados fundamentalmente y en principio al agua, aire, suelos y alimentos.

La localización de los laboratorios de medida se ha escogido de modo que cubra razonablemente todo el país, sin involucrar un número demasiado grande de ellos que dificulte el control de calidad de las medidas.

IAEA-SM-306/31P

SYSTEM FOR MONITORING RADIOACTIVE CONTAMINATION
OF FOODSTUFFS AND OTHER AGRICULTURAL PRODUCTS

T. BRUNCLÍK, H. PROCHÁZKA, V. JIRÁSEK
Veterinary Research Institute,
Brno, Czechoslovakia

A system proposed by the Veterinary Research Institute in Brno for monitoring environmental contamination by fission products arising from the nuclear power plant accident in Chernobyl has been accepted by the Federal Ministry of Agriculture and Foodstuffs. Monitoring has concentrated on investigation of radioactive contamination of foodstuffs and feedstuffs. Only a system of measurement at an accuracy level higher than 90% for samples weighing at least 250 g was useful. Therefore, a two step system of monitoring was employed. Because the most important radionuclides were ^{134}Cs, ^{137}Cs and ^{131}I, gamma radiation measurement was given priority.

Laboratories in the first monitoring step formed the base of this system because they enabled us to screen the contaminated samples by means of measurement of the total gamma radiation. Selected samples were measured in second step laboratories, where gamma spectrometric analysis was performed. As a result of experience gained after the Chernobyl accident the first step was expanded to include outdoor measurement of caesium contamination in living animals too. This might be important especially in the case of higher caesium contamination of beef, e.g. where contamination levels exceed a derived intervention level. In such a case decontamination could be achieved by feeding with uncontaminated fodder for a short time (up to ten days), leading to an acceptable hazard.

Monitoring results are collected in a central database organized by the reference laboratory of the Veterinary Research Institute. This database includes all the information about every sample, including its source (e.g. the organ of an animal), the place of origin (sorting is mostly by district but for selected commodities it may go down to the village level), the history of the sample and the kind and quantity of the contamination. These characteristics are represented in a numerical code which allows both extension of a basic file of commodities and its modification at any time. The construction of the database is kept consistent with the features of structural programming, which allows the design of special programs for future use with the database, which now includes information on more than 40 000 samples examined.

The database is used very extensively and has three main areas of application: (a) prognostics and provision of expert advice and information, (b) scientific research, and (c) data on safety measurements. The database is very valuable since it is based on high quality measurement and serious evaluation of results. Thus it is possible to predict for which regions only restricted monitoring will be sufficient or for which regions there is a high probability that agricultural production need no longer be monitored. The effect of such competent forecasting is evident in terms of both safety and the economy. This can be demonstrated by the following two examples.

(1)   A region considered acceptable for production of baby food was to be surveyed because the environmental caesium contamination was very inhomogeneous. First, complete screening expertise was employed so that suitable dairy farms could be chosen. On the basis of the results the transport routes to the dairies were changed periodically, which enabled a limited volume of baby food to be produced in the acute period after the contamination. Experiments with live animals were then performed in order to examine the influence of various feeding methods, types of feedstuff and types of caesium contamination of green fodder on the activity of the udder under prolonged ingestion. The results of both repeated application of expertise and experiment were used for making prognoses, which allowed a very reliable control of milk production.

(2)   A high accuracy level of the results from monitoring of deer in the first hunting season after the accident together with screening results (about ten animals from each district) made it possible to select Moravian districts as those where only wild animals, mainly red deer and roe deer, with contamination below the derived intervention level could be hunted in the second season after the accident. Both the ecological and economic effects of these restrictions were significant because there is no possibility of measuring wild animals in vivo in advance.

The contribution of the database to scientific research is irreplaceable. The high statistical accuracy of the database made it possible, for example, to develop on new aspects of transfer factors in food chains, which were derived mostly from model experiments in the past. Now, transfer factors seem to be a function of the internal (health) and external living conditions of the animal and may vary by up to a factor of 5.

The utilization of information from the database will also make it possible to prepare reliable data for future safety projects.

IAEA-SM-306/105P

SURVEILLANCE OF ENVIRONMENTAL RADIOACTIVITY IN THE FEDERAL REPUBLIC OF GERMANY BY AN INTEGRATED MEASUREMENT AND INFORMATION SYSTEM

R. DEHOS, A. BAYER, A. KAUL
Institute for Radiation Hygiene
  of the Federal Health Office,
Neuherberg,
Federal Republic of Germany

On 19 December 1986, the Federal German Parliament (Deutscher Bundestag) passed the Act on the Precautionary Protection of the Population against Radiation Exposure. For the radiation protection of the public this Act has two chief purposes:

— To survey environmental radioactivity, and
— To keep both radiation exposure of man and radioactive contamination of the environment as low as possible by introducing appropriate measures in the event of an incident of potential radiological impact.

For the purpose of monitoring the environmental radioactivity, the Act has laid the basis for an Integrated Measurement and Information System for the Surveillance of Environmental Radioactivity (IMIS), which is currently in preparation.

The concept provides for the following tasks of the IMIS to be acted upon:

— Continuous surveillance of environmental radioactivity;
— Early detection and assessment of events that are of potentially significant radiological impact (early warning);
— Upon occurrence of such an event, continuous acquisition of comprehensive information about the radiological situation and assessment of the consequences.

According to this scheme, the different environmental fields mainly concerned with the large scale transport of radioactive substances are monitored by Federal Measurement Networks:

— German Weather Service (Deutscher Wetterdienst, DWD):
  air and precipitation;
— Federal Environmental Office (Umweltbundesamt, UBA): air;
— Federal Office for Civil Defence (Bundesamt für Zivilschutz, BZS): gamma dose rate, ground surface deposition;

— Federal Institute for Hydrology (Bundesanstalt für Gewässerkunde, BfG): surface waters other than coastal waters;

— German Hydrographic Institute (Deutsches Hydrographisches Institut, DHI): surface waters of the North Sea and the Baltic Sea, including coastal waters.

All data collected by the Federal Measurement Networks are gathered by the Central Federal Agency for the Surveillance of Environmental Radioactivity (Zentralstelle des Bundes für die Überwachung der Umweltradioaktivität, ZdB) at the Institute for Radiation Hygiene (Institut für Strahlenhygiene, ISH) of the Federal Health Office (Bundesgesundheitsamt, BGA).

The surveillance of specific activity in other environmental fields such as:

— Foodstuffs, tobacco products, commodities, medical drugs and their basic substances;

— Animal feed;

— Drinking water, groundwater and surface waters;

— Sewage, sewage sludge and residual and waste materials;

— Soil and plants;

— Organic fertilizer

is carried out by 43 state measurement agencies linked together into 11 State Networks of the individual Federal states.

All data collected by the State Networks are again first gathered by the ZdB and then forwarded for checking and processing to so-called Guiding Agencies. These agencies are:

— Federal Institute for Hydrology (BfG): surface waters;

— Federal Research Institute for Nutrition (Bundesforschungsanstalt für Ernährung, BfE): foodstuffs;

— Federal Institute for Milk Research (Bundesanstalt für Milchforschung, BfM): milk, milk products, soil, plants, animal feed and organic fertilizer;

— Federal Research Institute for Fishery (Bundesforschungsanstalt für Fischerei, BfF): fish, fish products, Crustacea and Mollusca;

— Institute for Water, Soil and Air Hygiene of the Federal Health Office (Institut für Wasser-, Boden- und Lufthygiene des Bundesgesundheitsamtes, BGA/WaBoLu): drinking water, groundwater, sewage and residual and waste materials;

— Institute for Radiation Hygiene of the Federal Health Office (BGA/ISH): tobacco products, commodities and medical drugs.

After checking and processing, the data are returned to the ZdB for further processing.

The ZdB has been assigned the task of summarizing, processing and documenting all data for the Federal Minister for the Environment, Nature Conservation and Nuclear Safety (Minister für Umwelt, Naturschutz und Reaktorsicherheit, BMU).

The ZdB is additionally assigned:

— To provide the state authorities with data from the Federal Measurement Networks;
— To co-ordinate the international exchange of data between the Federal Republic of Germany and international organizations (Commission of the European Communities, International Atomic Energy Agency);
— To develop models permitting an early assessment of radiation exposure;
— To define and develop parameters and programmes required for the whole of the IMIS.

The Federal Minister for the Environment, Nature Conservation and Nuclear Safety makes the final evaluation of the radioactivity data and gives corresponding recommendations. The ZdB assists the minister in the performance of this duty. The minister submits a report on the state of environmental radioactivity to the Federal Parliament and the Federal Council (Bundesrat). He also informs the media and the public.

BIBLIOGRAPHY

BAYER, A., et al., Integrated Measurement and Information System for the Surveillance of Environmental Radioactivity in the Federal Republic of Germany, Nucl. Eng. Int. (in press).

IAEA-SM-306/93P

IAEA ENVIRONMENTAL AND FOOD RADIONUCLIDE REFERENCE MATERIALS

V. STRACHNOV, R. DEKNER, R. SCHELENZ
Agency's Laboratory,
International Atomic Energy Agency,
Seibersdorf

The International Atomic Energy Agency has been distributing calibrated radionuclide solutions and reference and intercomparison materials since the early 1960s. The purpose of this activity has been to assist laboratories in the Member States in assessing and, if necessary, in improving the reliability of their analytical work in the field of nuclear technology and isotope utilization. The Analytical Quality Control Services (AQCS) programme, co-ordinated through the Agency's Laboratory at Seibersdorf, was established to deal with this task. Other divisions of

TABLE I. IAEA ENVIRONMENTAL AND FOOD REFERENCE MATERIALS
FOR RADIONUCLIDES INCLUDING Ra, Th AND U

Recommended values available for:	Matrix	IAEA code
U	Feldspar	F-1
	Uranium ore (pitchblende)	S-7, S-8, S-13
	Uranium ore (phosphate)	S-17, S-18, S-19
U, Th	Lake sediment	SL-1
	Soil	Soil-7
	Thorium ore	S-14, S-15, S-16
K-40, Cs-137	Lake sediment	SL-2
	Fish flesh	MA-B-3/RN
Sr-90, Cs-137, Ra-226, Pu-239	Soil	Soil-6
Sr-90, Ra-226	Animal bone	A-12
Sr-90, Cs-137	Milk powder	A-14
K-40, Sr-90, Cs-134, Cs-137	Milk powder	IAEA-152 [a]
	Whey powder	IAEA-154 [a]
K-40, Mn-54, Co-60, Sr-90, Tc-99, Cs-137, Ra-226, Pu-239, Pu-240	Alga, marine	AG-B-1
K-40, Pb-210, Po-210, Ra-226, Ra-228, Th-228, Th-230, Th-232, U-234, U-235, U-238	Sediment, deep sea	SD-A-1
K-40, Cs-137, Th-232	Sediment, marine	SD-N-2

[a] Affected by radioactive fallout from the Chernobyl accident.

the IAEA, including the International Laboratory of Marine Radioactivity at
Monaco, also contribute to this programme.

Shortly after the Chernobyl nuclear power plant accident, the IAEA started
collecting many environmental and foodstuff bulk samples (300–1000 kg each) for
the AQCS programme in order for them to serve as materials for intercomparison
studies. About thirty of these materials were affected by radioactive fallout from the

accident. After they had been checked for homogeneity, the materials were distributed to 20–50 laboratories in about 20–30 countries for intercomparison. For 1989, the AQCS intercomparison programme included such materials as: soil, stream sediment, uranium ore, clover, milk powder and tuna homogenate (Mediterranean), the latter three having been affected by fallout from Chernobyl. The radionuclides of interest in these intercomparison studies are ^{40}K, ^{90}Sr, ^{134}Cs, ^{137}Cs, $^{239,240}Pu$, ^{226}Ra, Th and U.

Other materials have been collected for determination of these radionuclides and are now in preparation or being considered. In all cases the results submitted by the participants are evaluated by the IAEA using numerous statistical tests and acceptance criteria to derive recommended values of the radionuclides of interest. In most cases, results from an intercomparison study must meet preset statistical criteria before the samples can be released as reference material.

The reference materials IAEA-152 (milk powder) and IAEA-154 (whey powder) are the first 'post-Chernobyl' natural matrix radionuclide reference materials internationally available. Three intercomparison exercises with Baltic Sea sediment, Mediterranean seaweeds and the sea plant *Posidonia oceanica* (collected after the Chernobyl accident) are already completed and evaluation of the results is in progress.

The IAEA environmental and food radionuclide reference materials which are currently available are listed in Table I. More information is given in the IAEA's yearly AQCS catalogue [1].

REFERENCE

[1] INTERNATIONAL ATOMIC ENERGY AGENCY, Analytical Quality Control Services Programme 1989, IAEA, Vienna (1989).

IAEA-SM-306/13P

FAST RADIONUCLIDE DETERMINATION USING SPECTROMETRIC PROCEDURES AND FIELD MEASUREMENTS

R. MARTINČIČ, M. KORUN,
U. MIKLAVŽIČ, B. PUCELJ
Jožef Stefan Institute,
Ljubljana, Yugoslavia

1. INTRODUCTION

The ability to assess as soon as possible the potential hazard due to a release of radioactive materials to the environment is important for providing timely data on which to base necessary protective and remedial actions. An Ecological Laboratory with a Mobile Unit was established at the Jožef Stefan Institute within the framework of a project of the United Nations Development Programme [1]. Its prime aims in radiological emergencies are fast detection and determination of contamination of the environment and formulation of expert recommendations to those authorities and organizations responsible for implementation of protective measures. To achieve these goals a special emergency monitoring system based on spectrometric procedures and field measurements was set up.

2. MONITORING STRATEGY

In the early phase of an accident radiological data on the immediate hazard include measurements of gamma dose rates, and airborne concentrations of particulate radionuclides and radioiodine.

For assessing the nature and extent of protective measures relating to the longer term hazard, survey data are collected which arise from measurements of the dose rates from, and isotopic analysis and quantification of, the deposited radionuclides, demarcation of contaminated areas, and fast in situ measurements of the concentrations of radioiodine and other relevant radionuclides in water and agricultural products associated with potential exposure via food chain pathways.

3. EQUIPMENT IN MOBILE LABORATORY

The equipment of the mobile radiological laboratory consists of:

(a) Emergency monitoring instrumentation, which consists of portable survey instruments and several gamma spectrometers (one HPGe spectrometer mounted in the vehicle, one portable HPGe spectrometer and one portable NaI spectrometer);

(b) A computerized meteorological station for field determination of local wind speed, wind direction, temperature, humidity, precipitation and gamma dose rates;

(c) Sampling and sample preparation equipment;

(d) Personal protective equipment;

(e) Accessories (personal computer, power generator, etc.).

4. MONITORING METHODOLOGY

The HPGe spectrometer mounted in the vehicle is used for radionuclide determination in prepared samples, the portable one for in situ measurements of plume and/or ground contamination, and the NaI spectrometer mainly for fast monitoring of food once the isotopic composition of the contamination is known. In situ gamma spectrometry has proven to be a sensitive and rapid method for field radionuclide determination [2]. The dose rates due to unscattered and scattered radiation from different gamma emitters can be derived as well.

Procedures have been set up for various types of measurements as well as for representative sampling and sample preparation. A method for rapid calculation of the efficiency of measuring voluminous samples with semiconductor detectors has been developed [3]. All these procedures are tested on a continuous basis in the routine monitoring of the environment surrounding the Krško Nuclear Power Plant and in regular emergency preparedness activities.

5. SAMPLING PROCEDURES

Sampling procedures have been prepared for:

(1) Air contamination: portable air samplers are used for airborne particulate radioactivity and radioiodine concentration [4];

(2) Rapid assessment of deposition: lucite plates of 0.3 m^2 area, brush painted with a layer of pharmaceutical Vaseline — experience since the Chernobyl accident showed Vaseline plates to be very convenient and reliable passive samplers [5];

(3)   Soil and sediment sampling: soil (private communication, Institute for Occupational Medicine, Bratislava, 1988) and sediment sampling is done by special sampling devices, which make it easy to take well defined samples.

6.   CONCLUSIONS

In the exercises and drills, in the routine monitoring of the environment around the Krško Nuclear Power Plant and especially in the measurement of contamination from the Chernobyl accident, the developed monitoring strategy and methodology as well as the equipment in the mobile laboratory have proved to be efficient for dealing with the early and later phases of a nuclear accident.

REFERENCES

[1]   MARTINČIČ, R., Ecological Laboratory with a Mobile Unit, IAEA/UNDP-YUG-79-006-TR, internal report, IAEA, Vienna, 1983.

[2]   MARTINČIČ, R., PUCELJ, B., KORUN, M., RAVNIKAR, M., Intercomparison In Situ Gamma Spectrometry, Internal Report DP-5391, Jožef Stefan Inst., Ljubljana, 1988.

[3]   KORUN, M., MARTINČIČ, R., A Method for Rapid Calculation of the Efficiency for Voluminous Samples on Semiconductor Detectors, Internal Report DP-5392, Jožef Stefan Inst., Ljubljana, 1989.

[4]   ZUPAN, M., MIKLAVŽIČ, U., PUCELJ, B., "Calibration of filters for detection of airborne I-131 in the environment of a nuclear power plant", Proc. 26th Yugoslavian ETAN Symp., Subotica (1982) IV.25–IV.31.

[5]   MIKLAVŽIČ, U., KORUN, M., BYRNE, A.R., BRAJNIK, D., JUŽNIČ, K., "Rapid assessment of Chernobyl deposition by gamma spectrometry of Vaseline samples", Proc. 14th Regional Congr. of Int. Radiation Protection Assoc., Dubrovnik (1987) 279–282.

IAEA-SM-306/12P

ESPECTROMETRIA GAMMA DE ALTA RESOLUCION APLICADA AL CONTROL MEDIOAMBIENTAL DEL ECOSISTEMA TERRESTRE URUGUAYO

C.M. VERA-TARTAGLIA, R. SUAREZ-ANTOLA,
R. GOYENOLA, G. BERNASCONI
Dirección Nacional de Tecnología Nuclear,
Ministerio de Industria y Energía,
Montevideo, Uruguay

Se presentan los resultados y conclusiones obtenidos de las medidas de baja actividad efectuadas mediante espectrometría gamma de alta resolución, para determinar los niveles de productos de fisión —provenientes del "fall-out" atmosférico— presentes en los estamentos de la cadena ecológica constituída por suelo/pastura natural/aguas superficiales/leche vacuna, tomados como indicador de la contaminación radiactiva ambiental.

Se eligieron, sobre una grilla cartográfica aplicada al mapa del país [1], 28 sitios de muestreo distribuidos homogéneamente en la superficie del territorio uruguayo.

A fin de determinar el perfil actual de la contaminación radiactiva ambiental y de confeccionar el "primer mapa radiactivo" de referencia nacional, se tomó como trazador artificial el ^{137}Cs incorporado por vía atmosférica.

Este trabajo describe el uso de los datos obtenidos en el desarrollo de un modelo de transferencia del ^{137}Cs en los estamentos elegidos. Se han vinculado estos datos con los obtenidos a partir de un sistema independiente y de alta frecuencia de muestreos de carne vacuna, lana de ovejas y forrajes en zonas ganaderas bien definidas del país [2]. Esta comparación muestra una buena concordancia con el modelo desarrollado y permite inferir una adecuada validación del mismo. La Fig. 1 ilustra los sistemas de muestreo aplicados y su correlación geográfica.

En el trabajo, además, se hace especial énfasis en el uso del análisis gamma-espectrométrico casi puramente instrumental [3]. Para ello se hace potencial uso de un detector de Ge Hiperpuro (3 × 3 pulgs.), con el cual se tomaron los espectros de las muestras que fueron medidas durante 16 horas, con la ayuda de una tarjeta multicanal para PC de 8192 canales. El proceso de cálculo se efectúa con apoyo de microordenadores. El detector fue calibrado en energías y en eficiencia, empleando una solución con 9 radisótopos aplicada a matrices semejantes con geometrías idénticas a las muestras analizadas [4].

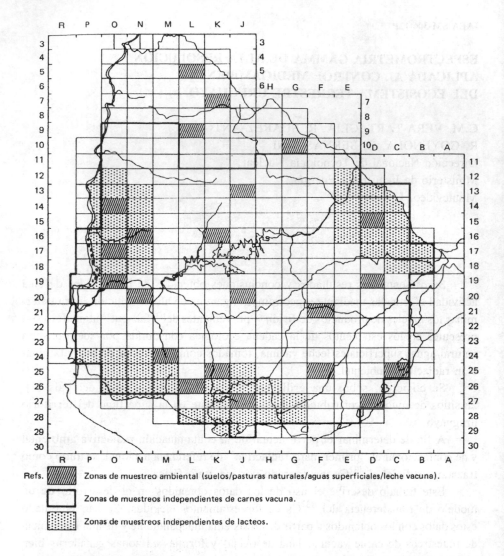

FIG. 1.   Sistemas de muestreos aplicados y su correlación geográfica (República Oriental del Uruguay).

Las líneas de los espectros obtenidos son comparadas con las del fondo radiactivo medido durante 16 horas, que sólo ha revelado las líneas correspondientes al ^{40}K y al decaimiento de las series radiactivas naturales del U y Th.

Los límites de detección se calcularon para el radisótopo estudiado, analizándose los resultados para diferentes geometrías de muestra [5].

REFERENCIAS

[1]   GARNER, A., Harmonizing Data and Estimates of the Radiological Impact of an
       Accident Using Extended European Grid Model, CEA, Inst. de protection et de sûreté
       nucléaire, Fontenay-aux-Roses (1986).
[2]   LINSLEY, G.S., HAYWOOD, S.M., DIONIAN, J., "Use of fall-out data in the
       development of models for the transfer of nuclides in terrestrial and freshwater sys-
       tems", Environmental Migration of Long-Lived Radionuclides (Proc. Symp. Knox-
       ville, 1981), IAEA, Vienna (1982) 615–634.
[3]   KNOLL, G., Radiation Detection and Measurement, McGraw-Hill, New York (1982).
[4]   IEEE Standard Techniques for Determination of Germanium Semiconductor Detector
       Gamma-Ray Efficiency Using a Standard Marinelli (Reentrant) Beaker Geometry,
       ANSI/IEEE Std 680, Inst. of Electrical and Electronics Engineers, New York (1978).
[5]   GONZALEZ, J.A., "Espectrometría gamma de muestras de agua ambiental",
       Methods of Counting and Spectrometry (Actas Simp. Berlín (Oeste), 1981), OIEA,
       Viena (1981) 59.

IAEA-SM-306/87P

UNIDAD MOVIL RADIOLOGICA
PARA ESTUDIO DE CONTAMINACION AMBIENTAL

A. CERVINI
Instituto Nacional de Investigaciones Nucleares,
México DF, México

1.   INTRODUCCION

Dadas las características del territorio mexicano: gran extensión, superficie
accidentada, dos cordilleras longitudinales y concentraciones de población con gran
densidad y distantes entre sí, se vió la necesidad de equipar una unidad móvil
radiológica para poder hacer estudios de las condiciones ambientales en puntos
específicos del territorio.

Asimismo, se pensó en esta unidad para poder dar apoyo en mediciones de
contaminaciones en caso de un accidente nuclear y/o radiológico.

Se realizó un estudio de las necesidades de la unidad para hacer frente a un
programa de caracterización de zonas en función de una posible deposición de
materiales contaminantes en caso de accidente.

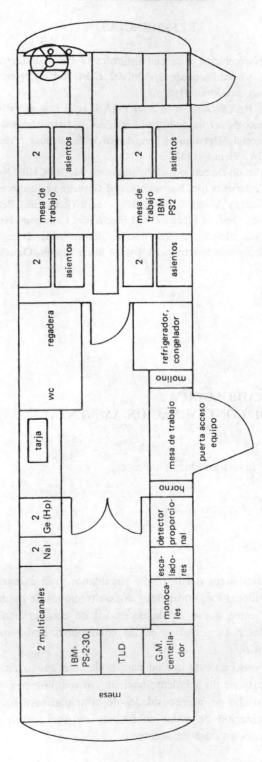

FIG. 1. Vista general de la Unidad Móvil Radiológica.

Se seleccionó un autobús existente en el Instituto, con capacidad de carga de 12,5 toneladas y un espacio útil de 11 m de largo por 2,40 m de ancho por 2,10 m de alto, y con un espacio de carga de 8 m³, y se dividió su interior en tres espacios de trabajo: el primero como zona de trabajo de escritorio y cálculo, con un baño equipado de regadera; el segundo, dedicado a la preparación y conservación de muestras; y el tercero para efectuar mediciones (véase la Fig. 1).

2. CARACTERISTICAS

2.1. Distribución de áreas y equipo

El espacio de escritorio y cálculo consiste en dos mesas con ocho asientos, librero y una computadora IBM-PS-2-30. El baño está equipado con WC y una regadera que se puede utilizar para descontaminación del personal. Se tienen conexiones para realizar descontaminación al exterior de la Unidad.

El espacio de preparación y conservación de muestras tiene una tarja y mesa para preparados radioquímicos, refrigerador, congelador, muestreadores de aire de alto y bajo volumen, horno y molino.

El espacio de medición está equipado con un detector de Ge(Hp), dos de NaI, uno proporcional semiautomático, dos sistemas multicanal independientes con capacidad de soportar cuatro detectores cada uno, dos monocanales portátiles, dos escaladores, tubos centelladores, tubos GM, un sistema multicanal portátil, un sistema lector TLD, y una computadora IBM-PS-2-30. Se tiene contemplado sustituir los NaI por Ge(Hp).

2.2. Sistema de climatización

Las tres áreas están separadas y con puertas para la comunicación. Se tiene un sistema de climatización por recirculación independiente para cada área, con una capacidad para garantizar un clima estable de 20°C ± 2°C en condiciones de trabajo en zonas tropicales o en zonas de clima continental. La alimentación del sistema es a 120 V.

2.3. Sistema de alimentación eléctrica

Se tiene contemplada una capacidad total de carga de 12 kV·A con una tensión de 120 V.

La alimentación puede ser directa de la línea comercial o por medio de dos autogeneradores independientes de 7,5 kV·A y 4,5 kV·A instalados en la parte inferior de la unidad.

Para cada circuito de carga se tiene un selector de la línea de alimentación.

El circuito de alimentación de los instrumentos de medición tiene un sistema de energía ininterrumpible por medio de baterías, regulador de tensión y tierra física independiente.

Asimismo, se dispone de un sistema de tierra general y de un sistema de pararrayos.

2.4. Sistema de drenaje

Se tienen a disposición tres sistemas de drenaje de las aguas: uno para aguas negras provenientes del WC; otro para aguas grises de la regadera y tarja; y un tercero para agua contaminada proveniente de la regadera o tarja, con sus respectivos selectores.

2.5. Preparación de muestras

La preparación de muestras se simplifica al máximo, considerando, por lo general, el tratamiento físico sin un tratamiento radioquímico. Las muestras que necesitan un tratamiento radioquímico complejo se conservan por refrigeración o congelación, realizándose posteriormente el tratamiento y la medición en el Laboratorio base.

2.6. Métodos de medición

Las geometrías que se tienen contempladas para la medición son del tipo Marinelli de medio litro y un litro, disco de 5 cm y puntual.

Los equipos están calibrados con estandares preparados en el Laboratorio base, y se verifica su estabilidad.

2.7. 'Software' disponible

Se encuentra a disposición un sistema de búsqueda y cuantificación de radionucleidos.

Para apoyo a emergencia, se tiene un programa de cálculo de exposiciones y blindajes para fuentes puntuales, lineales, superficiales y volumétricas.

Asimismo, se cuenta con un programa para dispersión atmosférica y proyección de dosis.

2.8. Sistema de comunicación

Se dispone de un sistema de radiocomunicación con una frecuencia de 169 MHz y conexión para teléfono.

IAEA-SM-306/7P

MODEL FOR THE CHOICE OF THE
ACCEPTABLE RISK TO THE POPULATION
IN THE CASE OF A 'SEVERE' NUCLEAR ACCIDENT

L. FAILLA
National Institute for Risk Prevention
  in Technological Activities,
Rome, Italy

Presented by L. Frittelli

Any anomalous situation involving nuclear materials can be considered as a 'nuclear accident', which can take place wherever nuclear materials are present. Any nuclear accident affecting the public could be considered as a 'severe nuclear accident'. Our purpose here is to compare the long term radiation risk incurred by the population in a severe nuclear accident with the other risks associated with the interventions undertaken in order to reduce the radiation risk.

Any nuclear accident causes to the people involved an increase, ΔD, in the absorbed ionizing radiation, which implies an increase, Δr, in the risk of damage. For evaluating Δr it is necessary to estimate ΔD by means of the usual methodologies. The evaluation of the increase Δr resulting from the increase ΔD is based on the well known 'linear hypothesis'. However, ΔD must be integrated with two variables which are particularly pertinent to our topic: the age and the physical state of the affected individuals. These variables are fundamental for our purpose.

The following expression can therefore be considered:

$$\Delta r = f(\Delta D, E, F) \tag{1}$$

where E is the age and F the physical state of the people involved. Reducing ΔD would not be sufficient to reduce Δr, since age and physical condition must also be taken into account.

In order to reduce ΔD, several interventions could be considered: (a) evacuation; (b) cut-off or reduction of the drinking water supply; and (c) cut-off or reduction of the food supply. These interventions, however, imply in turn other risks, R.

The appropriateness of the interventions and their characteristics can be determined from comparison of these new risks with the increase in the radiation risk, Δr. Therefore, in order to put the interventions into effect we must have:

$$R < \Delta r \qquad (2)$$

The risk R resulting from the interventions would be evaluated as follows.

(a)   *Evacuation*

The risk connected with evacuation (R_e) or with restricted use of a territory is linked to the age of the persons involved in this intervention, to their physical condition and to the circumstances under which the evacuation takes place. Therefore, with the adoption of the criterion expressed in inequality (2), the evacuation is justified only if:

$$R_e = f(E, F, c) < \Delta r \qquad (3)$$

where c is related to the evacuation circumstances.

(b)   *Intervention related to drinking water supply cut-off*

The health risk (R_a) related to cut-off of the drinking water supply is linked to the possibility of replacing the water subject to the ban. Whenever this cannot be done, the radiation risk doubtlessly is preferred to the risk arising from the intervention. However, this risk can also be evaluated on the basis of the modalities (timing and quantity) of water substitution. In this case, the age and physical state of the people involved should be taken into account. All these elements are expressed in a parameter, A. Therefore, the intervention may be undertaken only if:

$$R_a = f(A) < \Delta r \qquad (4)$$

(c)   *Restrictive measures on food supply*

The health risk (R_b) arising from this intervention is linked to the possibility of substituting the limited or forbidden food, taking into account the 'weight' (p) of this food in the diet. For instance, for a group of vegetarians a meat cut-off will give $R_b = 0$. In contrast, a similar intervention on milk could give a very high value of

R_b and would be even more serious than Δr. Therefore, the intervention will only be justified if:

$$R_b = f(E, F, B, p) < \Delta r \tag{5}$$

where B and p are related to the possibility of substitution and to the 'weight' of the food in the diet.

Conclusions

The increased health risk following an increase, ΔD, in absorbed ionizing radiation cannot be estimated in the case of a 'severe nuclear accident', nor can it be subjected to specific rules a priori. In contrast, the interventions to reduce ΔD can be controlled and subjected to regulations in the execution phase. Therefore, interventions to reduce the risk of damage from a severe nuclear accident cannot be evaluated according to the nuclear risk only, but must be compared with the risks arising from the interventions themselves.

This proposal is not easy to implement. In fact, we must consider that the risk cannot be physically quantified, and therefore cannot be measured in terms of a specific unit of measurement. It is based on probabilities which do not allow for protecting man from the probable damage.

The expressions given above, which link the different risks with the variables examined, call for knowledge of the population that may be involved in a severe nuclear accident. In particular, the age parameter appears in all the expressions, directly or indirectly through other variables. But other parameters, such as the physical state of the people involved, the possibility of supply, the relevant factors which characterize a diet, and the circumstances of the evacuation, are also important for taking the necessary decisions on the above cited interventions. With highly accurate knowledge of the above parameters, it will be possible to make timely decisions on interventions.

However, these decisions cannot be made only on the basis of the reduction of ΔD, even though it is an important element, because it does not exclusively identify the risk of damage to the involved population. Therefore, an increase, ΔD, in absorbed ionizing radiation can be considered 'acceptable' if it implies an increase in the nuclear risk, Δr, lower than the risk, R, arising from the interventions undertaken to reduce Δr, i.e. ΔD is acceptable if $\Delta r < R$. The risks associated with evacuation (R_e), drinking water supply cut-off (R_a) and food supply cut-off (R_b) must be evaluated separately with reference to the involved population on the basis of the previously examined parameters.

BIBLIOGRAPHY

FAILLA, L., "Statistics and risk philosophy in human activities", paper presented at 10th Congr. of Int. Radiation Protection Assoc., Avignon, 1982.

FAILLA, L., "Nuclear accidents with regard to human health in the work place", paper presented at 10th World Congr. on Prevention of Occupational Accidents and Diseases, Hull, Quebec, 1983.

Part II

MONITORING OF RADIOACTIVITY

(c) Methods and Techniques

Part B

MONITORING OF RADIOACTIVITY

(c) Methodology Technique

MOSSES AS MONITORS
OF ^{137}Cs FALLOUT

S. DARÓCZY, Z. DEZSŐ, A. BOLYÓS, Á. PÁZSIT, J. NAGY
Isotope Laboratory,
Institute of Chemistry and Institute of Physics,
Kossuth University

M. NAGY
Department of Botany,
Institute of Biology, Kossuth University

Debrecen, Hungary

Abstract

MOSSES AS MONITORS OF ^{137}Cs FALLOUT.
  The reactor accident at Chernobyl proved that in the event of a major nuclear accident very inhomogeneous contamination covering a large area may be encountered. For making proper decisions, the relevant authorities require a detailed fallout map of the affected area, which may be difficult to obtain using conventional techniques. A biomonitor of widespread occurrence may provide the solution to this problem. In 1986, the radioactive contamination in Europe from the Chernobyl accident provided good experimental grounds to find a bio-monitor and prompted an investigation of the radioactivity of mosses for this purpose. In the methodological studies conducted to the present, several species abundant in Hungary have been used. The measurement of samples collected from the same place showed that the standard deviations are within reasonable limits. The retention of ^{137}Cs in mosses has been investigated since the beginning of 1987. Data obtained show no decrease of activity concentration at least for the first 1000 days after the contamination at the end of April 1986. The investigation of caesium uptake of different species showed no systematic difference among species. In order to determine the size of the area for which the data measured in one particular point are representative, many samples were collected in Debrecen and evaluated for their ^{137}Cs activity. The results show that only samples collected from unprotected areas give acceptable results. It is also clearly demonstrated that local conditions may affect radioactive fallout, a factor which then has an influence on the number of samples to be collected from a given area.

1. INTRODUCTION

The very long half-life and the volatility of ^{137}Cs, together with its similar behaviour in biological systems to potassium and sodium, make ^{137}Cs one of the most important of those radionuclides released into the environment by the nuclear industry.

The Chernobyl reactor accident provided a unique possibility to test and improve further the models available for predicting the movement and fallout of atmospheric contamination. The accident has also shown that public concern about radioactive contamination is growing worldwide. As a consequence, the relevant authorities require a detailed and accurate fallout map, either national or regional, in order to determine the long term effects of contamination.

Several methods exist and are used routinely to measure ground surface contamination. As an alternative to these methods, a biomonitor of widespread occurrence can be employed. The application of a biomonitor has several advantages: it is cheap and its use can establish a network with a very fine grid.

Gorham [1] was probably the first who observed that lichens and mosses accumulate radioactive fallout much more efficiently than higher classes of plants. Because they are readily available and easy to collect, mosses seem to be more suitable for mapping. Other authors also investigated the radioactivity of mosses after the Chernobyl reactor accident [2, 3]. It has been proved that these plants accumulate all kinds of radioisotopes present both in the atmosphere and in rain water.

Although mosses have long been used to measure metallic pollutants of the atmosphere [4], their use as a biomonitor of ^{137}Cs contamination is not straightforward at all. In contrast to the ordinary pollutants which are permanently present in the atmosphere, radioactive contaminants resulting from a major nuclear accident occur abruptly in the atmosphere, with relatively rapid fallout. Questions on the uptake and retention efficiency and their dependence on species and on other parameters, such as the weather conditions, for example, are only a few of the many questions which must be answered.

The release of radionuclides from the reactor at Chernobyl in 1986 provided excellent conditions for investigating many of the above problems. Owing to the long half-life of ^{137}Cs, its study seemed to be most promising and prompted us to investigate the possibility of using mosses as biomonitors of ^{137}Cs. In this paper we summarize the results of the methodological studies obtained so far.

2. SAMPLE COLLECTION

Since the beginning of 1987 moss samples have been collected from those places where they were fully exposed to atmospheric contamination in 1986. Unprotected roofs of buildings seemed to be the best collecting sites. In some cases samples were also collected from the tops of brickworks and concrete surfaces as well as from streets, etc.

Ceratodon purpureus, Grimmia pulvinata, Hypnum cupressiforme, Tortula ruralis and *Bryum argenteum,* as the most abundant moss species, were collected in

as large a quantity as possible in order to study the behaviour of the different species
and to gather information on the reproducibility and accuracy of the method.

3.   SAMPLE PREPARATION

Simple mechanical procedures were used to remove larger pieces of contaminants. Since each sample consisted of hundreds of small plants, cleaning them
one by one was not feasible; instead the entire dried sample was rubbed through a
series of sieves and the Ø 0.5–1.4 mm fraction was used to prepare samples.
Although 20% of the material was lost in this way, a simple inspection of the fraction
below Ø 0.2 mm showed that it consisted mainly of sand and other mineral substances. Using a laboratory oil press, moss samples were then pressed into discs of
Ø 41 mm.

A detailed investigation of nearly 600 samples prepared in this way showed a
normal density distribution with 1.2 g/cm^3 and $\pm$ 0.17 g/cm^3 as average value and
standard deviation (SD), respectively. Both data are reasonable if we consider that
a certain but variable amount of inorganic material with very small particle size
strongly adheres to the plant and is almost impossible to remove.

4.   ACTIVITY MEASUREMENT

In moss samples the activity concentration (AC) of ^{137}Cs and ^{134}Cs isotopes
has been determined with Ge(Li) gamma spectrometry. In the early phase of the
investigations, ^{106}Ru could also be measured and in some cases hot particles in the
samples were also identified through the measurement of the activity of ^{144}Ce and
^{95}Zr isotopes.

The gamma spectrometer consisted of a 40 cm^3 active volume Ge(Li) detector
with 3.5 keV resolution (FWHM) for the 661.6 keV line of ^{137}Cs and with 4.9%
relative efficiency, a Canberra C-20 MCA and ATOMKI (Hungary) electronics.

All activity data have been corrected to 26 April 1986 using
$T_{1/2}(^{137}\text{Cs}) = 30.14$ a and $T_{1/2}(^{134}\text{Cs}) = 2.0685$ a; the final results were then given as
ACs in Bq/g for all moss samples.

5.   RESULTS

Data for moss samples collected at the same time from different parts of a
given roof are presented in Table I. The SD of ⟨AC⟩ is 3–15% for 90% of the 22
sample sites investigated. It is important to note that no systematic AC difference has

Text cont. on p. 401

TABLE I. DATA FOR MOSS SAMPLES COLLECTED FROM DIFFERENT PARTS OF A GIVEN ROOF

Sample sites	AC of moss samples (Bq/g)					⟨AC⟩ (Bq/g)	SD (%)
	1	2	3	4	5		
Debrecen							
Lefkovics u. 40	0.749	0.995	0.904	0.988		0.909	10.9
Simonyi u. 29	0.991	1.055	0.786	0.752		0.896	14.4
Péterfia u. 41	0.998	1.032	0.893	0.971	1.075	0.994	6.1
Bodaszöllö	0.914	0.961	0.796	0.945	0.977	0.919	7.1
Hajdúszoboszló							
Állami Gazdaság (State farm)	0.538	0.549	0.551	0.542	0.508	0.538	2.8
Állami Gazdaság	0.617	0.599	0.352	0.489		0.514	20.6
Szekszárd							
Csatár u. 44	0.858	0.877	0.774	0.724		0.808	7.7
Rákóczi u. 103	1.051	1.149	1.096			1.099	3.6
Szluka Gy. u. 16/b	1.404	1.338	1.359	1.901		1.500	15.5

Kunszentmiklós							
Tavasz u. 12	0.529	0.533	0.483			0.515	4.5
Bacsó B. u. 3	0.528	0.497	0.397			0.468	11.5
Kálvin tér 17	1.512	1.457	1.442			1.470	2.0
Ősz u. 5	0.507	0.412	0.492	0.515	0.557	0.497	9.4
Nyírbátor							
Rózsa F. u. 61/a	0.570	0.617	0.614	0.736		0.634	9.8
Rózsa F. u. 61/b	0.941	0.924	0.843			0.903	4.8
Kiskunhalas							
Fecske u. 7	1.481	1.483	1.304			1.423	5.9
Kiskundorozsma							
Tas u. 5	1.138	1.231	1.018			1.129	7.7
Jászdózsa							
Árpád u. 17	4.959	4.842	5.568			5.123	6.2
Boglárlelle							
Szabadság u. 19	6.138	5.850	4.903			5.630	9.4
Látrány							
Széchenyi u. 8	11.077	12.047	12.645	13.598	12.987	12.471	6.9

TABLE I. (cont.)

Sample sites	AC of moss samples (Bq/g)					⟨AC⟩ (Bq/g)	SD (%)
	1	2	3	4	5		
Dunakömlöd							
Szabadság u. 11	2.109	1.910	2.447			2.155	10.3
Hatvan							
Vörös Hadsereg (Army installation)	3.245	2.745	3.132	2.995		3.029	6.1
Ösz u. 5	0.507	0.412	0.492	0.515	0.557	0.497	9.4

Note: u.: street; tér: square.

TABLE II. ACTIVITY CONCENTRATIONS (Bq/g) FOR FIVE DIFFERENT MOSSES

Sample sites	Bryum argenteum	Ceratodon purpureus	Grimmia pulvinata	Hypnum cupressiforme	Tortula ruralis	⟨AC⟩ (Bq/g)	SD (%)
Egyházasdaróc							
Dózsa Gy. u. 1	–	4.456	3.939	–	3.205	3.883	13.3
Ménfőcsanak							
Győri u. 126	1.839	–	1.637	1.889	–	1.788	6.1
Győri u. 111	–	1.126	–	–	1.478	1.302	13.5
Gyomaendrőd							
Hídfő u. 15	0.514	–	–	–	0.525	0.519	1.0
Hídfő u. 11	0.445	–	–	–	0.502	0.473	6.0
Nagykölled							
Fö u. 59	–	2.963	–	2.566	2.462	2.664	8.1
Sopronhorpács							
Dózsa Gy. u. 6	–	–	11.911	8.875	–	10.393	14.6
Kőröstarcsa							
Petőfi u. 16	0.442	0.378	–	–	0.487	0.436	10.3

TABLE II. (cont.)

Csárdaszállás							
Kossuth u. 13	—	0.772	—	—	0.918	0.845	8.6
Tótszerdahely							
Béke u. 20	—	—	—	6.494	6.342	6.418	1.2
Kistelek							
Bercsényi u. 4	—	2.259	—	—	2.186	2.222	1.6
Debrecen							
Egyetem tér 1	—	—	—	1.306	1.622	1.464	10.8
Nyírbátor							
Rózsa F. u. 61	0.991	0.863	—	—	—	0.927	6.9
Süttő							
Vasútállomás (Railway station)	—	4.985	—	—	4.481	4.733	5.3
Szentes							
Futball pálya (Football stadium)	—	4.222	—	—	3.989	4.105	2.8
Komlóska							
Ady E. u. 1	—	1.580	—	—	1.757	1.668	5.3

Note: u.: street; tér: square.

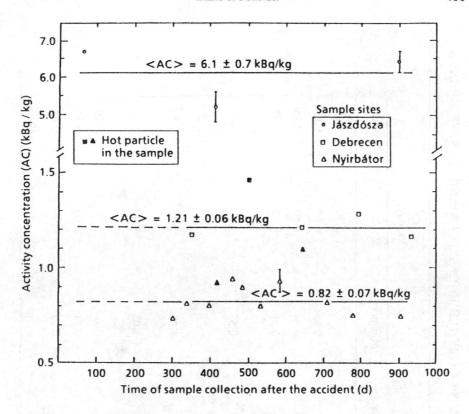

FIG. 1. *Data for mosses collected at different times from the same roof.*

been observed for mosses living at different heights of the roof. It seems that the flow of rain water down the roof does not influence the caesium uptake of mosses.

From a practical point of view it is interesting to study the caesium uptake of different species. For this reason as many species as possible were collected from one sample site (roof) and then processed separately for measurements. The results obtained are summarized in Table II. All the five mosses studied take up caesium to the same extent.

The retention of caesium in mosses is of primary importance if it is to be applied as a biomonitor. Therefore, a careful investigation has been carried out using four experimental sites where mosses cover an area large enough to facilitate such a study for many years. Samples have been regularly collected, their ^{137}Cs activity determined and calculated back to the time of the Chernobyl reactor accident. The results of three sites for the first 1000 days after the accident are shown in Fig. 1. No systematic decrease can be observed; furthermore, the SDs of the means are similar to those shown in Table I. It should be mentioned, however, that the most recent

TABLE III. ACTIVITY CONCENTRATIONS FOR MOSSES COLLECTED IN DEBRECEN AND ITS SURROUNDINGS

Sample sites	Features of sample sites	Number of samples	AC (Bq/g)			SD (%)
			Minimum	Maximum	Average	
Bodaszölö	Roof, 40°	7	0.80	0.99	0.93	7.0
Bodaszölö	Roof, 0°	2	0.85	0.93	0.89	
Apafa	Roof, 25°	2	0.64	0.69	0.66	
Izotóp Laboratórium	Asphalt	5	0.51	0.90	0.73	21.2
Izotóp Laboratórium	Shaded ground	1			0.25	
Köztemetö (Public cemetery)	Tombstone, 0°	6	0.53	0.79	0.65	17.1
Köztemeto	Shaded ground	1			0.17	
KLTE menza (University area)	Concrete	9	1.00	2.21	1.50	27.3
KLTE kémiai épület (University area)	Asphalt	2	1.02	1.36	1.19	
Thomas M. u. 41	Asphalt	1			0.70	
Lefkovics u. 40	Roof, 40°	8	0.71	1.00	0.85	13.1
Simonyi u. 29	Roof, 0°	3	0.75	0.99	0.84	15.3
Menyhárt J. tér 1	Concrete	1			0.50	

Location	Type					
Péterfia u. 41	Roof, 15°	5	0.89	1.08	0.99	6.9
Rézmetszö u. 21	Roof	1			0.41	
Csüry B. u. 20	Roof	2	0.77	0.92	0.85	
Hétvezér u. 85	Roof, 30°	3	0.51	0.60	0.54	8.6
Erzsébet u. 27	Roof, 0°	1			0.46	
Krúdy Gy. u. 6	Roof, 35°	4	1.16	1.28	1.21	4.5
Debrecen, Kondoros	Roof	1			0.82	
Balaton u. 9	Roof, 20°	2	0.71	0.94	0.82	
Balaton u. 62	Roof, 45°	2	0.90	1.36	1.13	
Kanális u. 37	Roof, 45°	3	0.76	0.93	0.83	10.8
Gázvezeték u. 59	Roof, 30°	2	0.65	0.76	0.70	
Kerekes F. u. 23	Shaded roof	1			0.16	
Áchim A. u.	Roof, 45°	2	0.67	0.70	0.69	
Mikepércsi u.	Roof, 30°	1			0.80	

Note: u.: street; tér: square.

data which are not included in Fig. 1 for one of our experimental sites show a weak decrease in ^{137}Cs concentration.

From the viewpoint of mapping, the detailed investigation of a relatively small area, such as a town, seemed to be useful in order to determine the number of sample sites required to produce a fallout map with reasonable accuracy. The area of Debrecen and its surroundings, covering some 36 km^2, has been used for this purpose, where samples were collected from 27 different places. The results summarized in Table III show the importance of the unprotected sample site (see the data for the three pairs of samples from unprotected and protected places). The mean value of the ^{137}Cs AC for the whole area is 0.84 Bq/g. The 32% relative SD of the mean is much more than expected from the previous experiments and indicates that the contamination of the area in question was not uniform. The above results suggest that at least 3–5 samples have to be collected from one site to obtain reasonable data on contamination.

The above as well as other methodological investigations not mentioned in this paper show that mosses can be used effectively as biomonitors of radiocaesium released in a major nuclear accident.

REFERENCES

[1]   GORHAM, E., A comparison of lower and higher plants as accumulators of radioactive fallout, Can. J. Bot. **37** (1959) 327.

[2]   SAWIDIS, T., Uptake of radionuclides by plants after the Chernobyl accident, Environ. Pollut. **50** (1988) 317.

[3]   LILJENZIN, J.O., et al., Analysis of the fallout in Sweden from Chernobyl, Radiochim. Acta **43** (1988) 1.

[4]   RICHARDSON, D.H.S., The Biology of Mosses, Blackwell Scientific Publishers, Oxford (1981).

RAPID DETERMINATION OF SOIL CONTAMINATION BY HELICOPTER GAMMA RAY SPECTROMETRY

I. WINKELMANN, S. SCHMERBECK, H.J. ENDRULAT
Institute for Radiation Hygiene
   of the Federal Health Office,
Neuherberg,
Federal Republic of Germany

Abstract

RAPID DETERMINATION OF SOIL CONTAMINATION BY HELICOPTER GAMMA RAY SPECTROMETRY.

The paper describes aerial nuclide specific measurements of surface contamination that were performed after the Chernobyl reactor accident in the southern region of the Federal Republic of Germany in August 1989. For these measurements, a helicopter equipped with a gamma ray spectrometer system including an HPGe detector with a relative efficiency of 50% was used. Soil contamination due to ^{134}Cs and ^{137}Cs was measured during a number of flights covering a total distance of about 300 km. The average flying altitude measured by a laser altimeter was about 70 m above ground level and the speed was about 130 km/h. The measuring time was chosen to be 60 s for each spectrum, corresponding to a flight path distance of about 2.2 km over which the average soil contamination was determined. The measured ^{137}Cs values of up to 25 kBq/m^2 are in good agreement with the results of measurements obtained by other methods. The values measured for ^{134}Cs were lower by a factor of 5.

1. INTRODUCTION

After the reactor accident in Chernobyl, nuclide specific soil contamination measurements were carried out in the southern part of the Federal Republic of Germany with an in situ gamma ray spectrometer system. By this method, a period of about two to three weeks was required to measure the soil contamination over the whole country [1]. To reduce the time necessary for such contamination measurements, a helicopter was equipped with a semiconductor spectrometer system and used for rapid aerial measurements. This method has the combined advantages of in situ gamma ray spectrometry and helicopter mobility. The primary goal in using an airborne gamma ray spectrometer system is to obtain a fast survey of the soil contamination of a large area due to a nuclear accident. In this way, widespread contamination can be assessed within a very short time [2].

2. METHODOLOGY

A gamma ray spectrometer installed in an aircraft permits the rapid measurement of the gamma radiation of radionuclides deposited on the surface. These measurements can be performed at different altitudes and flight speeds.

At an altitude of 70 m, a soil contamination of several kilobecquerels per square metre for ^{134}Cs and ^{137}Cs can be measured by a semiconductor detector having a relative efficiency of 50% and with a measuring time of 60 s. The activity of the individual radionuclides deposited on the surface is calculated, considering altitude and the intensity of the photopeaks measured in the spectra. Another significant parameter is the relaxation depth, which describes the distribution of the radionuclides in the soil [3]. It indicates the depth at which the activity has decreased to 37% of the surface value. Immediately after the deposition of radionuclides a 'pure' surface contamination can be assumed for data evaluation (dry deposition). In the case of wet deposition of radionuclides, a relaxation depth of 1–2 cm may be suitable, as was the case for the measurements performed in May 1986 in southern Bavaria after the Chernobyl reactor accident. The relaxation depth can be determined by soil sample measurements [1, 4].

3. MEASURING SYSTEM

A helicopter (type Alouette) was equipped with a gamma ray spectrometer consisting of an HPGe detector with a relative efficiency of 50%, and electronic components such as a high voltage supply and an amplifier. The detector signals were analysed in a microcomputer (multichannel analyser card) to evaluate the data during flight. The data were stored on a hard disk after each measurement (the required time for each spectrum was 6 s). Thus the radiation due to soil contamination could be measured continuously.

The whole system was mounted in a 19 in rack. The detector was fixed in a metal box lined with foam rubber to reduce the microphonic effects due to helicopter vibrations. This box was placed over an opening in the helicopter bottom.

The flying altitude was measured by a laser altimeter. The altitude data could be read by the pilot from an instrument display and subsequently stored on a hard disk. The flight route was continuously filmed with a video camera and stored on magnetic tape. The arrangement of the measuring system can be seen in Fig. 1.

4. MEASUREMENTS AND RESULTS

The soil contamination in the southern part of the Federal Republic of Germany after the Chernobyl accident had been measured with an in situ gamma ray

FIG. 1. Measuring system in helicopter.

spectrometer in May and June 1986 [1]. Further results from soil sample measurements were published in Ref. [5]. Detailed data were therefore available to test the method and the measuring system in the helicopter. For the first helicopter measurements, an area of about 2500 km² in the southeastern part of Bavaria was chosen.[1]

The routes of the test flights are shown in Fig. 2. The distance between routes was 10 km and the total distance flown was about 300 km. The average flying altitude was about 70 m above ground level and the speed was about 130 km/h. A total of 120 spectra were recorded in an energy range between 60 keV and 2 MeV. The measurement time was 60 s for each spectrum, corresponding to a distance of about 2.2 km over which the average contamination was determined.

The terrain surveyed consisted of small hills (with some deep valleys) covered occasionally by small wooded areas. The pilot followed the shape of the terrain as closely as possible. Sometimes, especially on routes I and II, cultivated fields surrounded by either woods or pastures were overflown, but most of the area consisted of grassland and pasture. In most cases the influence of fields and forests in this region could be neglected. The data were omitted where this influence could not be neglected and the activity distribution in the soil had been disturbed by soil treatment

---

[1] The flights were performed by the Federal Border Police (Bundesgrenzschutz).

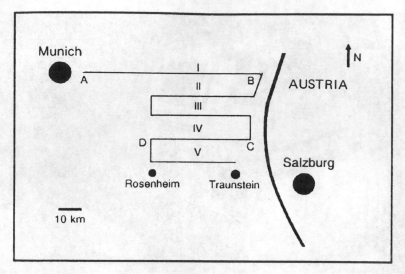

FIG. 2. Routes of the test flights in the southeast of the Federal Republic of Germany.

(fields). These effects are of minor interest for measurements performed immediately after the deposition of fresh fallout, the latter actually being the primary object of study by this rapid method.

For a measurement time of only 60 s and an altitude of about 70 m the ^{137}Cs photopeak could be clearly detected. For the same measurement time the ^{134}Cs photopeak could be seen in some spectra, but the statistical deviation of the net peak area was relatively high (up to 35%). To calculate the ^{134}Cs soil contamination, in general two or more spectra were added.

To evaluate the soil contamination, a relaxation depth of 3 cm was used. This value was determined on the basis of measurements of soil samples taken at some places in the region.

As examples, Figs 3 and 4 show the measured soil contamination by ^{137}Cs on routes I (from Munich to the border with Austria, total length about 80 km) and IV (from the Austrian border to the River Inn, total length about 50 km). On route I the measured values ranged from 5 to 17 kBq/m^2. Owing to the unknown activity distribution, one spectrum has not been evaluated. On route IV the soil contamination is slightly higher than on route I. Values up to 25 kBq/m^2 for ^{137}Cs were measured. The data agree with the results of measurements made directly after the Chernobyl accident. The low values on route IV (points a and b in Fig. 4) are due to the fact that a great part of the spectrum was taken over the Inn Valley and over the lake Wagingersee. The observed variation is in the range measured in this region by other techniques. More detailed analysis and data evaluation are being carried out.

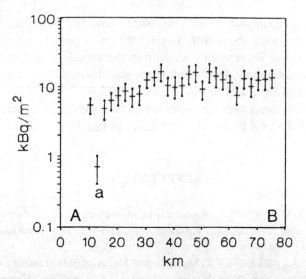

FIG. 3. Soil contamination due to ^{137}Cs between Munich and the border with Austria meas-ured with an airborne gamma ray spectrometer (flight route I from A to B in Fig. 2; total length about 80 km). a: not evaluated, activity distribution not known (cultivated fields).

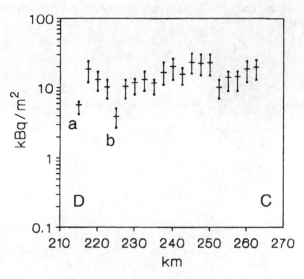

FIG. 4. Soil contamination due to ^{137}Cs between the Austrian border and the River Inn measured with an airborne gamma ray spectrometer (flight route IV from C to D in Fig. 2; total length about 50 km). a: Inn Valley; b: Wagingersee.

The measurement uncertainties, calculated according to standard error propagation procedures, were about 30%. Besides the uncertainty in determining the net peak area, additional uncertainties result from the variations in flying altitude and the inhomogeneous activity distribution in the soil. Nevertheless, such activity data can be very useful for a fast survey of contamination.

These briefly described test flights are to be continued over extended areas in the south of the Federal Republic of Germany in late 1989.

REFERENCES

[1]   WINKELMANN, I., et al., Radioactivity Measurements in the Federal Republic of Germany after the Chernobyl Accident, Rep. ISH-116, Inst. for Radiation Hygiene, Neuherberg (1987).

[2]   LINDÉN, A., MELLANDER, M., Airborne Measurements in Sweden of the Radioactive Fallout after the Nuclear Reactor Accident in Chernobyl, USSR, Rep. TFRAP 8606, Airborne Division, Swedish Geological Co., Uppsala (1986).

[3]   BECK, H.L., DeCAMPO, J., GOGOLAK, C., In-Situ Ge(Li) and NaI(Tl) Gamma-Ray Spectrometry, Rep. HASL-258, Health and Safety Lab., New York (1972).

[4]   MILLER, K.M., HELFER, I.K., "In situ measurements of Cs-137 inventory in natural terrain", Environmental Radiation '85 (Proc. 18th HPS Midyear Topical Symp. Colorado Springs, 1985), Proceedings Committee of Central Rocky Mountain Chapter, Health Physics Soc., Laramie, WY (1985).

[5]   Radioaktive Kontamination der Böden in Bayern, Bayerisches Staatsministerium für Landesentwicklung und Umweltfragen und für Ernährung, Landwirtschaft und Forsten, Munich (1987).

USE OF AIRBORNE RADIOMETRIC MEASUREMENTS FOR MONITORING ENVIRONMENTAL RADIOACTIVE CONTAMINATION

D.C.W. SANDERSON*, E.M. SCOTT**,
M.S. BAXTER*

* Scottish Universities Research
    and Reactor Centre,
East Kilbride

** Department of Statistics,
Glasgow University

Glasgow,
United Kingdom

Abstract

USE OF AIRBORNE RADIOMETRIC MEASUREMENTS FOR MONITORING ENVIRONMENTAL RADIOACTIVE CONTAMINATION.

The technique of airborne gamma spectrometry has many advantages over conventional ground based monitoring methods in the mounting of a time constrained response to an accident involving a release of radioactive material into the environment. The paper reviews the basis and background of the method, with an emphasis on the role that detector and nucleonics development plays in determining capabilities. The main features of detector response are briefly discussed, including nuclide specificity, circles of investigation and sensitivity. Identification of distributed and point sources of radioactivity is potentially of interest and leads to differing survey requirements. The relative merits and limitations of calibration methods are also reviewed. Practical examples of the use of aerial methods in the United Kingdom are given, including surveys conducted by the Scottish Universities Research and Reactor Centre using fixed wing and rotary wing aircraft in Scotland and west Cumbria in the aftermath of the Chernobyl accident. The results clearly show the potential of aerial methods for guiding ground based work in the recovery phase of an accident involving release of radioactivity to the environment. Further needs include baseline studies, technical developments and improved knowledge of nuclide mobility and transfer factors.

1. INTRODUCTION

Aerial radiometric survey provides an extremely rapid and economical means of locating high deposition areas in the event of an accident involving the release of significant quantities of gamma ray emitting radionuclides. This paper outlines the basis and background to the method and recent experience of it in the United Kingdom, and indicates the potential application to monitoring of environmental contamination following a major nuclear accident.

2. BACKGROUND

The gamma rays which accompany many nuclear transitions can penetrate up to several hundred metres in air, producing radiation fields which can be mapped with sensitive spectrometry equipment operated from aircraft. The distribution of terrestrial, or surficial marine, radionuclides, whether natural or anthropogenic, can be determined by constructing appropriate low altitude flight paths.

These principles have been applied to mineral exploration and geological mapping for over forty years. Gas filled detectors used in the late 1940s for mapping total radiation fields around uranium reserves [1] were quickly supplanted by more efficient solid scintillation detectors [2, 3]. Early British work included studies of total radiation mapping in mineral exploration [4, 5], a simple line survey of west Cumbria in 1957 following the Windscale fire [6, 7] and an appraisal of the use of helicoper mounted equipment in emergencies [8]. Equipment used in this early work involved detectors of modest area and volume, coupled to single channel ratemeters, with analogue chart recorder output. Despite the success of early studies, the inability to identify or quantify specific nuclides and adverse cost–benefit analyses were probably among the reasons why work of the United Kingdom Atomic Energy Authority in aerial survey methods was suspended in the early 1960s.

The need to resolve contributions from U, Th and K gamma radiation for geological mapping and particularly for reliable location of uranium resources [9–11] provided the impetus for development of spectral analysis systems [12, 13] coupled to high volume NaI detector arrays. A typical modern geophysical system may have one or two 255 channel multichannel analysers (MCAs) and up to 50 L of NaI frequently in the form of a summed array of 10 cm × 10 cm × 40 cm slab detectors. Equipment of this sort has been used for environmental purposes. Examples include searches conducted in 1978 when the Cosmos 954 satellite deposited the contents of a nuclear reactor over the Great Slave Lake in northern Canada [14, 15] and the rapid national mapping undertaken in Sweden after the Chernobyl accident [16].

Nevertheless the approach was not used in the United Kingdom at the time of the Chernobyl accident. With hindsight it is clear that aerial survey at an early stage could have contributed to fuller and more convincing knowledge of the fallout deposition pattern and would have helped to direct ground based resources to places of greatest need. Even more than three years after the event, the picture of deposition produced by ground based methods is less than adequate.

Awareness of the inherent difficulties of using conventional methods to map fallout on a national or regional scale prompted the Scottish Universities Research and Reactor Centre (SURRC) to examine the potential of aerial survey methods for mapping detailed local variations in anthropogenic nuclides. This was first undertaken in a short feasibility study conducted in southwest Scotland, and has been followed up by a continuing programme of research and survey flights to extend and apply the technique.

3.   THE METHOD

The basis of the method lies in the character of gamma ray emission and transport from terrestrially deposited radioactivity to an airborne gamma ray spectrometer. The gamma ray fluxes comprise a mixture of primary (unscattered) photons, whose energies are characteristic of individual nuclides, and secondary (scattered) radiation resulting from partial energy transfer interactions in the path from nuclide to detector [17]. Distributed radioactivity in, or on, water or soil can generate a surface gamma ray flux from depths of up to 20–30 cm (mineral rich soil or rocks) or 40–50 cm (organic or waterlogged matrix). This flux propagates through the air and can be detected up to several hundred metres above ground. With increasing height above ground, both the total gamma flux and the relative amount of primary information conveyed by unscattered radiation decrease, by attenuation and scattering in the air path. Typical half-depths for primary radiation in dry air range from 72 m (for ^{137}Cs at 662 keV) to 158 m (for 2.62 MeV photons from ^{208}Tl).

TABLE I. PRINCIPAL NUCLIDES OF INTEREST FOR AERIAL GAMMA SPECTROMETRY

Nuclide	Energy (keV)	Half-life	Origin	Comment
Am-241	59.5	432.2 a	Activation product	Special method needed
I-131	364	8.04 d	Fission product	Sensitivity to be determined
Cs-137	662	30.2 a	Fission product	Readily detected
Cs-134	796	2.06 a	Activation product	Readily detected
Ar-41	1295	110 min	Activation product	Readily detected
K-40	1462	1.28×10^9 a	Primordial nuclide	Readily detected
Bi-214	1764	19.9 min	U-238 daughter	Readily detected (radon daughter)
Tl-208	2615	3.05 min	Th-232 daughter	Readily detected

A gamma ray spectrometer raised above the ground therefore records a spectrum comprising photopeaks, escape peaks and scattered radiation, degrading progressively with increasing height. Table I summarizes some of the gamma rays and nuclides which may be of interest for aerial radiometric survey.

The other important feature of raising a detector above ground is that the detection geometry opens up so that the area on the ground being sampled increases very rapidly. Typical areas of investigation are such as to give 90% of the detected signal from a circle of diameter four to five times the height above ground [18–20].

Sensitivity depends strongly on aircraft altitude and speed, detector size and energy resolution, integration time and extent of spectral interferences (especially for low energies). Effective use of aircraft in aerial survey can be achieved by matching the integration time to the time taken to cross a circle of investigation at safe flying speeds, and choosing a detector volume and altitude which will give satisfactory sensitivity under these conditions. Minimum detectable levels of some kilobecquerels per square metre are readily achievable.

Considerations of this type lead to selection of large volume (16–48 L) NaI detectors for rapid fixed wing survey at high resolution, and smaller (4–16 L) arrays for use from light helicopters, if 10 cm crystal thickness is adopted. Choice of crystal thickness is debatable, and there are arguments that thinner detectors than those conventionally selected for geophysical survey would be better suited to detection of anthropogenic nuclides. Future use of hybrid detector arrays incorporating solid state elements is also an attractive possibility.

Survey procedures involve flying along search paths while recording detector spectra in conjunction with navigational and altitude data through a data logging computer. In the absence of prior knowledge of deposition patterns, the adoption of systematic flight paths such as creeping line or box and grid surveys is appropriate. It may be possible under some conditions to take other indications, such as weather patterns or topographic predispositions to elevated deposition, into account in designing strategies. This is an area for further attention.

Data analysis comprises collation and validation of results, extraction of integrated and spectrally deconvoluted photopeak counts, correction for in-flight altitude variations and other minor influences, and calibration to the response under standardized conditions.

Whereas for geological work it makes sense to standardize results to the response of a hypothetically homogeneous radiation field, and calibration pads have been prepared in a number of countries for this purpose [21], no consensus has yet been reached on the basis for calibrating anthropogenic sources. Standardization to layered sources, homogeneous sources and implicit vertical distributions from typical sites are all possible approaches. Associated questions also arise on the appropriate methods for practical realization of calibrations where a number of approaches, ranging from Monte Carlo simulation to the use of large pads for many nuclides, are possible.

TABLE II. ENVIRONMENTAL AERIAL SURVEYS CONDUCTED BY THE SURRC

Location	Date	Area (ha)	Line spacing	Detector volume (L)	Flying time (h)	Type of aircraft
Whithorn, Mull of Galloway (southwest Scotland)	Feb. 1988	30 000	1 mile	7	5	Cessna Titan[a]
North and South Uist, Benbecula (Western Isles)	Mar. 1988	30 000	1 mile	7	5	Cessna Titan[a]
West Cumbria	Aug. 1988	45 000	500 m	7	36	Bell helicopter[a]
Upper Clyde Valley (southwest Scotland)	Dec. 1988	8 000	1 km	7	2	Aérospatiale Squirrel[a]
Central Highlands[b]	Dec. 1988	5 000	1 km	7	5	Aérospatiale Squirrel[a]
Eaglesham Moor (southwest Scotland)	Jan. 1989	8 000	1 km	7	1	Aérospatiale Squirrel[a]
North Wales	Jul. 1989	32 000	500 m	20	20	Aérospatiale Squirrel[c]
Southwest England	Sep. 1989	225 000	1 km/500 m	24	51	Aérospatiale Squirrel[c]

Note: 100 m altitude adopted throughout.

[a] Integration time 30 s.

[b] This survey included two 40 km transects and transport time from southern Scotland.

[c] Composite method with 10 and 15 s integration times in separate multichannel analysers.

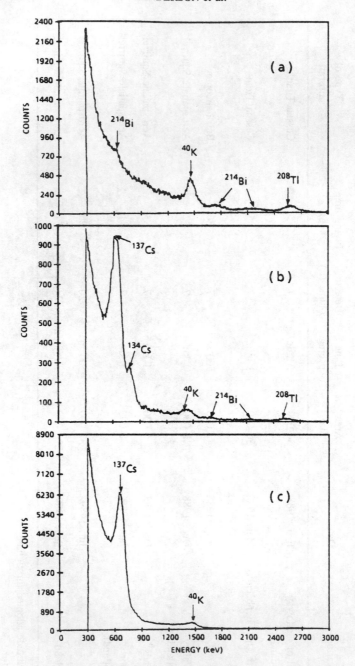

FIG. 1. (a) Natural radioactivity recorded at 100 m altitude in north Cornwall, September 1989. (b) Chernobyl fallout in west Cumbria (Birker Fell) recorded at 10 m altitude in August 1988. (c) Environmental contamination from the Sellafield site recorded in west Cumbria (Muncaster) in 1988 at 30 m altitude with a 7 L detector (note the absence of ^{134}Cs).

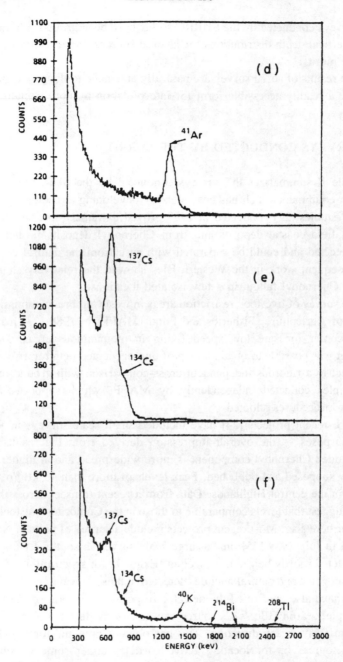

FIG. 1.   (d) Argon-41 in spectrum recorded at 100 m altitude 2 km downwind of a Magnox power station. (e) Ground level spectrum recorded in the central Highlands of Scotland (Meall Glas) in December 1988, showing extensive contributions from anthropogenic nuclides. (f) Typical spectrum from a survey at 100 m altitude in the central Highlands of Scotland (Lochan na Laing, north of Loch Tay), recorded in December 1988.

Surveys conducted by the SURRC so far have been ground calibrated to comparable vertical depth distributions for layered sources; however, other approaches also have merits.

The results of aerial survey are naturally amenable to the production of maps, which are a readily accessible form for interpretation by non-specialists.

4.   SURVEYS CONDUCTED BY THE SURRC

Table II summarizes the surveys conducted by the SURRC since February 1988 with equipment which has been steadily developing at each stage. The initial feasibility studies [22] conducted in 1988 proved extremely successful; ^{137}Cs both from Sellafield coastal deposits and from Chernobyl deposits further inland was readily detected and could be separated with ease from the natural components.

Subsequent work in the Western Isles showed the relatively slight extent of deposited Chernobyl fallout in a few isolated locations.

The survey of livestock restriction areas in west Cumbria commissioned by the Ministry of Agriculture, Fisheries and Food (MAFF) in 1988 [23] made use of a light helicopter for safe low speed flying in mountainous areas. Despite poor weather, it was possible to obtain a map of unprecedented detail in two weeks. This was subject to a rigorous independent cross-comparison with ground based results from samples collected independently by MAFF, which supported the overall calibration procedures adopted.

More recent projects [24] have included small scale surveys in Scotland for various purposes — the overall impression derived from the results is that the recently added Chernobyl component is more widespread and at higher levels than previously supposed and published. Peak levels of more than 40 kBq/m^2 have been observed in the central Highlands. Data from a recent transect across the Southern Uplands suggest that levels comparable to those in the Cumbrian livestock restriction zone occur here also. Most recent projects include a survey of parts of North Wales, conducted in July 1989 [25], and a large scale survey in southwest England which was completed shortly before this meeting. Examples of spectra illustrating some of the anthropogenic and natural nuclides detected in these surveys are shown in Fig. 1.

We have also confirmed the detector response over point sources of activity. For example, a 5.68 GBq ^{137}Cs source can be readily detected using aerial survey up to 1 km away in any direction. This has obvious potential for searches for lost industrial sources or for locating active particles under some possible accident conditions.

It has been possible to extend both detector specifications and methodology at each stage. Our first experiments were conducted using a simple 7 L detector with a portable multichannel analyser and laptop computer, with successful results. Since then, extension of the equipment and method has taken place sequentially with each survey. A methodology for asynchronous recording of spectra from multiple computer based MCAs, coupled to a multicrystal array of 8% resolution NaI detectors, was adopted in 1989.

Practical experience has shown that high volume summed detectors may be prone to serious dead time limitations if the whole spectrum from 100 keV to 3000 keV is analysed. Similarly summed arrays inevitably produce poorer energy resolution than each individual component — which is of considerable importance for accurate spectral stripping of low energy components. We believe that extension of our decentralized, multi-MCA acquisition approach to the extent of having one ADC memory segment per detector would greatly enhance the operational reliability and ability to handle unusually high signal levels such as may be encountered in a nuclear emergency. This would also provide the development environment for assessment of the potential of solid state detectors in a hybrid array.

5. FURTHER NEEDS AND POTENTIAL APPLICATIONS

The importance of aerial survey in the aftermath of a major nuclear accident lies in its unique potential for rapid location of areas of high deposition. Even using relatively simple equipment, it is possible to conceive of broad national mapping of a country the size of the United Kingdom within a time-scale of a few days. Detailed surveys could, if necessary, be conducted over a period of weeks rather than years. It is very unlikely that ground based measurements, whether by vehicle or by collection of samples for remote analysis, could be achieved for large areas on such time-scales.

The sensitivity of established methods allows ready detection of 10 Bq/L on water, or about 1 kBq/m^2 on ground, of many fission products, including ^{137}Cs.

Incorporation of provision for such emergency surveys into contingency plans could make a vital contribution, both to constructing effective countermeasures and reassuring the public. It is rather hard to imagine how reliable countermeasures could be enacted without a detailed knowledge of deposition; and aerial survey appears to be the only possible method to obtain sufficient information in time for such countermeasures to be effective.

Some further research and development is certainly needed to improve sensitivity and to resolve some of the finer details of response analysis and calibration. Further knowledge of transfer factors under different environmental conditions is also desirable. Nevertheless we feel that the technique is sufficiently useful in its present form to encourage Member States of the International Atomic Energy Agency to give serious consideration to the use of this method both for baseline studies and in the event of an emergency. At the SURRC we have committed major equipment and staff resources to furtherance of the science and technology of aerial radiometrics.

REFERENCES

[1] STEAD, F.W., Airborne radioactivity surveying speeds for uranium prospecting, Eng. Min. J. (1950) 74–77.

[2] PRINGLE, R.W., ROULSTON, K.E., BROWNELL, G.W., LUNBERG, H.T.F., The scintillation counter in the search for oil, Min. Eng. (1953) 1255–1261.

[3] PEIRSON, D.H., PICKUP, J., A scintillation counter for radioactivity prospecting, J. Br. Inst. Radio Eng. 14 (1954) 25–32.

[4] WILLIAMS, D., BISBY, H., The aerial survey of terrestrial radioactivity, Rep. AERE-R-3469, Atomic Energy Research Establishment, Harwell, UK (1960).

[5] WILLIAMS, D., CAMBRAY, R.S., Environmental survey from the air, Rep. AERE-R-2954, Atomic Energy Research Establishment, Harwell, UK (1960).

[6] WILLIAMS, D., CAMBRAY, R.S., MASKELL, S.C., An airborne radiometric survey of the Windscale area, October 19–22nd, Rep. AERE-R-2890, Atomic Energy Research Establishment, Harwell, UK (1957).

[7] CHAMBERLAIN, A.C., GARNER, R.J., WILLIAMS, D., Environmental monitoring after accidental deposition of radioactivity, Reactor Sci. Technol. 14 (1961) 155–167.

[8] PEIRSON, D.H., CROOKS, R.N., Survey of environmental radioactivity: A preliminary assessment of the use of helicopters in an emergency, Rep. AERE-M-927, Atomic Energy Research Establishment, Harwell, UK (1961).

[9] DARNLEY, A.G., "Airborne gamma ray survey techniques — Present and future", Uranium Exploration Methods (Proc. Panel, Vienna, 1972), IAEA, Vienna (1973) 67–108.

[10] INTERNATIONAL ATOMIC ENERGY AGENCY, Recommended Instrumentation for Uranium and Thorium Exploration, Technical Reports Series No. 158, IAEA, Vienna (1974).

[11] INTERNATIONAL ATOMIC ENERGY AGENCY, Radiometric Reporting Methods and Calibration in Uranium Exploration, Technical Reports Series No. 174, IAEA, Vienna (1976).

[12] DICKSON, B.H., BAILEY, R.C., GRASTY, R.L., Utilizing multichannel airborne gamma-ray spectra, Can. J. Earth Sci. 18 (1981) 1793–1801.

[13] GRASTY, R.L., GLYNN, J.E., GRANT, J.A., The analysis of multichannel airborne gamma-ray spectra, Geophysics 50 (1985) 2611–2620.

[14] BRISTOW, Q., "The application of airborne gamma-ray spectrometry in the search for radioactive debris from the Russian satellite Cosmos 954", Current Research, Part B, Geol. Surv. Can., Pap. 78-1B (1978) 151–162.

[15] GRASTY, R.L., "The search for Cosmos 954", Search Theory and Applications (HEVLEV, K.B., STONE, L.D., Eds), Plenum Press, New York (1980) 211–220.

[16] LINDEN, A., MELLANDER, H., Airborne Measurements in Sweden of the Radioactive Fallout after the Nuclear Accident in Chernobyl, USSR, Rep. TFRAP 8606, Swedish Geological Co., Uppsala (1986).

[17] LOVBORG, L., KIRKEGAARD, P., Numerical Evaluation of the Natural Gamma Radiation Field at Aerial Survey Heights, Rep. Risø-R-317, Risø Natl Lab., Roskilde, Denmark (1975).

[18] DUVAL, J.S., COOK, B., ADAMS, J.A.S., Circle of investigation of an airborne gamma-ray spectrometer, J. Geophys. Res. **76** (1971) 8466–8470.

[19] GRASTY, R.L., KOSANKE, K.L., FOOTE, R.S., Fields of view of airborne gamma-ray detectors, Geophysics **44** (1979) 1447–1457.

[20] LOVBORG, L., The Calibration of Portable and Airborne Gamma Spectrometers — Theory, Problems and Facilities, Rep. Risø-R-2456, Risø Natl Lab., Roskilde, Denmark (1984).

[21] GRASTY, R.L., The Design, Construction and Application of Airborne Gamma-Ray Spectrometer Calibration Pads — Thailand, Geol. Surv. Can., Pap. 87-10 (1987).

[22] SANDERSON, D.C.W., SCOTT, E.M., BAXTER, M.S., PRESTON, T., A feasibility study of airborne radiometric survey for UK fallout, Rep. 88-03, Scottish Universities Research and Reactor Centre, East Kilbride, 1988.

[23] SANDERSON, D.C.W., SCOTT, E.M., Aerial Radiometric Survey in West Cumbria in 1988, Food Science Report N611, Ministry of Agriculture, Fisheries and Food, London (1988).

[24] SANDERSON, D.C.W., SCOTT, E.M., BAXTER, M.S., "The use and potential of aerial radiometrics for monitoring environmental radioactivity", Proc. Conf. on Water Resource Consequences of a Nuclear Accident, Glasgow, 1989, Instn of Civil Engineers, London (in press).

[25] SANDERSON, D.C.W., EAST, B.W., SCOTT, E.M., Aerial Radiometric Survey of Parts of North Wales in July 1989, Survey Report, Scottish Universities Research and Reactor Centre, East Kilbride (1989).

METHODS FOR RADIONUCLIDE DETERMINATION IN THE ENVIRONMENTAL MONITORING SYSTEM OF THE GERMAN DEMOCRATIC REPUBLIC

H.-U. SIEBERT, J. THIELE, M. LÖNNIG,
K. GROCHE, E. ETTENHUBER
National Board for Atomic Safety
   and Radiation Protection,
Berlin

Abstract

METHODS FOR RADIONUCLIDE DETERMINATION IN THE ENVIRONMENTAL MONITORING SYSTEM OF THE GERMAN DEMOCRATIC REPUBLIC.

The National Board for Atomic Safety and Radiation Protection is responsible for the environmental surveillance of man-made and natural sources of radiation in the German Democratic Republic. Therefore, the Board performs monitoring programmes itself and controls monitoring programmes performed by other institutions. The aim of the programmes is to determine the radioactive contamination of environmental media under normal and accidental radiation conditions. To achieve comparable measuring results from all institutions involved in environmental monitoring the Board establishes the radionuclide determination procedures for environmental monitoring. In the paper a survey is given of the sample preparation, non-nuclide-specific enrichment and nuclide specific enrichment methods used. For low activity concentrations in samples as well as for radionuclide determination for which the production of special measuring samples is necessary, separation and preparation methods are described. The measuring methods used to determine radionuclide concentrations in environmental media include the following: the determination of α emitting radionuclides by use of integral and spectrometric α activity measurements; the determination of β emitting radionuclides by use of β anticoincidence arrangements as well as liquid scintillation spectrometers; and the determination of γ emitting radionuclides using highly effective NaI(Tl) scintillation detectors, high resolution p and n type HPGe detectors with well type, on-top and Marinelli geometries, and anti-Compton spectrometers. Lower detection limits for the activity measuring devices are presented and the specific uses of these devices under normal and accidental radiation conditions are discussed.

1. ENVIRONMENTAL MONITORING SYSTEM

For the purpose of determining contamination levels in the environment and assessing the radiation dose to man due to man-made sources of radioactivity, as well as for comparing man-made radiation with natural background radiation, a set of co-ordinated surveillance programmes has been developed in the German Democratic Republic (GDR). On the basis of the Atomic Energy Act and the Radia-

423

TABLE I. ENVIRONMENTAL MONITORING TASKS

Institution	Kind of measurement
Meteorological Service of the GDR	Fallout sampling and determination of gross β activity and γ activity
	Sampling of rain and determination of gross β activity and γ activity
	Continuous measurement of γ dose rate
	Continuous measurement of radioactive aerosol contamination
	Continuous measurement of radioactive iodine contamination
Ministry of Environmental Protection and Water Management	Continuous sampling of surface water for determination of gross β activity
	Sampling of surface water for determination of γ activity concentration
	Continuous measurement of γ dose rate
	Continuous measurement of γ activity concentration in surface waters
Ministry of Agriculture, Forestry and Food	Determination of γ activity concentration in samples of agricultural products and food
Nuclear facilities	Continuous monitoring of gaseous and liquid radioactive effluents
	Fallout sampling and determination of gross β activity
	Continuous measurement of γ dose rate
	Continuous measurement of radioactive aerosol contamination
	Continuous measurement of air contamination by iodine
	Determination of radionuclide concentration in samples of environmental media
Civil Defence Forces of the GDR	Dose rate measurements
	Measurement of γ activity concentration in environmental samples

tion Protection Ordinance the National Board for Atomic Safety and Radiation Protection (hereinafter called 'the Board') is responsible for the surveillance of the territory of the GDR. It establishes the scope and methods of environmental monitoring. The Board performs monitoring programmes itself for the purpose of governmental supervision and controls monitoring programmes carried out by other institutions. The contamination of environmental media has to be measured by:

— The Meteorological Service of the GDR (MS), in the ground level air;
— The Ministry of Environmental Protection and Water Management (MEPWM), in surface water and groundwater resources;
— The Ministry of Agriculture, Forestry and Food (MAFF), in agricultural and animal products;
— The operators of nuclear facilities, in the surroundings of nuclear power stations and other nuclear facilities;
— The Civil Defence Forces of the GDR (CDF), in the event of a nuclear accident.

Environmental monitoring includes:

— Monitoring of radionuclide concentrations in radioactive fallout, ground level air, surface water, soil, biological media and foodstuffs;
— Monitoring of local dose or local dose rate; and
— Monitoring of radioactive effluents and their release conditions according to GDR regulations.

The tasks of environmental monitoring which have to be performed by the above mentioned institutions in normal situations and in the early and intermediate stages of a nuclear accident are summarized in Table I.

In the case of a nuclear accident, the results of all measurements recorded have to be reported to the Board for the purpose of assessing the general radiological situation on the territory, supporting the emergency control centre and determining the mean annual effective radiation exposure of the population.

The environmental monitoring system is designed to meet the requirements of the Convention on Early Notification of a Nuclear Accident.

2.   REQUIREMENTS ON METHODS OF DETERMINING RADIONUCLIDE CONCENTRATION

Within the framework of the environmental monitoring system radionuclide determination is performed by:

— The environmental monitoring laboratory of the Board;
— Laboratories of
  — nuclear facilities,

— the Meteorological Service,
— the Ministry of Environmental Protection and Water Management,
— the Ministry of Agriculture, Forestry and Food,
— the Civil Defence Forces;
— Mobile laboratories set up by
— nuclear facilities,
— the Civil Defence Forces,
— the Board.

To obtain comparable measurement results from all institutions involved in environmental monitoring, the radionuclide determination methods to be used are established by the Board.

Depending on the special monitoring tasks of the laboratories involved, demands on the radionuclide determination methods differ according to:

— The nature and quantity of samples to be analysed at different contamination levels of the environment;
— The radionuclide spectrum to be determined;
— The qualifications and training of the laboratory staff;
— The special analytical arrangements in the different types of laboratories;
— The required lower detection limit or the actual activity level of the sample;
— The chemical and physical properties of the radionuclides to be determined;
— The radiation properties of the radionuclides to be determined; and
— The required accuracy of analysis.

These general requirements imply special demands on the laboratories and the availability of radionuclide determination methods at the institutions involved in environmental monitoring (Table II).

The methods of radionuclide determination applied in the laboratories of the GDR under the existing nuclear environmental monitoring programmes allow the analysis of about 10 000 samples per year, of which about 7000 are processed by the Board. In the event of an accident the capacity can be increased to about 2000 samples per day.

3.   METHODS OF RADIONUCLIDE CONCENTRATION
      DETERMINATION

Determination of the activity concentration in environmental samples requires a system of:

— Sample preparation and non-nuclide-specific or nuclide specific enrichment;
— Chemical methods of separation and isolation of the radionuclides to be determined; and
— Activity and analytical measuring techniques.

3.1. Sample preparation and radionuclide enrichment

For a large proportion of samples from nuclear environmental surveillance programmes in normal situations direct determination of radionuclide concentration is not possible because of the low activity concentration. In such cases, activity determination has to be preceded by sample preparation, which consists of mechanically converting the rough sample into a form and consistency more appropriate for analysis and sample enrichment, and reducing the sample mass and volume by removing most of the mainly organic sample matrix.

Figure 1 gives a survey of the sample preparation and enrichment methods used.

As a rule, dry ashing at a maximum of 450°C is the simplest and most effective method of removing the organic sample matrix and thus of non-nuclide-specific enrichment. A special type of ashing, in particular for samples of small mass and volume, is 'low temperature ashing', which precludes losses of particularly volatile elements. Instead of this method, which is rather expensive, the predried and crushed sample can also be decomposed by acid. This is also the starting point of the separation and preparation methods to be used.

Depending on the type of sample, various steps of preparation such as evaporation, cutting up, breaking, filtration and drying are necessary before ashing in order to make the ashing process more effective. After ashing, mixing and homogenization of sample residues is necessary in almost all cases. The enrichment factors are about 10–100 for biomedia and foodstuffs and up to 3000 for water samples.

In some cases, particularly in large water samples, nuclide specific enrichment of radionuclides can be achieved by ion exchange. This applies to the enrichment of ^{137}Cs by ammonium-12-molybdatophosphate under pressure from sea or surface water samples having a volume of up to 100 L. Here, enrichment factors of up to 10 000 can be obtained. Another nuclide specific enrichment procedure is electrolytic decomposition for enriching of ^{3}H from surface and sea water samples, which leads to enrichment factors of about 20.

After sample preparation it has to be decided whether the sample residues can be subjected to an integral or spectrometric activity measurement or whether analytical chemical separations are necessary to remove interfering elements or radionuclides.

In the case of increased contamination levels of environmental media, radionuclide enrichment can be largely omitted and mechanical conversion of the rough sample becomes the crucial step of sample preparation. If for radionuclide determination the production of a special measuring sample is necessary, enrichment cannot be omitted. Therefore, rapid sample enrichment, especially by optimized dry ashing procedures as well as by precipitation or co-precipitation from liquid samples, can be used.

TABLE II. DEMANDS ON LABORATORIES OF INSTITUTIONS INVOLVED IN ENVIRONMENTAL MONITORING

Institution	Requirements	Radionuclide determination methods for samples	Equipment
National Board for Atomic Safety and Radiation Protection	Full scale environmental laboratory	Sample preparation and enrichment	Fully equipped radiochemical laboratories
	High and low level activity determinations	Radiochemical separation and preparation procedures	Highly effective low level α, β and γ spectrometers
	Non-urgent and rapid determinations	α, β and γ activity spectrometry	Low level α/β counters
	Simple mobile methods for sampling and activity screening	Gross α, β and γ activity measurement	Atomic absorption spectrometer
	Special measurements	Direct gross β activity measurement	Scintillation γ spectrometers
Meteorological Service of the GDR	Simple measuring procedures	Gross β activity measurement of dried fallout	Simple sample preparation equipment
	Sensitive methods for early notification	Direct γ spectrometric activity measurement of fallout and rain	β counters
	High throughput in accident situations		Scintillation γ spectrometers
Ministry of Agriculture, Forestry and Food	Capability for detection of 1% of DIL[a]	Direct γ spectrometric activity measurement	High resolution γ spectrometers
	High throughput in accident situations		

Ministry of Environmental Protection and Water Management	High throughput in accident situations Simple measuring procedures	Gross β activity measurement of dried samples Direct γ spectrometric activity measurement	Simple sample preparation equipment β counters Scintillation γ spectrometers
Civil Defence Forces of the GDR	Simple measuring procedures for direct activity determination High throughput in accident situations Capability for detection of 1% of DIL	Gross γ activity measurement Direct γ spectrometric activity measurement	β counters High resolution γ spectrometers Scintillation γ spectrometers
Nuclear facilities	Sensitive measuring procedures for radiologically relevant radionuclides in effluents and releases and environmental media Capability for detection of 1% of ALI[b] or DAC[c] High throughput in accident situations Simple mobile methods for sampling and activity screening	Gross β activity measurement Sample preparation and enrichment Radiochemical separation β and γ activity determination	Radiochemical laboratory β counter High resolution γ spectrometer Scintillation γ spectrometer

[a] DIL: derived intervention level. [b] ALI: annual limit of intake. [c] DAC: derived air concentration.

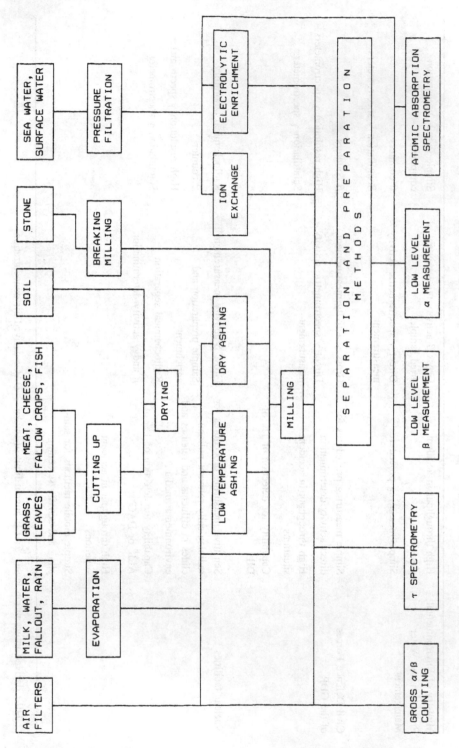

FIG. 1. Sample preparation and enrichment methods.

3.2. Separation and preparation methods

Although a number of highly sensitive and selective measuring methods for direct determination of the radionuclide concentration in environmental samples are available, in cases of low activity concentration and the presence of pure α or β emitters in samples, separations and preparations are inevitable. The first step in processing ashing residues is decomposition. Depending on the sample medium, the sample amount and the radionuclides to be determined, acid decomposition, acid leaching or alkaline melt is used. With highly efficient pressurized bomb digestion techniques the time consuming traditional acid decomposition can be omitted for small samples of ash of less than 10 g.

To isolate the radionuclide of interest from interfering radionuclides and matrix elements, various separation methods are employed:

— Precipitation and co-precipitation are performed mainly at the beginning of analytical chemical separation runs to remove interfering trace or matrix elements or for group separation preceding an element separation ($BaSO_4$ precipitation in Ra determination; $Fe(OH)_3$ precipitation in Cs determination; $Ca(COO)_2$ precipitation in Th determination).
— Extraction and ion exchange are used mainly for selective separation of the radionuclides to be determined (Alamine/Aliquat for the extraction of U, Pu, Th; HDEHP to separate Y from Sr; AMP to separate Cs).

For the radiometric measurement of isolated radionuclides by α/β low activity measurement methods, the preparation of a measuring sample of minimum as well as defined basic weight is necessary. For this purpose,

— Precipitation of the radionuclides to be determined along with an inactive carrier substance is used to prepare measuring samples, particularly of β emitting nuclides (^{89}Sr, ^{90}Sr, ^{137}Cs);
— Electrolytic deposition of the radionuclides to be determined is used to prepare 'mass free' measuring samples, mainly of α emitting nuclides. Both spontaneous electrolytic deposition (^{210}Bi/^{210}Po) and electrolyses from inorganic solutions (U, Pu, Th) are employed.

Table III presents a list of separation and preparation methods used for individual radionuclide determinations.

Special efforts are being made to modify and shorten routine work in radiochemical separation and preparation procedures to facilitate rapid determination under accident conditions. Work is going on to shorten the routine radiochemical procedures for determining ^{90}Sr and Pu nuclide concentration. This work is focused on:

TABLE III. SEPARATION AND PREPARATION METHODS USED FOR INDIVIDUAL RADIONUCLIDE DETERMINATIONS

Radionuclide	Method of separation and determination	Kind of sample	Lower detection limit
H-3	Electrolytic enrichment, LSC[a]	Water	0.2 Bq/L
C-14	$BaCO_3$ precipitation, LSC	Air	1 mBq/m^3
Sr-89	$SrCO_3$ precipitation, β counting,	Water	5 mBq/L
	subtraction of Sr-90 activity	Plant	0.2 mBq/g
		Soil	0.05 mBq/g
Sr-90	Y-90 extraction with HDEHP, $Y_2(COO)_3$	Water	0.5 mBq/L
	precipitation, β counting	Plant	0.2 mBq/g
		Soil	0.04 mBq/g
I-131	AgI precipitation, γ spectrometry	Water	5 mBq/L
Cs-137	Ion exchange with AMP, $CsPtCl_6$	Water	0.1 mBq/L
	precipitation, β counting	Plant	0.2 mBq/g
		Soil	0.5 mBq/g
Pb-210,	Spontaneous electrolytic deposition	Water	10 mBq/L
Bi-210,	on Ni discs, β counting, α counting	Plant	1 mBq/g
Po-210		Soil	10 mBq/g
Ra-224	Ra co-precipitation with $BaSO_4$, α counting,	Water	10 mBq/L
	Rn-222 emanation, α counting,	Plant	0.1 mBq/g
	subtraction of Ra-226 activity	Soil	2 mBq/g
Ra-226	Ra co-precipitation with $BaSO_4$,	Water	10 mBq/L
	Rn-222 emanation, α counting	Plant	0.1 mBq/g
		Soil	2 mBq/g
Ra-228	Ra co-precipitation with $BaSO_4$, Ac-228	Water	20 mBq/L
	extraction with DTPA, Ac co-precipitation with		
	$Ce_2(COO)_3$, β counting		
Th-230,	Th co-precipitation with $Ca(COO)_2$, Th extrac-	Water	1 mBq/L
Th-232	tion with Aliquat, electrolytic deposition,	Plant	0.01 mBq/g
	α spectrometry, Th-229 tracer	Soil	0.5 mBq/g
Uranium	Extraction with ethyl acetate,	Water	5 μg/L
	Na_2CO_3–K_2CO_3–NaF fusion, fluorimetric	Plant	0.01 μg/g
	determination	Soil	1 μg/g
U-234,	Ion exchange on Dowex 1×8, electrolytic	Water	0.6 mBq/L
U-235,	deposition, α spectrometry,	Plant	2 μBq/g
U-238	U-232 tracer	Soil	15 μBq/g
Pu-238,	Ion exchange on Dowex 1×4, electrolytic	Water	0.6 mBq/L
Pu-239–240	deposition, α spectrometry,	Plant	2 μBq/g
	Pu-242 tracer	Soil	15 μBq/g

[a] LSC: liquid scintillation counting.

— ^{89}Sr and ^{90}Sr determination in samples using a time reduced Ca/Sr separation procedure based on extraction techniques with crown ethers, precipitation of SrCO$_3$ and a β absorption filter technique;
— Pu concentration determination in samples using a modified selective extraction method based on Alamine combined with a rapid and quantitative electrolytic precipitation method for measuring samples.

These procedures will allow a sample analysis to be carried out within 24 hours.

Radionuclide losses incurred by several separation steps during analysis are corrected by yield determination after addition of inactive carrier material or radioactive tracers. The necessary yield determination is made by atomic absorption spectrometry (Sr, Cs), by titration (Y) or by activity measurement (U, Pu, Th).

3.3. Determination of radionuclide concentrations

Figure 2 gives a survey of measuring methods used to determine radionuclide concentrations.

3.3.1.   Determination of α emitting nuclides

For integral α activity measurement, ZnS(Ag) scintillation detectors are employed. Here, the precondition for determining the radionuclide concentration is a nuclide-pure sample (e.g. ^{224}Ra, ^{226}Ra, ^{210}Po) as a result of chemical separation and preparation methods. For the measuring equipment employed, the minimum detectable activity (MDA) is 1.5 mBq for a measuring time of 100 min. At high selectivity and low background, the use of Si semiconductor spectrometers allows the sensitive and simultaneous detection of several α emitting nuclides. Here, particularly high demands are made on the homogeneity and 'mass freedom' for the measuring sample. In activity determination of Th, U and Pu isotopes with Si surface barrier detectors, having an effective area of 4 cm^2, for a measuring time of 1000 min an MDA of 0.5 mBq is attained.

To determine uranium, measurement of fluorescence radiation in uranium compounds excited by UV light is used as an alternative method with a lower detection limit of 1 ng.

3.3.2.   Determination of β emitting nuclides

For gross β activity determination β plastic scintillation counters having an active area of 20 cm^2 are used together with 5 cm Pb shielding. Integral measurements are carried out from 100 keV on. An MDA of 0.3 Bq is attained for a 100 min measuring time. Efficiency and self-absorption calibration is done according to KCl

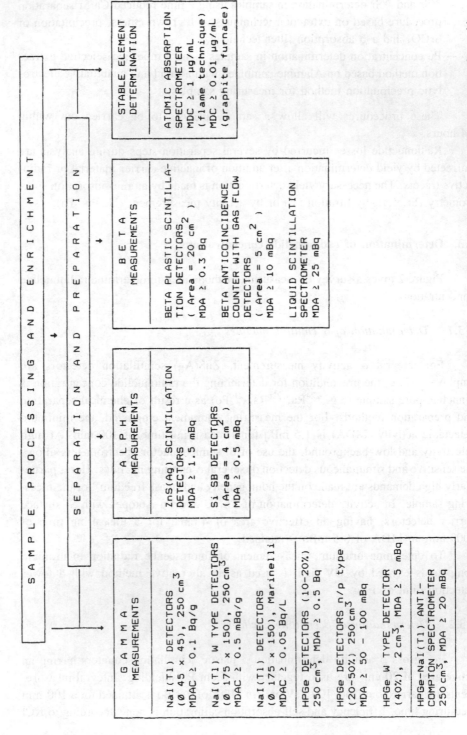

FIG. 2. Measuring methods used to determine radionuclide concentrations. All ø dimensions in millimetres; MDA: minimum detectable activity; MDC: minimum detectable concentration.

standards. Owing to the different radionuclide compositions of the samples in different accident situations, corrections of the counting results are necessary.

The low activity determination of ^{89}Sr, ^{90}Sr, ^{137}Cs, ^{210}Bi and ^{228}Ac samples of high radionuclide purity is carried out mainly by integral measuring methods. For this purpose, β anticoincidence arrangements are employed, consisting of an end window gas flow counter with an effective area of 5 cm^2 equipped with an integrated shield counter. Owing to the high efficiency of the counting equipment (29–35% for ^{137}Cs, depending on the area mass of the measuring sample; 43% for ^{90}Y) and the extremely low background rate of 0.4 count/min, an MDA of 10 mBq for ^{137}Cs is attained for a 100 min measuring time.

The content of β emitting nuclides with a low β maximum energy, such as ^{3}H and ^{14}C, is determined by liquid scintillation spectrometry. If a Tricarb 2250 CA device is used, the MDA for ^{3}H amounts to 3 Bq/L for a measuring time of 200 min.

3.3.3. Determination of γ emitting nuclides

The majority of radionuclides of significance in environmental surveillance can be determined by γ spectrometric measuring methods. The main advantages of γ spectrometric activity determination, such as:

— Simultaneous measurement of radionuclides, and
— Omission of expensive enrichment, separation and preparation,

are associated with the disadvantage of a higher detection limit as compared with β activity measurements. By use of highly effective detector equipment sufficient for a large number of samples, γ spectrometric activity determination has become the dominant method of activity measurement in nuclear environmental surveillance. In conditions of increased environmental contamination its importance is growing considerably.

For samples containing radionuclides with γ energies so different that not much can be expected from the energy resolution, or with only one main radionuclide, high detection efficiency NaI(Tl) scintillation detection units in Pb–Cu radiation shielding are employed:

— NaI(Tl) detectors (ϕ 45 mm × 45 mm) with 5 cm Pb shielding for screening analysis of samples in 250 cm^3 plastic bottles, giving a minimum detectable activity concentration (MDAC) of 100 mBq/g for ^{137}Cs and a 100 min measuring time. These simple detection arrangements are used mainly in laboratories for a rough analysis of environmental samples.
— NaI(Tl) well type detectors (ϕ 175 mm × 150 mm) in 20 cm Pb–Cu shielding equipped with automatic sample changers for sensitive analysis of samples in 250 cm^3 geometry with an MDAC of 0.5 mBq/g for ^{137}Cs and a measuring time of 100 min.

— NaI(Tl) detectors (ϕ 175 mm × 150 mm) for water samples having a
  Marinelli volume of 20 L, giving an MDAC of 0.05 Bq/L for ^{137}Cs and a
  measuring time of 100 min.

A major application of this large detector equipment is, for example, in the determination of ^{226}Ra, ^{232}Th and ^{40}K by the three channel method and of ^{131}I and ^{134,137}Cs activity concentration in environmental samples.

Samples containing a radionuclide mixture are analysed at high energy resolution with Ge semiconductor spectrometers. In view of the different demands on the detector efficiency and measuring geometry for sample analysis in the different types of laboratories, the following equipment is used:

— Portable HPGe detectors with a relative efficiency of 10–20% in 5 cm Pb
  shielding for measurement of samples with a volume of up to 250 cm^3 in local
  and mobile laboratories, giving an MDA of 0.5 Bq for ^{137}Cs over a measuring time of 100 min.
— Portable HPGe detector arrangements with a relative efficiency of 20% used
  in mobile laboratories for special in situ measurements of surface contamination on the ground, giving a minimum detectable area activity of 40 Bq/m^2
  for ^{137}Cs over a measuring time of 100 min.
— HPGe detectors of p and n types with a relative efficiency of 20–50% and with
  automatic sample changers in 10–20 cm Pb shielding. For 250 cm^3 samples
  an MDA of 50–100 mBq for ^{137}Cs and a measuring time of 1000 min is
  attained.
— HPGe well type detectors with a relative efficiency of 40% for small samples
  (fallout and vegetable ashes) with a volume of 2 cm^3, achieving an excellent
  MDA of 5 mBq for ^{137}Cs and a measuring time of 1000 min.

When highly efficient Ge detectors are used, corrections due to sum-coincidence effects have to be carried out.

On the one hand, a reduction of the background to improve the lower detection limit can be achieved by effective radiation shielding. On the other hand, the lower detection limit is improved by electronic measures employing an anti-Compton spectrometer consisting of a 50% efficiency HPGe detector and a special NaI(Tl) well type detector surrounding the semiconductor detector. By means of an anti-coincidence circuit, the Compton fraction in the γ spectrum of the Ge detector is reduced by a factor of 2–3, and the lower detection limit improved as compared with single detector arrangements.

3.3.4. Determination of stable elements

To determine a great number of stable elements relevant for assessing the ecological exposure pathway and to determine the yield of analytical chemical separation steps, atomic absorption spectrometry is used. Concentrations are measured for elements such as Ca and K and for Cs, Sr and Co, whose concentrations are in the milli- or microgram per gram range. Depending on the element to be determined, the sample medium and the necessary minimum detectable concentration (MDC), the sample material is atomized in a flame (≥ 1 μg/mL) or in a graphite furnace (≥ 0.01 μg/mL).

4.   CONCLUSIONS

The environmental monitoring system of the German Democratic Republic provides a number of well established methods for the determination of radionuclide concentrations in environmental media. Based on the determination methods summarized in this paper, they are modified for use in the different types of environmental monitoring laboratories. The establishment of radionuclide determination procedures for environmental monitoring by the National Board for Atomic Safety and Radiation Protection serves to ensure comparable measuring results from all institutions involved in environmental monitoring.

GROSS BETA AND GAMMA ACTIVITY MEASUREMENTS FOLLOWING A NUCLEAR ACCIDENT: METHODS, BENEFITS, LIMITATIONS

R. MAUSHART
Laboratorium Prof. Dr. Berthold,
Wildbad, Federal Republic of Germany

Abstract

GROSS BETA AND GAMMA ACTIVITY MEASUREMENTS FOLLOWING A NUCLEAR ACCIDENT: METHODS, BENEFITS, LIMITATIONS.

On the basis of experience since the Chernobyl event it is shown that, in the case of major environmental contamination, gross beta and gamma activity measurements can be a valuable complement to spectrometric radionuclide determinations. The types, purposes and advantages of gross activity measurements at various phases after an accident are surveyed. The performance and use of some newly developed instruments, particularly designed for field use and ease of operation by intervention teams, are described and typical calibration factors given. Modern contamination monitors may assess surface radioactivity of the order of kilobecquerels to megabecquerels per square metre, and bulk activity in water or milk of several kilobecquerels per litre. Battery powered, portable foodstuff monitors can measure gross gamma activities down to 10 Bq/kg on untreated samples within 30 min and to 100 Bq/kg within 2–5 min. The methods described are particularly useful for making many screening measurements in order to separate samples with insignificant activity concentration from high level specimens that need further evaluation.

1. INTRODUCTION

In the case of a nuclear accident with widespread distribution of radioactivity, it is one of the foremost tasks to determine the radionuclide concentration in foodstuffs, compare the results with predetermined intervention levels and take appropriate actions.

This necessitates measurements on a large number of samples. Practical experience since Chernobyl has clearly demonstrated that the selection of samples — and thereby the limitation of their numbers — is not left exclusively to the government authorities. Producers, processors and vendors of foodstuffs, as well as individual consumers, request the measurement of their own particular products. This puts a heavy work-load on all institutions and persons involved in emergency radioactivity measurements.

It also requires the availability of a multitude of field instruments that are easy to operate, to avoid lengthy time delays due to transportation of samples to a few well equipped laboratories.

It seems therefore appropriate to investigate the possibilities and limitations of comparatively simple and rapid measurement techniques, such as the assessment of gross beta and gamma activity, for extensive application following a nuclear accident.

2.   PURPOSE OF GROSS ACTIVITY MEASUREMENTS AFTER A NUCLEAR ACCIDENT

There is no doubt that, first of all, the individual radionuclides in the fallout mixture must be identified. On the one hand, dose and risk estimates can only be based on knowledge of the kinds of isotopes present. On the other hand, any instrument for determination of activity concentrations can only be calibrated properly if the radionuclide, or mixture of radionuclides, to be assessed is known.

TABLE I. PURPOSES AND ADVANTAGES OF GROSS ACTIVITY MEASURE-MENTS AT VARIOUS PHASES AFTER A NUCLEAR ACCIDENT

Accident phase	Type of measurement	Purpose	Advantages
Immediately after spread of radioactivity	Screening measurements	Selection of low, medium and high activity samples to make first survey	Fast results with simple instruments Large sample numbers Unskilled staff
Intermediate phase (decay of short lived nuclides)	Satellite measurements in co-operation with central analysing laboratories	Survey of extended areas, e.g. by measurement of milk samples on farms; assessment of samples exceeding limits	Immediate, decentral- ized decisions on fur- ther use of samples possible
Late phase (only long lived nuclides left)	Evaluation measurements on samples with known nuclides	Identification of the few remaining high activity samples	Measurements can be made by all people affected

For identification of radionuclides in a given sample, and for measurement of their individual activity concentrations, high resolution gamma spectrometry will be the prevailing kind of measurement [1]. However, the general use of this method is limited for three reasons:

— Gamma spectrometers are still rather expensive and therefore not available in sufficiently large numbers to handle the vast quantities of samples to be measured following an accident.
— The reliable operation of gamma spectrometers requires skilled and experienced personnel; they could hardly be used by emergency teams.
— With few exceptions, gamma spectrometers are laboratory equipment and not suited for field measurements.

Because of the necessity for extensive in situ measurements by untrained people on the one hand and numerous measurements of milk and other foodstuffs on the other, simple gross activity measurements become unavoidable.

Methods for measuring gross activity are particularly suited for making many screening measurements to separate samples with insignificant activity concentration from high level specimens that need further evaluation. But that is not the only advantage. Table I gives a survey of the purposes and further advantages of gross activity measurement at various phases after an accident [2].

3. TYPES OF INSTRUMENTS FOR GROSS ACTIVITY MEASUREMENTS

Beta radiation measurements on solid or liquid samples require instruments having flat, thin-window detectors with large sensitive areas, possibly more than 100 cm^2. As it happened, contamination monitors with integrated large area proportional counter probes proved to be in rather widespread use in the Federal Republic of Germany. They were commonly in use before the Chernobyl accident for routine radiation protection measurements in clinical and research radionuclide laboratories. Equally, fire brigades and civil defence teams were equipped with similar instruments, which were instantly applied to the emergency situation after Chernobyl.

Consequently, seemingly innumerable measurements of gross beta activity in open areas were performed in the first weeks of May 1986 with proportional detector contamination monitors, both for surface contamination (Fig. 1) and for contamination of vegetables and other foodstuffs (Fig. 2).

Portable scintillation based systems, however, using NaI crystals for gross gamma measurements were practically not available at the time of the accident, although they would have been much better suited to the purpose of foodstuff activity measurement, as will be shown below.

FIG. 1. *Measurement of radioactivity deposited on the lawn of a public swimming pool, performed by fire brigade staff members.*

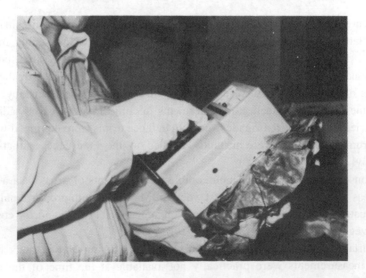

FIG. 2. *Measurement of lettuce at a market in Stuttgart on 5 May 1986, performed with a contamination monitor by a health inspector on duty.*

In the mean time, new instruments have been developed, taking into account the experience gained since Chernobyl. These are particularly designed for field use and ease of operation by intervention teams. Main features include automatic background subtraction, automatic activity calculation by means of preset calibration factors for specific nuclides or typical post-accident nuclide mixtures, and indication of statistical error margins. These new techniques have been described in detail elsewhere [3, 4].

In the following sections, some practical aspects of the various gross activity measurement techniques will be discussed, and performance specifications given.

4.   SURFACE ACTIVITY MEASUREMENTS WITH A CONTAMINATION MONITOR

When, after Chernobyl, extended radiation surveys had to be made by rather inexperienced people, much confusion was caused by the fact that there are three distinctly different categories of measurements that can be performed with a contamination monitor (Table II): assessing the increase in overall radiation, as measured in counts per second and expressed as 'times background'; assessing the activity deposited on surfaces, primarily on the soil, as expressed in becquerels per square centimetre or metre by converting the count rate to such a quantity; and assessing the activity in water or foodstuffs, as expressed in becquerels per litre or kilogram

TABLE II. TASKS OF FIELD MEASUREMENTS WITH A CONTAMINATION MONITOR

Counts/s	Detection of increased radiation levels:
	— Locate and demarcate 'radioactivity infested' areas
	— Find and isolate contaminated objects and persons
Bq/cm^2	Measurement of surface contamination:
	— Identify where established contamination limits on soil, vehicles and persons have been exceeded
Bq/L or Bq/kg	Measurement of activity concentrations:
	— Identify where established activity limits in cleaning water, drinking water, vegetables and other foodstuffs have been exceeded

also by converting the count rate. Both the surface activity and volume activity determinations are gross beta measurements, whereas the assessment of overall radiation is based primarily on gamma measurement.

Just expressing the instrument readings as a multiple of the background level seems to be the easiest thing to do. However, to arrive at a meaningful interpretation of the results, one must know the contributions of the various sources that are causing the background reading, such as cosmic radiation, terrestrial gamma radiation, and beta radiation from natural radioactivity in the soil. The background count rate, therefore, varies according to the positioning of the monitor. This is a fact that might become important when making a measurement in the field [5].

Another problem when assessing the deposited activity per unit area on soil is the structure of the surface, which is usually far from smooth. The calibration factors given in the instrument handbooks, or stored in the microprocessor of the instrument, are related to even, smooth surfaces and can therefore be applied only with great care. Some typical ones, nevertheless, are given in Table III.

Taking into account also the said difficulties of estimating natural background readings, which may vary by a factor of 2 from place to place, one arrives at realistic values of lower detection limits for artificial beta emitting radionuclides on the soil of 2 to 5 kBq/m^2.

TABLE III. SURFACE ACTIVITY CALIBRATION FACTORS FOR VARIOUS BETA EMITTING RADIONUCLIDES (counts/s per kBq/m^2)
(*xenon filled detector, window area 100 cm², window thickness 5 mg/cm² titanium, average background count rate 8 counts/s*)

Co-60	Sr-90/Y-90	I-131	Cs-137
1.7	3.6	2.6	2.5

5. MEASUREMENTS OF BETA ACTIVITY IN FOODSTUFF SAMPLES WITH A CONTAMINATION MONITOR

The determination of radioactivity content in foodstuffs requires suitable preparation of the vegetables, meat, etc., prior to measurement, in order to provide for a defined sample geometry.

The material to be measured has to be chopped or ground and then packed firmly into a brick shaped dish that offers a surface of at least the same area as the detector window of the monitor. The proportional detector has a beta/gamma efficiency ratio of better than 100 and therefore assesses almost exclusively the beta component of the radiation emitted by the sample.

It should be noted that a measuring arrangement such as that shown in Fig. 2 is not suited for determining the specific activity in becquerels per kilogram but can only be used for screening purposes owing to the undefined sample geometry.

Calibration factors for homogenized samples of foodstuffs and radionuclides such as ^{137}Cs or ^{131}I (beta maximum energies of 514 and 600 keV, respectively) are typically 500 Bq/kg per count/s, leading to a lower detection limit of about 1 to 2 kBq/kg.

As a first approximation, the calibration factor for beta radiation is not affected by the density of the sample material, the beta particles always arising from a layer of the same mass per square centimetre, e.g. 230 mg/cm^2 for ^{131}I. This gives a thickness of the 'active' layer — the layer whose activity can be detected — of 2.3 mm for a density of 1, or 4.6 mm for a density of 0.5. In both cases, however, the activity concentration per unit mass is the same. Only if the density is very small and the saturation layer therefore very deep may the change of the geometric factor influence the efficiency.

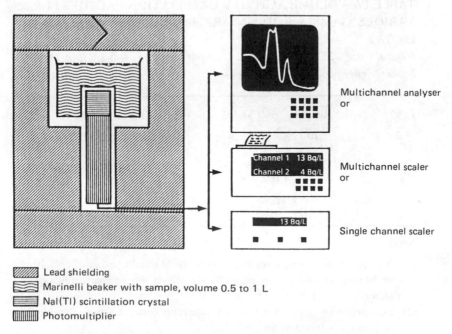

Multichannel analyser
or

Multichannel scaler
or

Single channel scaler

Lead shielding
Marinelli beaker with sample, volume 0.5 to 1 L
NaI(Tl) scintillation crystal
Photomultiplier

FIG. 3. Scintillation detector system for gamma activity measurements in foodstuffs and drinking water.

6.   MEASUREMENTS OF GAMMA ACTIVITY IN FOODSTUFF
     SAMPLES WITH A SCINTILLATION BASED SYSTEM

Much lower detection limits for radioactivity in foodstuff samples may be
achieved by assessing the gamma radiation with scintillation detectors. The usual
arrangement for measuring gamma emitting radionuclides in foodstuff samples is a
Marinelli type beaker (Fig. 3).

Most laboratory set-ups use crystal sizes of 5–7.5 cm and multichannel
analysers for evaluation. Portable field units, as developed in the aftermath of the
Chernobyl accident, have 2.5–5 cm crystals and electronics with either preset energy
windows or broad energy band single channel evaluation devices.

The broad energy band type seems preferable because it will assess all gamma
emitting nuclides. Thus it can be used for any possible event without a change of
window settings. If the nuclides that are present have been identified previously in
comparable samples, then calibration factors can be determined for gross activity
measurement. Anyhow, in most cases one or two radionuclides will be so
predominant that calibrations can be related to only these nuclides and still yield useful results. For example, in the first weeks after Chernobyl, the 'reference nuclide'
was ^{131}I, while later on it was the ^{137}Cs–^{134}Cs mixture.

TABLE IV. VOLUME ACTIVITY CALIBRATION FACTORS FOR
VARIOUS GAMMA EMITTING RADIONUCLIDES (counts/s per
kBq/L)
(crystal size 25 mm × 25 mm, sample volume 0.5 L (Marinelli
beaker), energy window >40 keV)

Zr-95	Ru-103	I-131	Cs-137	K-40
4.2	10	14	8	1
^{137}Cs/^{134}Cs	1:0.4	1:0.3	1:0.2	
	12	11	10.5	

Notes

(1) Because of the sample layer thickness in the Marinelli beaker of only 20
     to 30 mm, sample density only slightly affects the volume calibration
     factor.

(2) For gamma energies above 250 keV, the error between sample densities
     of 1 and 0 is less than 5%.

(3) To convert the volume activity (Bq/L) to mass activity (Bq/kg), the result
     must be divided by the sample density.

Some typical calibration factors are given in Table IV. It can be seen that for key radionuclides, such as ^{131}I and ^{137}Cs, the differences are to within a factor of 2.

Activity calculation may be done by such a scintillation based instrument using an internally set calibration factor, and the approximate value of the activity will be indicated from the beginning of a measurement, along with the statistical accuracy reached at any moment. Thus, the user obtains an immediate, approximate result, and, on the basis of the error displayed, can decide how long to continue counting the sample. This procedure considerably shortens the necessary measuring time in accordance with the amount of activity present. Practical detection limits are about 10 Bq/kg for a 30 min count and 100 Bq/kg for a counting time of 2–5 min.

It is advisable to make regular checks of instrument performance with suitable calibration sources. A typical source applied after Chernobyl was a mixture of ^{137}Cs and ^{134}Cs in the same ratio as emitted, moulded as a firm plastic material into a Marinelli beaker. In this way, the faster decay of the ^{134}Cs is automatically accounted for.

One particular cause of error with gross gamma measuring systems might be their inability to discriminate against the natural activity of ^{40}K as contained in most foodstuffs. A typical potassium content, however, of 2–3 g/kg will result in a reading equivalent to less than 10 Bq/kg of ^{137}Cs, so the error will be negligible when measuring artificial radioactivities higher than some tens of becquerels per kilogram.

7. OUTLINE OF A STRATEGY FOR MEASUREMENT OF RADIOACTIVITY IN FOODSTUFFS AFTER A NUCLEAR ACCIDENT

As already mentioned, contamination monitors are frequently in use in radionuclide laboratories and nuclear medicine departments. In some countries they are also part of the emergency equipment of fire brigades, rescue organizations and civil defence units, or have become so since Chernobyl. They are therefore likely to be easily available and should be used to obtain a first estimate of the situation, by determining 'radioactivity infested' regions, screening for high or low activity samples and selecting those which necessitate further measurements.

To measure the true activity concentration in foodstuffs, however, contamination monitors should be used with proper care and knowledge, and when no other means are readily available.

Selected foodstuff samples representative of a certain kind of victual, e.g. milk, and for a certain region should be analysed by gamma spectrometric methods. This having been done, however, similar samples may then be measured with sufficient accuracy by gross gamma measurement set-ups, as described, the advantages being that such measurements can be performed:

— In the field or at the point of origin of the victuals, because the instruments are easily transportable and even battery operated;

— On a large number of samples because the instruments are inexpensive;
— By comparatively untrained people because operation is simple and no further calculations or evaluations of results are required.

It must be stressed, however, that all field measurements are to be supervised and guided by master laboratories, as described in Ref. [1].

REFERENCES

[1]   INTERNATIONAL ATOMIC ENERGY AGENCY, Measurement of Radionuclides in Food and the Environment: A Guidebook, Technical Reports Series No. 295, IAEA, Vienna (1989).

[2]   MAUSHART, R., "Measuring the radioactivity in foodstuffs — A comparative assessment of common methods", Radiation Protection — Theory and Practice (Proc. 25th Anniv. Symp. Malvern, 1989) (GOLDFINCH, E.P., Ed.), Inst. of Physics, Bristol and New York (1989) 389–392.

[3]   KIRSCH, H., "New, enlarged operational possibilities for contamination monitors in environmental monitoring", The Radioecology of Natural and Artificial Radionuclides (Proc. 15th IRPA Regional Congr. Visby, 1989) (FELDT, W., Ed.), TÜV Rheinland, Cologne (1989) 535–541.

[4]   MAUSHART, R., The measurement of radioactivity in foodstuffs — A re-evaluation of methods after Chernobyl, Radiat. Prot. Manage. 5 3 (1988) 45–50.

[5]   MAUSHART, R., Überwachung der Radioaktivität in der Umwelt, GIT, Darmstadt (1989) 200.

SPECTROMETRIC METHOD FOR DETERMINING ^{90}Sr ACTIVITY IN ENVIRONMENTAL SAMPLES

V.A. KNAT'KO
Institute of Physics,
Byelorussian Academy of Sciences

I.I. UGOLEV, M.E. ZHAMOZDIK, A.S. SOKOLOVSKIJ
Institute of Physical and Organic Chemistry,
Byelorussian Academy of Sciences

Minsk,
Union of Soviet Socialist Republics

Abstract

SPECTROMETRIC METHOD FOR DETERMINING ^{90}Sr ACTIVITY IN ENVIRON-
MENTAL SAMPLES.

The paper considers an instrumental method for determining ^{90}Sr in environmental
samples. The method is based on measurement of the beta spectrum of a sample followed by
special processing of the high energy portion of the spectrum in order to obtain the
contribution of ^{90}Y. A procedure for calculation of activities from better spectrometry data
obtained in measurements with calibrated samples is discussed.

The radioactive contamination of the environment which followed the accident
at the Chernobyl nuclear power plant made it essential to carry out prolonged large
scale measurements of the concentrations of long lived radionuclides in samples of
various kinds. Among these radionuclides ^{90}Sr occupies an especially important
place because of its powerful radiobiological toxicity. This being so, the need for
faster and more reliable methods of measuring ^{90}Sr was noted in Ref. [1]. A review
of methods available for ^{90}Sr analysis may be found in Ref. [2].

Knat'ko et al. [3] proposed an instrumental method for quick, routine determinations of ^{90}Sr, based on measurement of the beta spectrum of the sample followed
by special processing of the high energy portion of the spectrum. The end point
energy ϵ_0 of the beta spectrum of ^{90}Y (daughter nucleus of ^{90}Sr) is 2.27 MeV. The
analysis presented in Ref. [3] showed that only a small number of radioactive fission
products with half-lives $T_{1/2} > 0.5$ year and their daughter nuclides have
$\epsilon_0 > 1.5$ MeV. Apart from ^{90}Y, the radionuclides in question are ^{106}Rh(^{106}Ru),
^{110}Ag(^{110}Ag^m), ^{144}Pr(^{144}Ce) and ^{154}Eu (the parent nucleus is indicated in brackets).

TABLE I. CHARACTERISTICS OF THE RADIONUCLIDES ^{90}Y, ^{106}Rh, ^{110}Ag, ^{144}Pr AND ^{154}Eu [4, 5]

	Y-90	Rh-106	Ag-110	Pr-144	Eu-154
$T_{1/2}$	64.0 h	29.6 s	24.6 s	17.28 min	8.8 a
ϵ_0 (MeV)	2.27	3.54	2.89	3.00	1.89
I_0 (%)[a]	98.5	78.8	95.2	97.7	11.5
P (%)[b]	100	100	1.33	98.2	

[a] I_0 is the intensity of the beta transition with beta particle end point energy ϵ_0.
[b] P is the yield of the daughter nucleus indicated in the table.

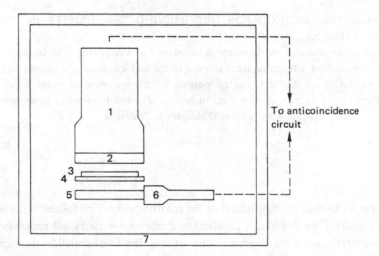

FIG. 1. *Detector part of the beta spectrometer: 1: photomultiplier tube, Ø 150 mm; 2: scintillator, 10 mm thick, Ø 150 mm; 3: sample, Ø 140 mm; 4: screen, Ø 150 mm; 5: scintillator, 150 mm × 150 mm × 10 mm; 6: photomultiplier tube, Ø 25 mm; 7: lead shield, 50 mm thick.*

The data for these nuclei are given in Table I. Given the differences in ϵ_0 values, it should be possible to resolve the high energy range of the sample spectrum (1.5–3.5 MeV) into components attributable to the radionuclides mentioned above, and in this way to determine the activity of ^{90}Y and, accordingly, of ^{90}Sr.

To investigate the possibilities of this method, we designed an automated beta spectrometer (a detailed description of this instrument is given in Refs [6, 7]). A block diagram of the detector arrangement used for this spectrometer is shown in Fig. 1. The beta particle detector consists of a plastic scintillator (polystyrene + p-terphenyl + POPOP) with an area of 100 cm^2. The background is reduced by a 50 mm thick lead shield and by anticoincidence shielding. The latter reduces the background count rate by about one half in the energy range 1.5–3.5 MeV. The background spectrum for this range is shown in Fig. 2. The total number of background pulses in the 1.5–3.5 MeV range is 2420 for a measuring time t = 10^4 s.

The results of ^{90}Sr measurements depend not only on the background level but also on the content of the other radionuclides in the sample. The high energy portions of the beta spectra for ^{90}Y, ^{106}Rh and ^{144}Pr obtained after subtraction of

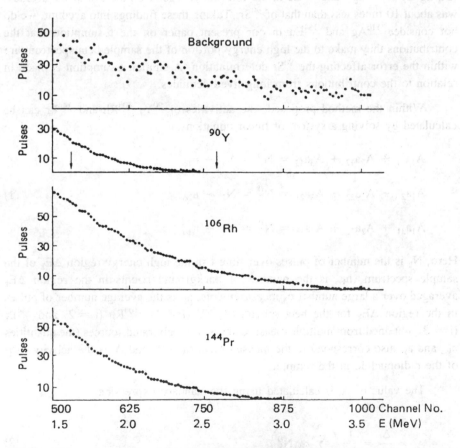

FIG. 2. *High energy portions of background and radionuclide spectra. The spectra shown correspond to a measurement time of 10^4 s and an activity of 10 Bq. The radionuclide spectra were processed by a five channel smoothing procedure: each point in the figure corresponds to the sum of the pulses in five channels of the smoothed spectrum. Arrows indicate the boundaries (X$_1$ and X$_2$) of the energy intervals (see text).*

the background are shown in Fig. 2. In measuring these spectra we used calibrated samples simulating soil, with a surface density of 0.1 g/cm^2, containing $^{90}\text{Sr} + {}^{90}\text{Y}$, $^{106}\text{Ru} + {}^{106}\text{Rh}$ or $^{144}\text{Ce} + {}^{144}\text{Pr}$ (parent and daughter nuclei were present in the samples in equal proportions). According to our results, over a measuring time $t = 10^4$ s, the number of pulses, N, in the high energy region of the beta spectra for ^{90}Y, ^{106}Rh and ^{144}Pr, for an activity of 10 Bq, was 458, 2148 and 1170, respectively. For the radionuclides ^{110}Ag and ^{154}Eu, the value of N was substantially lower (<15 pulses): this was because of the low yield (1.33%) of ^{110}Ag in the decay of the parent nucleus $^{110}\text{Ag}^m$, and in the case of ^{154}Eu it was due to the low intensity of the beta transition ($I_0 = 11.5\%$) with $\epsilon_0 = 1.89$ MeV (Table I). According to the evaluations cited in Ref. [6], the specific activity (Bq/kg U) of $^{110}\text{Ag}^m$ and ^{154}Eu in the Chernobyl reactor on the eve of the accident was about 10 times less than that of ^{90}Sr. Taking these findings into account, we do not consider ^{110}Ag and ^{154}Eu in our present paper on the assumption that the contributions they make to the high energy portion of the sample beta spectrum are within the error affecting the ^{90}Sr determination. A similar assumption is made in relation to the contributions from long lived actinides.

Within the method proposed, the activities of ^{90}Y, ^{106}Rh and ^{144}Pr can be calculated by solving a system of linear equations:

$$A_1 a_{11} + A_2 a_{12} + A_3 a_{13} = N_1^0 = N_1 - n_{1\phi}$$

$$A_1 a_{21} + A_2 a_{22} + A_3 a_{23} = N_2^0 = N_2 - n_{2\phi} \tag{1}$$

$$A_1 a_{31} + A_2 a_{32} + A_3 a_{33} = N_3^0 = N_3 - n_{3\phi}$$

Here, N_k is the number of pulses over time t in the high energy region ΔE_k of the sample spectrum, $n_{k\phi}$ is the number of background counts in the region ΔE_k averaged over a large number of measurements, a_{ki} is the average number of pulses in the region ΔE_k for the beta spectra of ^{90}Y ($i = 1$), ^{106}Rh ($i = 2$) and ^{144}Pr ($i = 3$), obtained from multiple measurements with calibrated sources (the quantities $n_{k\phi}$ and a_{ki} also correspond to the measurement time t), and A_i is the activity (Bq) of the radionuclide in the sample.

The value of A_i is calculated using the standard expression:

$$A_i = \frac{\Delta_i}{\Delta} = \frac{1}{\Delta} \left(\sum_{k=1}^{3} \alpha_{ki} N_k^0 \right) \tag{2}$$

where Δ is the determinant of the system in (1), and Δ_i is a determinant obtained from Δ when we replace the elements of the i-th column with free terms. The

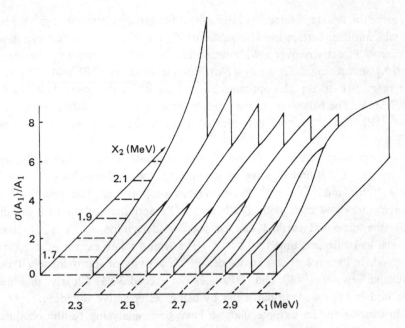

FIG. 3. Behaviour of the relative error $\sigma(A)_1/A_1$ in the measurement of ^{90}Y activity as a function of the boundaries of the energy intervals.

quantity α_{ki} is a cofactor of the element a_{ki} in the determinant Δ. To evaluate the mean square deviation of the quantity A_i, we use the expression:

$$\sigma(A_i) = (D(A_i))^{1/2} \approx \frac{1}{\Delta} \left(\sum_{k=1}^{3} \alpha_{ki}^2 N_k \right)^{1/2} \tag{3}$$

In deriving expression (3) it was assumed that the dispersions of N_k^0 and a_{ki} have the forms:

$$D(N_k^0) = D(N_k) + D(n_{k\phi}) \approx N_k \quad \text{and} \quad D(a_{ki}) \approx 0$$

In using the procedure proposed here for calculating activities, it is important to select the optimum boundaries, X_1 and X_2, for the energy intervals (Fig. 2). To elucidate this problem we performed calculations of $\sigma(A_i)$ using (3) and varying X_1 and X_2, as well as the value of A_i. The A_i values lay in the range 0.1–100 Bq. The values of N_k needed for calculating $\sigma(A_i)$ were specified with the help of the expression:

$$N_k = \sum_{i=1}^{3} A_i a_{ki} + n_{k\phi}$$

From the results obtained, it follows that the error in determining ^{90}Y activity has a minimum for certain specific values of X_1 and X_2, and also that variations in ^{106}Rh and ^{144}Pr activity over a wide range have virtually no effect on the values of X_1 and X_2 minimizing the quantity $\sigma(A_i)$. For variations in ^{106}Rh and ^{144}Pr activity in the range 0.1–10 Bq, the optimum X_1 and X_2 lie in the intervals 2.5–2.8 and 1.6–1.7 MeV. The behaviour of the relative error $\sigma(A_1)/A_1$ as a function of X_1 and X_2 for ^{106}Rh and ^{144}Pr with activities of 1 Bq is shown by way of example in Fig. 3.

The optimum values of X_1 and X_2 thus obtained were used in analysing the beta spectra of calibrated samples containing the radionuclides $^{90}Sr + ^{90}Y$, $^{106}Ru + ^{106}Rh$ and $^{144}Ce + ^{144}Pr$ in various proportions. The results of these measurements show that, with ^{106}Rh and ^{144}Pr activities of up to 3 Bq in the sample, the proposed method can be used to determine a ^{90}Y (^{90}Sr) activity equivalent to 10 Bq in a time t = 2.5 × 10^4 s with a relative error of $\leq 50\%$ (this corresponds in the present case to a specific ^{90}Sr activity of approximately 1 Bq/g). The relative error for ^{106}Rh and ^{144}Pr does not exceed 30%. Clearly, this method can be further improved, in particular by using better active shielding.

In conclusion, let us note that we have been analysing results obtained by solving a system of linear equations. It would be interesting to examine other procedures for calculating activities from beta spectrometry data. This question and also the problem of the influence of sample preparation on the accuracy of results will be the subjects of future investigation.

REFERENCES

[1]   Report of the Expert Consultation on Recommended Limits for Radionuclide Contamination of Food, FAO, Rome (1986).

[2]   HELLMUTH, K.H., Rapid Determination of ^{89}Sr and ^{90}Sr — Experiences and Results with Various Methods after the Chernobyl Accident in 1986, Rep. STUK-A70, Finnish Centre for Radiation and Nuclear Safety, Helsinki (1987).

[3]   KNAT'KO, V.A., UGOLEV, I.I., ZHAMOZDIK, M.E., Using the Special Characteristics of the ^{90}Y Beta Spectrum, Preprint No. 474, Inst. of Physics, Byelorussian Academy of Sciences, Minsk, 1987 (in Russian).

[4]   INTERNATIONAL COMMISSION ON RADIOLOGICAL PROTECTION, Radionuclide Transformations: Energy and Intensity of Emissions, Publication 38, Pergamon Press, Oxford and New York (1983).

[5]   GUSEV, N.G., BELYAEV, V.A., Releases of Radiation in the Biosphere, A Guide, Ehnergoatomizdat, Moscow (1986) (in Russian).

[6]   ZHAMOZDIK, M.E., KNAT'KO, V.A., SOKOLOVSKIJ, A.S., UGOLEV, I.I., "On beta-spectrometric analysis of environmental samples", Physics of the Biosphere, No. 13, Mokslas, Vilnius (1989) (in Russian) (in press).

[7]   ZHAMOZDIK, M.E., KNAT'KO, V.A., UGOLEV, I.I., "Beta-spectrometric method for determination of ^{90}Sr in soil", Documents of All-Union Conf. on Radiation Aspects of the Chernobyl Accident, Obninsk, 1988, USSR State Committee for Hydrometeorology, Moscow (in Russian) (in press).

ANALYSES OF Pu IN SOME NATURAL MATRIX MATERIALS

J. LaROSA, A. GHODS-ESPHAHANI,
J.C. VESELSKY, R. SCHELENZ
Agency's Laboratory,
International Atomic Energy Agency,
Seibersdorf

M. MATYJEK
Institute of Chemistry and Nuclear Technology,
Warsaw, Poland

Abstract

ANALYSES OF Pu IN SOME NATURAL MATRIX MATERIALS.

Under the Supplementary Programme on Nuclear Safety of the International Atomic Energy Agency, analytical procedures are being developed for the reliable determination of various radionuclides in a wide range of materials. The isotopes of Pu are of particular interest because they are very important for a total dose assessment. The Agency's Laboratory at Seibersdorf is therefore developing and testing appropriate analytical methods for the determination of Pu in environmental samples such as vegetation and soil as well as in major food groups. A radiochemical procedure developed for the determination of Pu at Seibersdorf is described in the paper. The procedure was applied to three marine substances and one type of soil for the purpose of assessing its accuracy and reliability. Analytical results are presented for the activity concentrations of ^{238}Pu and of total ^{239}Pu and ^{240}Pu in these natural matrix materials. Advantages, difficulties and limitations of the method are discussed.

1. INTRODUCTION

In 1986, in response to the many requests from Member States for radionuclide analysis of a large number of food and environmental samples following the Chernobyl accident, the International Atomic Energy Agency (IAEA) established a new programme entitled Fallout Radioactivity Monitoring in Environment and Food (MEF). The major objective of the MEF programme was to provide the Member States with reliable analytical methods for the determination of radionuclides in various food and environmental samples following a nuclear accident. In 1989, after three years of intensive work by laboratories participating in the MEF programme, the IAEA published an MEF guidebook [1] which contained recommended procedures for taking of representative samples as well as sample handling, preparation and dissolution and also listed detailed radiochemical procedures. This publication contributed significantly towards meeting the original objective of the programme and requirements of Member States.

Under the MEF programme a small group was formed in the Chemistry Unit of the Agency's Laboratory at Seibersdorf to implement and improve radioanalytical procedures for the determination of the actinides (Pu, Am, Cm), with particular emphasis on plutonium. To achieve this goal and gain experience in actinide analysis, this group recently participated in an intercomparison exercise organized by the IAEA's International Laboratory of Marine Radioactivity (ILMR) in Monaco. The Seibersdorf group tested its method to determine plutonium, in particular [238]Pu and [239,240]Pu, in environmental samples by radiochemical separation of the plutonium in combination with alpha spectrometry measurements. This paper presents the radiochemical procedure utilized at Seibersdorf and compares the plutonium results obtained by this procedure with the recommended values from the intercomparison study.

2.   EXPERIMENTAL

The separation procedure for the determination of plutonium is outlined below and shown schematically in Fig. 1. For the intercomparison exercise three types of marine material which had been processed (dried, ground, sieved and homogenized) by the ILMR were analysed for Pu concentration. These materials were designated IAEA-306 (Baltic Sea sediment), IAEA-307 (sea plant *Posidonia oceanica*) and IAEA-308 (mixed Mediterranean seaweeds). In addition, the Pu content of one terrestrial soil sample designated as IAEA-Soil-6, which was available as a reference material from the Analytical Quality Control Services at Seibersdorf, was also determined. This latter material was derived from topsoil (to a depth of 10 cm) collected near Ebensee in Upper Austria at an altitude of 1100 m above sea level.

Before dissolution, the samples were dried for two days at 80°C and ashed at 600°C for 6–9 h to remove any carbonaceous material. Ashing temperatures above 600°C should be avoided to minimize conversion of Pu to intractable PuO_2 [2]. From the ash, 8–10 g of sample were accurately weighed out and spiked with a known amount of [242]Pu (0.036–0.16 Bq) to determine the chemical recovery of Pu. The spiked ash in a polytetrafluoroethylene beaker was treated with 70 mL of concentrated HF (40 wt%) and strongly heated to dissolve silica and to expel volatile SiF_4. The HF mixture was evaporated nearly to dryness. The HF treatment was usually followed by addition of 50 mL of concentrated HCl (32 or 37 wt%) and evaporation of the mixture to a thick paste (10–20 mL in volume). Then 50 mL of concentrated HNO_3 (65 wt%) were combined with the paste and the mixture was again evaporated nearly to dryness to remove free HF and to convert insoluble fluorides to soluble nitrates. This HNO_3 evaporation was repeated two more times. The

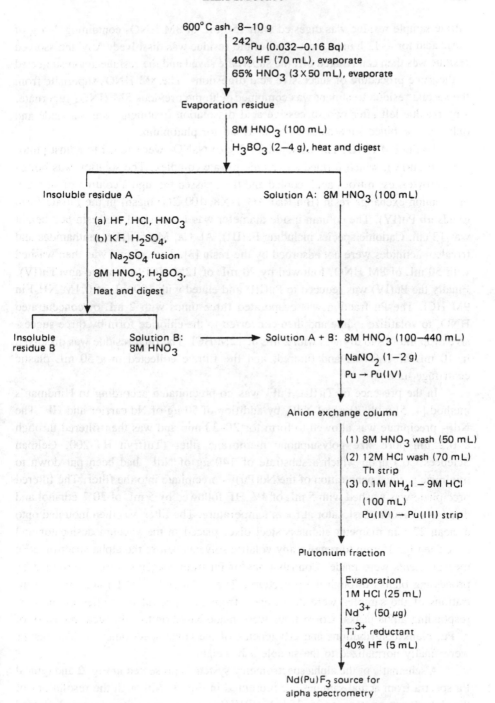

600°C ash, 8—10 g
242Pu (0.032—0.16 Bq)
40% HF (70 mL), evaporate
65% HNO3 (3 × 50 mL), evaporate

Evaporation residue

8M HNO3 (100 mL)
H3BO3 (2—4 g), heat and digest

Insoluble residue A   Solution A: 8M HNO3 (100 mL)

(a) HF, HCl, HNO3
(b) KF, H2SO4,
    Na2SO4 fusion
8M HNO3, H3BO3,
heat and digest

Insoluble   Solution B: ────────▶ Solution A + B: 8M HNO3 (100—440 mL)
residue B   8M HNO3
                                        NaNO2 (1—2 g)
                                        Pu → Pu(IV)

Anion exchange column

(1) 8M HNO3 wash (50 mL)
(2) 12M HCl wash (70 mL)
    Th strip
(3) 0.1M NH4I — 9M HCl
    (100 mL)
    Pu(IV) → Pu(III) strip

Plutonium fraction

Evaporation
1M HCl (25 mL)
Nd3+ (50 µg)
Ti3+ reductant
40% HF (5 mL)

Nd(Pu)F3 source for
alpha spectrometry

FIG. 1. General chemical scheme for the dissolution of inorganic sample ash and separation of Pu.

nitrate sample residue was digested with 150 mL of 8M HNO_3 containing 2–4 g of boric acid for 4–12 h until most or all of the residue was dissolved. Any undissolved residue was then centrifuged off, the supernate saved and the residue again subjected to the same procedure of successive acid digestions. The 8M HNO_3 supernate from the second residue treatment was combined with the previous 8M HNO_3 supernate. Any residue left after two successive acid dissolution treatments was set aside and only the combined supernates were analysed for plutonium.

To the 8M HNO_3 sample solution, 2 g of $NaNO_2$ were added to adjust plutonium to Pu(IV), which formed an anionic nitrate complex. The solution was boiled to destroy excess nitrous acid, cooled and then passed through a column of strongly basic anion exchange resin (Bio-Rad AG 1-X8, 100–200 mesh) in the nitrate form to adsorb Pu(IV). The column inside diameter was 1.0 cm and the resin bed height was 13 cm. Cationic species including Fe(III), Al, Ca, Mg, Ti(IV), lanthanides and trivalent actinides were not adsorbed by the resin [3]. The column was then washed with 50 mL of 8M HNO_3 followed by 70 mL of 12M HCl to remove any Th(IV). Finally the Pu(IV) was reduced to Pu(III) and eluted with 100 mL of 0.1M NH_4I in 9M HCl. The Pu fraction was evaporated three times with 2 mL of concentrated HNO_3 to volatilize iodine and then converted to the chloride form by three successive evaporations to dryness with 2 mL of 12M HCl. The final residue was dissolved in 10 mL of 1M HCl and filtered, and the filtrate collected in a 50 mL plastic centrifuge tube.

In the presence of Ti(III), PuF_3 was co-precipitated according to Hindman's method [4, 5] with 70 μg of NdF_3 by addition of 50 μg of Nd carrier and HF. The NdF_3 precipitate was allowed to form for 20–30 min and was then filtered through a 0.2 μm pore size polysulphone membrane filter (Tuffryn HT-200, Gelman Sciences Co.) upon which a substrate of 140 μg of NdF_3 had been put down to minimize surface penetration of the $Nd(Pu)F_3$ precipitate into the filter. The filtered precipitate was washed with 5 mL of 4% HF followed by 5 mL of 20% ethanol and dried in a vacuum desiccator at room temperature. The filter was then mounted onto a clean 27 mm diameter stainless steel disc, placed in the vacuum desiccator and evacuated for 15 min to remove any volatile solvents before the alpha spectrometric measurements were made. Contributions of Pu from reagents were determined by processing blanks through the procedure. The [238]Pu and [239,240]Pu activity concentrations of the samples were determined from the spectral intensities of the corresponding alpha peaks. Corrections were made based on the chemical recovery of [242]Pu, blank contributions and efficiencies of the alpha spectrometers. The results were finally normalized to the sample ash weight.

A schematic of the alpha spectrometry system is presented in Fig. 2 and typical Pu spectra from actual samples are presented in Fig. 3. Although the resolutions of the alpha spectrometers are 25–35 keV (FWHM), the co-precipitated sources yielded alpha spectra with typical resolutions ranging from 50 to 80 keV (FWHM). These were adequate for the determination of Pu in the samples. The efficiencies of the

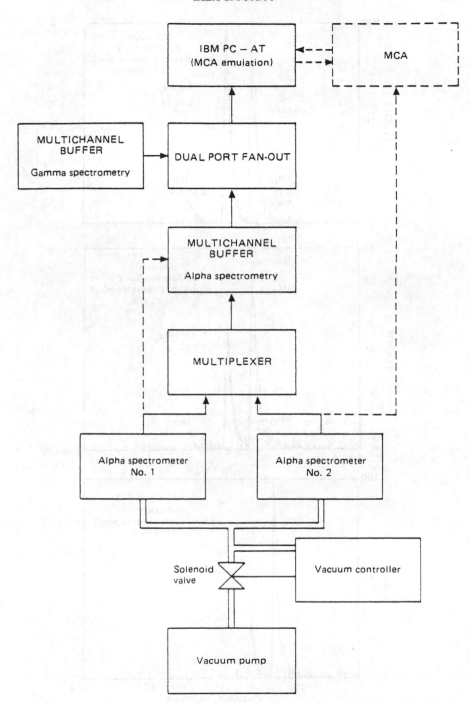

FIG. 2. Block diagram of the alpha spectrometry system. The dashed lines and box depict the electronics configuration used in the absence of the multiplexer unit.

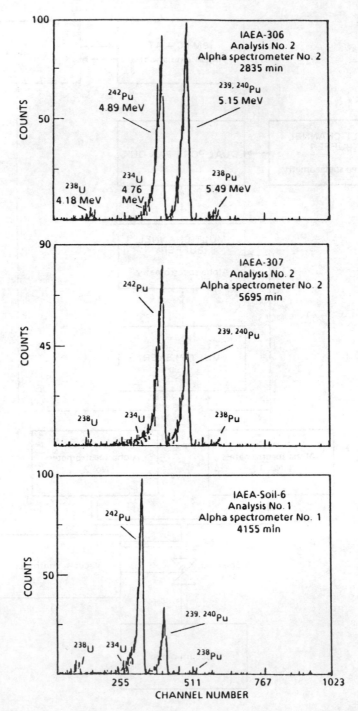

FIG. 3. *Representative alpha spectra of NdF₃ co-precipitated sources prepared from radio-chemically separated Pu fractions of marine and soil samples.*

TABLE I. RESULTS OF Pu ANALYSES OF MARINE AND SOIL MATERIALS[a]

| Material | Analysis | | | Pu-242 recovery (%) | Pu activity concentration (Bq/kg) | | | |
| | No. | Dry wt (g) | Days[b] | | Agency's Laboratory, Seibersdorf | | References | |
					Pu-238	Pu-239, 240	Pu-238	Pu-239, 240
IAEA-306	1	11.0	3	100	0.159 ± 0.016	6.30 ± 0.09	0.17 (0.14–0.19)[c]	5.7 (5.5–6.3)[c], 6.59 ± 0.22[d]
	2	11.7	3	102	0.173 ± 0.015	6.41 ± 0.06		
IAEA-307	1	24.0	8	91	0.022 ± 0.005	0.94 ± 0.02	0.025 (0.022–0.028)[c]	0.72 (0.66–0.79)[c]
	2	24.0	8	87	0.024 ± 0.005	0.95 ± 0.02		
	3	24.0	8	100	—	0.94 ± 0.02		
IAEA-308	1	14.6	12	100	<0.01	0.59 ± 0.02	0.017 (0.015–0.023)[c]	0.50 (0.46–0.52)[c]
	2	14.5	8	36	0.04 ± 0.03	0.53 ± 0.03		
IAEA-Soil-6	1	10.4	6	85	0.033 ± 0.009	1.04 ± 0.04		1.04 (0.96–1.11)[e]
	2	10.5	6	92	0.019 ± 0.008	0.94 ± 0.05		
	3	10.4	5	70	0.018 ± 0.008	0.97 ± 0.04		
	4	10.6	5	73	0.032 ± 0.008	0.99 ± 0.04		

[a] All activity concentrations are expressed per kilogram of dry material. All uncertainties are expressed at the 1σ confidence level and are derived from counting statistics unless otherwise indicated.

[b] Time in working days from the beginning of sample ash dissolution to preparation of the final source for alpha spectrometry.

[c] Results of the plutonium measurements obtained from intercomparison exercises. The median value and the 95% confidence interval (in parentheses) are reported here and are from a private communication with the ILMR.

[d] The mean Pu-239, 240 result is from Table 1 of Ref. [8] for marine sediment from the Gulf of Finland, 1974.

[e] This is the reference value and the 95% confidence interval (in parentheses) for IAEA-Soil-6 [9].

alpha spectrometers were determined by preparing several co-precipitated (^{241}Am–NdF$_3$) sources containing known quantities of ^{241}Am (0.22 Bq) and measuring the ^{241}Am activity with source–detector geometries similar to those in the case of Pu.

3.   RESULTS

The results for the Pu analyses on the marine and soil samples are presented in Table I along with the intercomparison exercise results for the marine samples and the reference value for the soil. In general the agreement is quite satisfactory considering the small number of analyses, indicating that the procedure can be used for analysis of Pu in samples of these types. The uncertainties in the reported results were calculated by taking into consideration only counting statistics, blank contributions and corrections due to the presence of low levels of U, Th and ^{228}Th daughter activity. The reproducibility of the method is typically $\pm 3\%$ at the 1 Bq/kg level for a 10 g sample and a counting period of 4000 min.

4.   DISCUSSION

Although the reliability and accuracy of our procedure for the determination of Pu in various environmental samples were confirmed, the present method requires an average of seven working days to obtain a source suitable for alpha spectrometric measurements. A large fraction of the time is spent dissolving the sample to ensure the complete chemical exchange of ^{242}Pu tracer with the Pu in the sample [6, 7]. At least one full working day is necessary to complete the normal loading, washing and stripping steps associated with the operation of the anion exchange column. If post-precipitation occurs in the feed solution during loading of the column, it may result in drastically reduced flow rates, poor column performance and extended elution times. Nevertheless, the analytical results indicate that the method is applicable to diverse materials (soils, sediments, etc.) containing at least 1 Bq/kg of ^{239,240}Pu activity. This sensitivity is more than adequate as a basis for safety decisions and for dose assessment calculations based on current guidelines. The radiochemical purity of the Pu fraction is very good, and the alpha spectral resolution of the easily prepared NdF$_3$ source permits accurate nuclide identification and quantitative determination. A detection limit of approximately 4 mBq of ^{239,240}Pu per analysis is achievable for a measurement time of 4000 min.

Future work is directed towards investigating methods for more rapid and complete dissolution of sample ash (e.g. molten salt fusion) and for improved (faster, better resolution and efficiency) column chromatographic separations using conventional anion exchange, high performance liquid chromatography and supported liquid phase extraction techniques.

REFERENCES

[1] INTERNATIONAL ATOMIC ENERGY AGENCY, Measurement of Radionuclides in Food and the Environment: A Guidebook, Technical Reports Series No. 295, IAEA, Vienna (1989).

[2] HÖLGYE, Z., Influence of ashing temperature on plutonium separation from bones, Fresenius' Z. Anal. Chem. **331** (1988) 827–828.

[3] KRESSIN, I.K., WATERBURY, G.R., The quantitative separation of plutonium from various ions by anion exchange, Anal. Chem. **34** (1962) 1598–1601.

[4] HINDMAN, F.D., Neodymium fluoride mounting for alpha spectrometric determination of uranium, plutonium and americium, Anal. Chem. **55** (1983) 2460–2461.

[5] HINDMAN, F.D., Actinide separations for alpha spectrometry using neodymium fluoride coprecipitation, Anal. Chem. **58** (1986) 1238–1241.

[6] VESELSKY, J.C., Problems in the determination of plutonium in bioassay and environmental analysis, Anal. Chim. Acta **90** (1977) 1–14.

[7] SILL, C.W., Some problems in measuring plutonium in the environment, Health Phys. **29** (1975) 619–626.

[8] MIETTINEN, J.K., JAAKKOLA, T., JÄRVINEN, M., "Plutonium isotopes in aquatic foodchains in the Baltic Sea", Impacts of Nuclear Releases into the Aquatic Environment (Proc. Symp. Otaniemi, 1975), IAEA, Vienna (1975) 147–155.

[9] INTERNATIONAL ATOMIC ENERGY AGENCY, Analytical Quality Control Services, Information Booklet, IAEA, Vienna (1989) 24.

POSTER PRESENTATIONS

IAEA-SM-306/142P

PHYSICAL, TECHNICAL AND METHODICAL PROBLEMS OF EXPOSURE RATE MEASUREMENTS IN THE TERRITORIES CONTAMINATED AS A RESULT OF THE CHERNOBYL NUCLEAR POWER PLANT ACCIDENT

V.S. REPIN, I.P. LOS', A.V. ZELENSKY,
O.A. BONDARENKO, M.G. BUSINNY, D. NOVAK
All-Union Scientific Centre for Radiation Medicine,
Kiev,
Union of Soviet Socialist Republics

Instruments for measurement of exposure are the main instruments employed in a system of environmental monitoring in the case of a large radiation accident. Different types of detectors, however, give different results at the same point of measurement. A comparative analysis of different types of instruments based on the results of measurements in the territories contaminated after the Chernobyl accident is summarized here. Table I presents the main technical characteristics of the instruments used. They are divided into three groups according to the type of detector: Geiger–Müller tube (GMT), scintillation detector (NaI) and thermoluminescent dosimeter (TLD).

Open spaces situated within the limits of the 30 km zone around the Chernobyl nuclear power plant were chosen for the measurements. They are characterized by different types of dispersion of gamma decay (forest; lands ploughed and not ploughed after the accident; apartments; decontaminated highways). All the dosimeters were exposed at a level of 1 m above the ground. The TLDs were exposed for a time interval required for the cumulation of an absorbed dose ten times higher than the standard value of the minimum detectable dose for a dosimeter based on LiF. In addition, the gamma spectrum was measured with an HPGe spectrometer and an Ortec 7500 analyser. Sampling of different ground layers was done in order to estimate the distribution of radioactive fallout with depth.

Comparative analysis was based on the readouts of the Soviet DRG-01T device. The data analysis showed the following:

— The maximum measurement value exceeding the true value is observed for devices with scintillation detectors in the case of a large contribution from the low energy component (Fig. 1);

TABLE I. COMPARISON OF EXPOSURE RATES AS MEASURED BY DIFFERENT DEVICES IN DIFFERENT RADIATION FIELDS
(*values normalized to those for the USSR GMT*)

	DP-5B	S2000 Servior	NC482B + SWGM	DRG-01T	TLD (LiF)	TLD (CaSO$_4$)	SRP-68-01	Micro-Analyst
Country:	USSR	USA	USA	USSR	Finland	Finland	USSR	USA
Firm:		Bicron	Bicron		Alnor	Alnor		Bicron
Detector:	GMT	GMT	GMT	GMT	TLD	TLD	NaI	NaI
Point of measurement								
1	1.5	1.3	1.3	1.0			5.5	5.5
2	1.4	1.0	1.2	1.0	0.3	0.9	3.4	3.5
3	1.4	1.0	1.1	1.0	1.3	1.3	2.8	2.8
4	1.3	1.2	1.1	1.0	1.3	1.8	2.7	2.8
5	1.4	1.0	1.0	1.0	1.4	1.5	2.5	2.7
6	0.6		1.0	1.0	1.5	0.9	2.1	2.3
7			0.9	1.0				1.8

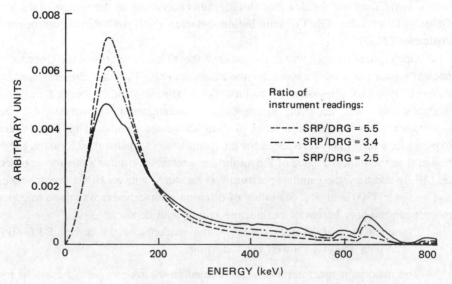

FIG. 1. Smoothed normalized gamma spectra for three different points of measurement and various ratios of instrument readings.

— The divergence in the measurements with scintillation detectors is determined by the contribution from the low energy (dispersed) spectrum component;
— The different readouts of GMT devices are within the limits of standard errors.

Thus the main reasons for the variable relative readouts of the instruments are the following:

— The shape of the gamma spectrum which falls on the detector;
— The shape of the gamma detection efficiency curve, which is different for each detector;
— The different energy (instrumental) thresholds of gamma detection for the various detectors;
— The different principles of exposure rate measurement: the absorbed gamma energy is measured in one case (TLD), and the integrated rate of counting of the interactions of gamma quanta and detector is measured in the other cases (GMT, NaI).

It was concluded that the devices for measuring exposure rate based on detectors with large Z (e.g. NaI) are not suitable for environmental monitoring. The most accurate device is the TLD. The possible errors in measurements with GMT dosimeters are compensated by their rapidity as compared with TLDs.

IAEA-SM-306/146P

IN SITU METHODS FOR GAMMA SPECTROMETRY OF CHERNOBYL FALLOUT ON URBAN AND RURAL ROUGH SURFACES

O. KARLBERG
National Institute of Radiation Protection,
Stockholm, Sweden

In situ gamma spectrometry and ionization chamber measurements were started about twenty days after the Chernobyl accident in the high fallout area of Gävle in middle Sweden. The purpose of the measurements was to study the retention of radionuclides and to estimate the environmental half-lives on different types of urban surfaces [1].

The uncollided fluences from ^{137}Cs showed a relatively large decrease with time, corresponding to a half-life of 80–100 d. This decrease, however, could not be interpreted only as a removal of the radionuclides from the surfaces, since it is not possible to differentiate between removal and increased attenuation due to migration into the surfaces.

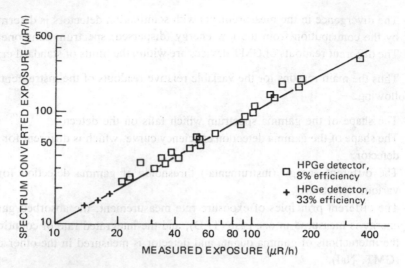

FIG. 1. Spectrum converted exposure plotted against exposure measured with an ionization chamber (cosmic component of 4 µR/h subtracted) (1 R = 2.58 × 10⁻⁴ C/kg).

Two methods to estimate this attenuation and to establish correction factors (here called surface attenuation factors) with respect to the ideal surface geometry were evaluated. One method, known as the energy dependent attenuation (EDA) approach, is based on the different attenuations for different energies of multienergy nuclides such as ¹⁴⁰La and ¹³⁴Cs. The differences in the measured attenuations of different peaks are compared with theoretical values, and estimates of the correction factor as well as the equivalent relaxation length for the surface are made.

In the other method, called the exposure–uncollided fluence comparison (EUC) approach, the exposure is calculated from the measured uncollided fluences for different values of the relaxation length and compared with the measured or spectrum converted exposures. The background exposures were estimated with the use of the measured uncollided fluences of the naturally occurring nuclides, with the assumption of a uniform distribution in the ground. Since only a few ionization chamber measurements were carried out, a spectrum–exposure conversion method [2] was used, where the spectra are converted with a so-called G function. The G function is found by the least squares fitting of spectra with known corresponding exposures. Figure 1 shows a comparison between measured and spectrum converted exposures.

Both methods were validated against measurements on different soil surfaces, where the reference values were obtained from soil samples. Both methods gave dis-

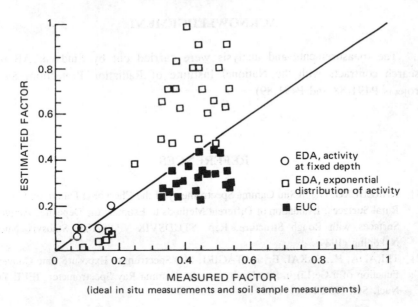

FIG. 2. Validation of both methods for determining surface attenuation factors using soil samples.

tinct responses and fairly reliable results for the deeply buried radionuclides, such as on ploughed or harrowed soils with surface attenuation factors below 0.1, but more uncertain results for undisturbed soils with factors larger than 0.3–0.5. Figure 2 shows the results of the validation.

The differential energy approach is only useful on surfaces with a deposition of the order of 10–100 kBq/m^2 and requires an accurate calibration with respect to the energy dependence. The EUC approach is sensitive to errors in the background exposure estimate and will improve considerably if background exposure measurements, prior to the fallout, are available.

The methods were applied to urban surface measurements, and indicated a decreasing attenuation factor from near unity shortly after the Chernobyl accident to typically 0.3–0.5 after a year or two. The latter factors were also confirmed in measurements with a 15 cm × 15 cm lead shield.

The decrease of these factors is similar to the decrease of the uncollided fluences, thus leading to the conclusion that none, or only small fractions, of the radionuclides are actually removed from urban surfaces after the initial removal by runoff and street cleaning.

ACKNOWLEDGEMENT

The measurements and analysis were carried out by Studsvik AB under research contracts with the National Institute of Radiation Protection, Sweden (projects P491.88 and P491.89).

REFERENCES

[1] KARLBERG, O., In Situ Gamma Spectrometry of the Chernobyl Fallout on Urban and Rural Surfaces. Evaluation of Different Methods to Estimate the Deposited Activity on Surfaces with Rough Structure, Rep. STUDSVIK-NP-89/108, Studsvik Nuclear, Nyköping (1989).
[2] TERADA, H., SAKAI, E., KATAGIRI, M., Spectrum to Exposure Rate Conversion Function of a Ge(Li) in Situ Environmental Gamma Ray Spectrometer, IEEE Trans. Nucl. Sci. **NS-24** 1 (1977).

IAEA-SM-306/20P

ENVIRONMENTAL SURVEILLANCE FOR RADIONUCLIDE CONTAMINATION UTILIZING A HIGH RESOLUTION INTRINSIC GERMANIUM DETECTOR

A.S. MOLLAH, M.M. RAHMAN
Institute of Nuclear Science and Technology,
Atomic Energy Research Establishment,
Dhaka, Bangladesh

In recent years the determination of artificial radionuclides in environmental and biological samples has become more and more important, owing to the world-wide distribution of radioactive fallout due to nuclear weapons tests and nuclear accidents as well as the development of the nuclear power industry in many countries [1]. Radioactivity from environmental contamination reaches the human population primarily through inhalation and ingested food. Therefore, it is necessary to identify and determine artificial radionuclides (particularly, long lived nuclides) in environmental and biological samples so that their impact on public health can be assessed and remedial actions taken.

A common problem in determining environmental radioactivity is the identification and measurement of several components in a complex mixture of radionuclides where concentrations may differ widely. As most of the radionuclides emit one or more gamma photons during the course of their decay, low level gamma spectrometry can be applied for the determination of the radionuclides. The advantages of a low level gamma spectrometry system for rapid determination are as follows: lowest possible background; efficiency; sharp resolution; longer counting time; and a large sample. HPGe detectors now constitute the basic instruments for direct analysis of a mixture of radionuclides.

In the study described here, the determination of the different gamma emitting radionuclides in environmental samples was achieved by gamma spectrometric investigation using a high resolution intrinsic HPGe detector coupled to a 4096 channel analyser. The detector (54 cm^3) has an efficiency of 19% at 1.33 MeV for ^{60}Co relative to a NaI(Tl) detector. The resolution at the same energy was 1.91 keV FWHM with a peak to Compton edge ratio of 31:1. A 10 cm thick lead enclosure was used to cut off cosmic rays. In order to use this detector, efficiency and minimum detectable activity (MDA) were measured for different radionuclides and different sample geometries [2]. The MDA that can be achieved with an assumed statistical accuracy can be calculated following an approach outlined in Ref. [1].

The sensitivity of the system permits detection of many radionuclides in environmental samples at levels of 0.01 Bq/kg, depending on the gamma energy, for a counting time of 20 000 s without any chemical separation and with a counting error of $\pm 8\%$ (1σ). Solid samples of plant, biological and inorganic origin are prepared by drying and grinding into fine powder, whereas liquid samples are measured without physical pretreatment in a standard plastic container in proportion to the system's geometric set-up. A more detailed description of the method will be found elsewhere [2, 3].

To determine the capacity of the system for rapid and accurate determination of environmental radioactivity, different environmental samples were analysed. The photopeaks of ^{137}Cs (due to fallout) and ^{40}K and daughter products of ^{238}U and ^{232}Th (due to natural background) were found for the samples and qualitatively analysed with the 4096 channel analyser. The levels of radioactivity were found to be in the range of 0.02 to 20 Bq/kg, depending on the type of sample. The accuracy and precision of this analytical technique were checked by radionuclide determination of the International Atomic Energy Agency's standard reference sediment No. SL 2. The ^{137}Cs results showed good agreement (to within 5%) with the certified values.

The system described is capable of providing qualitative and quantitative analysis of complex mixtures of gamma emitting radionuclides with high resolution and accuracy.

REFERENCES

[1] INTERNATIONAL ATOMIC ENERGY AGENCY, Measurement of Radionuclides in Food and the Environment: A Guidebook, Technical Reports Series No. 295, IAEA, Vienna (1989).

[2] MOLLAH, A.S., RAHMAN, M.M., HUSSAIN, S.R., Distribution of γ-emitting radionuclides in soils at the Atomic Energy Research Establishment, Savar, Bangladesh, Health Phys. 50 (1986) 835–838.

[3] MOLLAH, A.S., RAHMAN, M.M., Measurement of gamma activity from fallout Cs-137 in the environmental samples at AERE in Bangladesh, Bull. Radiat. Prot. 10 (1987) 3–8.

IAEA-SM-306/23P

IN SITU GAMMA SPECTROMETRY IN ACCIDENT SITUATIONS: POTENTIAL AND LIMITATIONS

L. BRAMATI, C. MAFFEI, G. PANDOLFI
Ente Nazionale per l'Energia Elettrica,
Rome, Italy

Within the framework of the Radiation Protection Research Programme of the Commission of the European Communities, the Ente Nazionale per l'Energia Elettrica, working in co-operation with the Centro Italiano Studi e Esperienze, has carried out a research and development programme on an in situ gamma spectrometry (ISGS) system appropriate for use in accident situations. The objectives were the following:

— Selection of instrumentation for high resolution ISGS,
— Preparation of a specifically designed computer program for obtaining and analysing spectra in situ.

The detector chosen is an HPGe detector (47 mm × 53 mm) with 1.76 keV resolution (FWHM) and 20% efficiency at 1332 keV. The cryostat permits measurements to be made within a period of three to four days without the need for refilling. The analyser has 4096 channels; the PC has 640 kbytes of random access memory, a 3.5 in floppy disk and a 10 Mbyte hard disk. The detector is mounted on a telescopic tripod. As a power supply we use a 400 W generator. The complete set of equipment is contained in two cases with a total weight of 60 kg. Each component has batteries; an external power supply is used only for the printer.

A linear relationship exists between the area of the photopeak and the exposure rate or concentration due to a radioisotope in/on the soil; the parameters entering the equation are the intrinsic efficiency of the detector and the theoretical fluence of photons for a defined geometry. A calculation code has been developed to transfer the spectrum from the multichannel analyser to the PC and to analyse it. The analysis of the spectrum is based on the identification of the isotopes and on calculation of the photopeak area.

In a one hour counting period, the numbers of net pulses for the following isotopes are:

- 2 per Bq/m^2 for ^{134}Cs and ^{137}Cs
- 3 per Bq/m^2 for ^{131}I
- 1.5 per Bq/m^2 for ^{60}Co
- 60 per Bq/kg for U (over three peaks)
- 45 per Bq/kg for Th (over three peaks)
- 7 per Bq/kg for ^{40}K.

Therefore, a surface concentration of 10 000 Bq/m^2, corresponding to a food contamination below any action level, can be measured in a few minutes, at 3% confidence level.

To verify the in situ responses of the HPGe detector and the related software, ad hoc campaigns were carried out in areas with uniform levels of radioactivity that had not been cultivated for many years. The results were compared with the exposure rates measured with ionization chambers and with the concentrations obtained from soil samples measured in the laboratory, and the agreement was quite good.

Two campaigns were then performed in Friuli and in Alto Lazio, in different environmental conditions. Friuli is the Italian region most affected by the Chernobyl fallout and has a very low natural background. In Alto Lazio the caesium deposition was generally low in an area of volcanic origin with high background radiation levels from natural radioisotopes.

The work has confirmed ISGS as a suitable method to map radioactive contamination over a wide area following an accident and to guide emergency actions in the intermediate period. It can usefully be substituted for a campaign of sample collection and measurement, mainly in regions far from a central laboratory. Owing to the complex instrumentation and the specialized operators required, it will be less effective in the initial period of an emergency.

IAEA-SM-306/92P

DETERMINATION OF URANIUM IN BIOLOGICAL
MATERIALS BY LASER FLUORIMETRY

A. GHODS-ESPHAHANI, J.C. VESELSKY
Agency's Laboratory,
International Atomic Energy Agency,
Seibersdorf

The uranium content of biological materials and foodstuffs is of the order of micrograms per gram. For the determination of these low uranium concentrations radiochemical methods and specific techniques are used, mainly alpha spectrometry, neutron activation and fission track methods. In general these procedures are time consuming (of the order of days for one determination).

TABLE I. DETERMINATION OF URANIUM IN SOME NATURAL MATRIX REFERENCE MATERIALS BY LASER FLUORIMETRY

Material	NIST[a] number	Certified value (ng/g)	This study (ng/g)
Oyster tissue	1566	116 ± 6	107 ± 3
Pine needles	1575	20 ± 4	22 ± 1
Tomato leaves	1573	61 ± 3	53 ± 2
Citrus leaves	1572	< 150	27 ± 1
Urban particulate matter	1648	5500 ± 100	5400 ± 200
		(µg/g)	(µg/g)
Spruce twigs[b]	CLV-1	86.8	84.6 ± 1.15
Spruce needles[b]	CLV-2	3.6	3.66 ± 0.04

[a] National Institute of Standards and Technology, Gaithersburg, Maryland, USA.
[b] Available from Vegetative Radionuclide Reference Materials (Canada).

In order to reduce time and manpower a method for uranium determination was developed using laser fluorimetry and minimal chemical operations. After dry ashing (200°C, then 650°C for 4 h) and wet ashing ($HClO_4$) of about 1 g of the material the uranium is extracted with methylisobutylketone from a $Ca(NO_3)_2$–EDTA solution. The element is stripped from the organic solvent into 0.001M HNO_3, which can be directly introduced into the laser fluorimeter. Analyses of natural matrix reference materials indicate agreement with the certified values (Table I). The method appears to be reliable and less time consuming than others presently in use.

IAEA-SM-306/91P

DETERMINATION OF ^{90}Sr AND ^{91}Y IN VEGETATION SAMPLES CONTAMINATED BY FALLOUT DEBRIS FROM THE CHERNOBYL ACCIDENT

A. GHODS-ESPHAHANI, J.C. VESELSKY, S. ZHU,
A. MIRNA, R. SCHELENZ
Agency's Laboratory,
International Atomic Energy Agency,
Seibersdorf

An improved method for the determination of ^{90}Sr and ^{91}Y in grass ash and other materials was developed, based on tributylphosphate (TBP) extraction of radioyttrium from a nitric acid sample extract solution. An important advantage of this method over the commonly used ^{90}Sr separation is that it is unnecessary to wait for the growing in of the ^{90}Y prior to its separation from the mother substance. Yttrium extraction has already been used for ^{90}Sr analysis, but the fission product ^{91}Y accompanying ^{90}Y in every chemical separation process was almost always considered negligible. This was not possible with relatively fresh Chernobyl fission products and consequently methods had to be found to distinguish the radiations of the two nuclides. Graphic analysis of the decay curve of the yttrium fraction isolated from the TBP extract was compared with the energy separation of the Cerenkov radiations of ^{90}Y and ^{91}Y, respectively. The latter method worked much faster and has therefore been adopted for routine analyses.

TABLE I. ^{90}Sr DETERMINATION IN MATERIALS USED IN IAEA INTER-COMPARISON STUDIES

Reference material	This study (Bq/kg)	IAEA recommended value (Bq/kg)	IAEA confidence interval (Bq/kg)
IAEA-152 (milk powder)	7.3 ± 0.6	7.7	7.3 − 8.2
IAEA-154 (whey powder)	7.1 ± 0.5	7.2	6.4 − 8.0
IAEA-Soil-6	30.2 ± 0.8	30.34	24.2 − 31.67

This method is also suitable for rapid determination of ^{91}Y in grass material. As a test of the methodology three IAEA standard samples (IAEA-152 (milk powder), IAEA-154 (whey powder) and IAEA-Soil-6) were analysed for ^{90}Sr. The results, as seen in Table I, were in good agreement with the certified values.

After ashing, the analysis takes only about six hours.

IAEA-SM-306/78P

^{242}Cm DETERMINATION IN ATMOSPHERIC FALLOUT AFTER A NUCLEAR ACCIDENT BY NON-DESTRUCTIVE ALPHA SPECTROMETRY ON AN AIR FILTER

M.L. DABURON, D. BULLIER,
C. PITIOT, M. VERRY
Institut de protection et de sûreté nucléaire,
Commissariat à l'énergie atomique,
Fontenay-aux-Roses, France

1. INTRODUCTION

Non-destructive alpha spectrometry was used to study an air filter exposed from 28 to 30 April 1986 at Grindsjön in southern Scandinavia immediately after the Chernobyl accident. As soon as the filter was received, on 20 May, gamma spectrometry was carried out in our laboratory. The spectrum clearly showed about twenty radionuclides; however, ^{239}Np was no longer detectable. Calculations from the ^{140}Ba level determined by gamma spectrometry and from the composition of the nuclear fuel of a French PWR [1] chosen as a model in the absence of other information at this time suggested a ^{242}Cm level sufficient for direct alpha spectrometry to be performed on the filter.

The alpha spectrometry was performed in a gridded low background ionization chamber. This detector permits measurement of discs up to 7 cm in diameter and is sensitive to alpha particles only. The resolution for electrolytic deposition is 40 keV for ^{241}Am. The filter, made of fibreglass, had an area of 10 cm × 10 cm.

2. RESULTS

Figure 1 shows the spectrum directly obtained on a disc cut from the filter. In spite of imperfect resolution due to autoabsorption, the following spectrum interpretation was possible:

— The energy drop at 6.10 MeV was typical of ^{242}Cm;
— A less well defined peak at about 5.3 MeV suggested the presence of ^{238}Pu, ^{239}Pu, ^{240}Pu and ^{241}Am, added to natural ^{210}Po;
— The absence of alpha activity beyond 6.10 MeV excluded the presence of Rn daughters at the time of measurement.

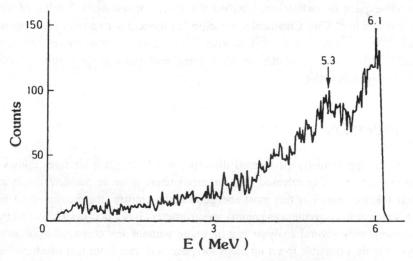

FIG. 1. *Alpha spectrum directly obtained on 20 May 1986 from air filter exposed at Grindsjön (south of Stockholm) from 28 to 30 April.*

TABLE I. RESULTS OF CHEMICAL SEPARATIONS

Radionuclide	Alpha activity for 410 m^3 of air (mBq)		Pu-(239+240) / Cm-242	Pu-238 / Pu-(239+240)
	20 May 1986	26 April 1986		
Cm-242	107	119	0.076	0.53
Cm-244	3.7	3.7		
Am-241	0.85	0.85		
Pu-(239+240)	9	9		
Pu-238	4.8	4.8		
Po-210	45	51		
Total	170	188		

Note: The ratios in the last two columns are very close to those measured by the USSR State Committee on the Utilization of Atomic Energy [3].

For a 28.3 cm^2 disc area corresponding to the crossing of 410 m^3 of air in 44 h, quantitative determination by graphic analysis of the spectrum gave a total alpha activity of 165 mBq, of which about 50% was due to ^{242}Cm.

Subsequent radiochemistry verified the spectrum results: 170 mBq, of which 60% was due to ^{242}Cm. Chemical separation [2] made it also possible to determine ^{239}Pu, ^{240}Pu, ^{238}Pu, ^{241}Am, ^{244}Cm and ^{210}Po. Table I summarizes the results obtained. Other portions of the air filter were analysed and the results showed a homogeneous deposition.

3.   DISCUSSION

Alpha spectrometry performed directly on a fibre glass air filter allows the determination of ^{242}Cm released into the atmosphere after an accident involving a nuclear reactor core. For this most energetic alpha emitter, the measurement technique is almost as convenient as gamma spectrometry. It is a non-destructive method, and consequently several analyses can be made without any chemical separation or delays. It is thus possible to set up maps in space and time from real measurements.

During the weeks after the Chernobyl accident, while attempts were made to evaluate the source term, the determination of the released actinides was of great importance. In this case, ^{242}Cm can be a good tracer for confirming ^{239}Np measurements or for replacing them when they cannot be performed any longer. At the time of the accident, the literature concerning gamma emitters was huge except in the case of ^{239}Np, for which emphasis was put on the difficulty of determination [4, 5]. Few publications dealt with alpha measurements and in most cases radiochemistry was involved [6–9].

REFERENCES

[1]   L'accident de Tchernobyl, Rev. 2, Rep. No. 2186, IPSN, Centre d'études nucléaires de Fontenay-aux-Roses (1986).

[2]   WILLEMOT, J.M., VERRY, M., JEANMAIRE, L., Détermination analytique du plutonium et des transplutoniens dans l'environnement, Rep. CEA-R-5460, Centre d'études nucléaires de Saclay (1988).

[3]   USSR STATE COMMITTEE ON THE UTILIZATION OF ATOMIC ENERGY, The Accident at the Chernobyl Nuclear Power Plant and its Consequences, information compiled for IAEA Experts Meeting, Vienna, 1986.

[4]   BYRDE, F., et al., "Mesures d'aérosols à la suite de l'accident de Tchernobyl et leur interprétation", Mesures de la radioactivité en Suisse après Tchernobyl et leur interprétation scientifique, Berne (1986).

[5]   ERLANDSON, B., ASKING, L., SWIETLICKI, E., Detailed early measurement of the fallout in Sweden from the Chernobyl accident, Water, Air, Soil Pollut. 35 (1987) 335.

[6]   Chernobyl — Its Impact in Sweden, Rep. 86-12, Natl Inst. of Radiation Protection, Stockholm (1986).

[7]   HENNIES, H.H., Radiation measurement in Germany resulting from the Chernobyl accident, Nucl. Eur. 7/8 (1988) 22.

[8]   HOLM, E., et al., "Fallout deposition of actinides in Monaco and Denmark following the Chernobyl accident", Proc. 4th Int. Symp. on Radioecology, Cadarache (1988).

[9]   POVINEC, P., et al., Aerosol radioactivity monitoring in Bratislava following the Chernobyl accident, J. Radioanal. Nucl. Chem. 126 (1988) 467.

IAEA-SM-306/47P

APPLICATION OF A HIGH PURITY GERMANIUM DETECTOR FOR RAPID IN SITU DETERMINATION OF ENVIRONMENTAL RADIOACTIVITY

F. STEGER, E. LOVRANICH, E. URBANICH
Institute for Radiation Protection,
Austrian Research Centre Seibersdorf,
Seibersdorf, Austria

I. NÉMETH, P. ZOMBORI, A. ANDRÁSI
Central Research Institute for Physics
  of the Hungarian Academy of Sciences,
Budapest, Hungary

The method of in situ gamma spectrometric determination of environmental radioactivity was introduced in 1987 in the environmental monitoring system of the Austrian Research Centre Seibersdorf. The semiconductor detector (HPGe) (Fig. 1) was calibrated and tested for field measurements of natural and artificial radionuclides distributed on/in the soil, as well as for determination of dose rates.

The calibration is based on a technique developed by Beck et al. [1]. The count rates of the full energy peaks in the spectrum are proportional to the concentrations of the corresponding gamma emitters in the soil (Fig. 2) and to the exposure rates from these nuclides.

Simple conversion factors are calculated for every peak in the spectrum. These energy dependent factors take account of the angular distribution of the uncollided gamma rays, of the detector efficiency and of the uncollided gamma flux from the sources in the soil.

Intercomparisons were made and good results were obtained.

Measurements after the Chernobyl accident were performed with this method in some of the more highly contaminated areas of Austria, and ^{137}Cs depositions of up to 100 kBq/m^2 were found. The determination of dose rates 1 m above the ground showed that in most cases the dose rates from the artificial (Chernobyl) radionuclides exceed the dose rates from the natural radionuclides or are at least of the same order of magnitude.

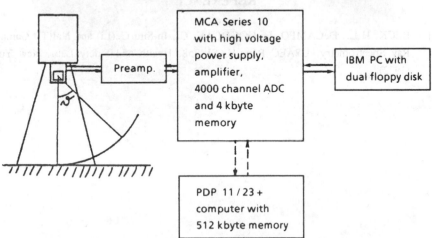

FIG. 1. Portable germanium spectrometer used by the Institute for Radiation Protection of the Austrian Research Centre Seibersdorf.

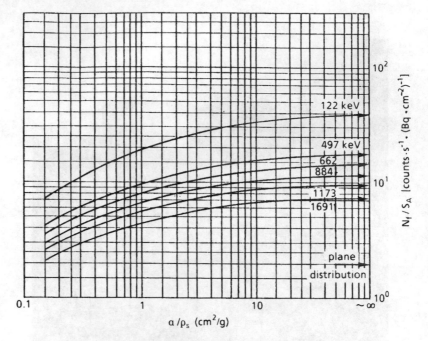

FIG. 2. Calibration factors for some gamma lines as a function of α/ρ_s.

REFERENCE

[1]   BECK, H.L., DeCAMPO, I., GOGOLAK, C., In-Situ Ge(Li) and NaI(Tl) Gamma-Ray Spectrometry, USAEC Rep. HASL-258, Health and Safety Lab., New York (1972).

IAEA-SM-306/37P

RAPID METHOD FOR LIVE MONITORING OF
CAESIUM ACTIVITY IN SHEEP, CATTLE AND REINDEER

P. STRAND
National Institute of Radiation Hygiene,
Østerås

L.I. BRYNILDSEN
Division of Veterinary Services,
Ministry of Agriculture,
Oslo

Norway

After the Chernobyl accident, the fallout differed considerably from area to area in Norway and soil samples showed that especially the mountain pastures in the south and middle of Norway were heavily contaminated. These natural ecosystems are important in several nutrition pathways, notably those involving reindeer, sheep and cattle.

In June 1986 the Directorate of Health imposed action levels for the nuclides ^{134}Cs and ^{137}Cs. The products (meat and dairy) from animals grazing on the mountain pastures were most severely contaminated. From about 20 to 35% of the sheep had activity levels above the action levels. Because of this a programme for mitigating actions was initiated which necessitated live monitoring and a high number of measurements in many parts of Norway.

It was necessary to estimate the activity levels of ^{137}Cs and ^{134}Cs by in vivo measurements in sheep, cattle and reindeer. The results from the in vivo measurements made it possible to sort out the animals with activity levels below the action levels, i.e. that could be slaughtered at once. The animals with higher levels than the action levels were given special feed for a certain period, depending on the measured activity level.

A fast and simple method for in vivo measurements was developed by using a 2 or 3 in NaI scintillation detector coupled to a multichannel analyser. The detector was placed on different parts of the animal, depending on the species and practical considerations. On sheep, the detector was placed on the back of the animal. On cattle, the detector was placed on the croup of the animal. For measuring reindeer, the animal was placed on its side and then the detector was firmly applied to the pelvis, between and parallel to the hind legs. The counting time varied from 20 to 60 s (normally 20 s) (Fig. 1).

FIG. 1. Live monitoring of caesium activity levels in sheep.

The shielding factor for each species was estimated by monitoring animals with very low activity levels in different geographic areas with different background levels. The equipment was calibrated for in vivo monitoring by estimating the ratio between the registered number of net impulses in the energy intervals for ^{134}Cs and ^{137}Cs of the live animal and the activity of muscle tissue samples from the same animal after slaughtering. In the first experiments we used 40 reindeer, 20 cattle and 15 sheep.

We have also developed a method for direct measurement on the carcasses of the same species.

Activity levels of tissue from the heart, kidney and different muscles and of wool of the same animals were also investigated and compared.

The uncertainty in calculating the activity level in meat from in vivo measurements was found to be less than 8% at the 95% confidence level for sheep, 7% for reindeer and 11% for cattle. The detection level depended on the background activity in the geographic area and the species. The detection level has normally been less than 50 Bq/kg.

IAEA-SM-306/55P

EFFICIENCY CALIBRATION OF GERMANIUM DETECTORS WITH ONE REFERENCE STANDARD GEOMETRY FOR RAPID ESTIMATION OF RADIONUCLIDE ACTIVITY IN ENVIRONMENTAL SAMPLES OF NON-STANDARDIZED GEOMETRY

A. PIETRUSZEWSKI
Central Laboratory for Radiological Protection,
Warsaw, Poland

1. INTRODUCTION

In rapid instrumental methods for analysis of gamma emitting radionuclides that are used for monitoring of food and environmental samples, determination of these radionuclides is required with an accuracy of $\pm 50\%$ within one day or of ± 200–300% within several hours of receipt of the sample at a concentration at least one order of magnitude below the derived intervention levels (DILs) established for foods by the World Health Organization and International Atomic Energy Agency, and at least at the lower level of 100 Bq/kg.

The accuracy with which the radionuclide activity is known, calculated (in becquerels per kilogram) as:

$$a = \frac{n}{fwtm}$$

is determined by the following factors:

n   is the number of photons registered by the detector: this can be high (> 1000) or low (< 100). To obtain better counting statistics one has to prolong the time of measurement or measure the sample under the best geometric conditions (usually with a Marinelli container).

t   is the measurement time, which is always determined with good accuracy (maximum error: $< 1\%$).

m   is the mass of sample available for the measurement: this can be small (several grams) or relatively large (~ 1000 g). The accuracy of mass determination is usually good (maximum error: ~ 3–5%).

w   is the number of photons per decay: this factor is also determined with good accuracy from tabulated data for radionuclide decay (error: $\sim 2\%$).

d   is the type of detector: typical detectors used for environmental sample radio-
      activity measurements have efficiencies of 7–40%.

f   is the detection efficiency: calibration can be performed with high accuracy —
      this is usually a time consuming procedure. If performed quickly, the accuracy
      is poor (±50% for a rapid method, ±200–300% for a very rapid method).

It is always preferable to have good efficiency calibration of the detector and
to use standard sample volume, mass and geometry. However, this optimum is not
always possible.

2.   DETECTOR EFFICIENCY CALIBRATION WITH ONE REFERENCE VOLUME STANDARD

Four germanium detectors of different sizes, efficiencies and resolution
parameters were each calibrated for 31 geometries. Cylindrical containers of
diameters 1.7, 3.5, 5.6, 7.5 and 9.5 cm filled to different heights of up to 12 cm
with radioactive standard solution were used for this experiment. These data were
used for calculation of the activity of selected radionuclides (^{141}Ce, ^{132}Te, ^{131}I,
^{103}Ru, ^{137}Cs, ^{134}Cs, ^{140}La) using one reference standard geometry (e.g. 1 cm
height standard for each container) for containers filled to different heights.

TABLE I. SUMMARY ERROR DATA FOR ALL ISOTOPES AND CONTAINERS

Max. container vol. (mL)	Dia. (cm)	Height range (cm)	Error range (%)	
			Detector with 30% efficiency	Detector with 15% efficiency
20	1.7	1–5	11–117	12–136
70	3.5	1–7	17–186	12–201
150	5.6	1–6	16–144	13–158
500	7.5	1–11	17–293	15–291
1000	9.5	1–12	26–333	25–339

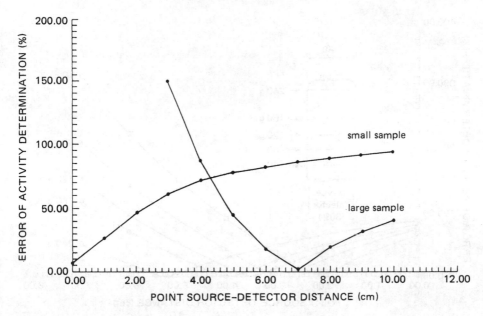

FIG. 1. *Comparison between errors of* ^{137}Cs *activity determination for small (~2 g) and large (~1000 g) samples, where point source efficiency calibration curves are used for calculations (HPGe detector with 30% efficiency).*

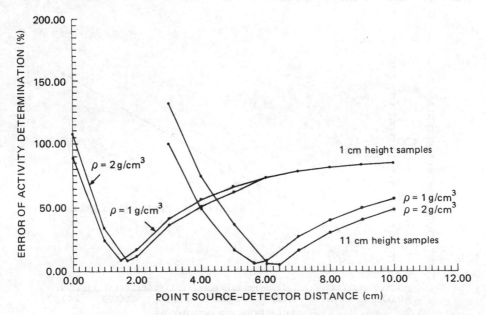

FIG. 2. *Errors of activity determination for samples of different heights and densities and a 500 mL container, with point source detector efficiency calibration. This calibration result shows that sample density does not significantly influence the shape or position of the error curve.*

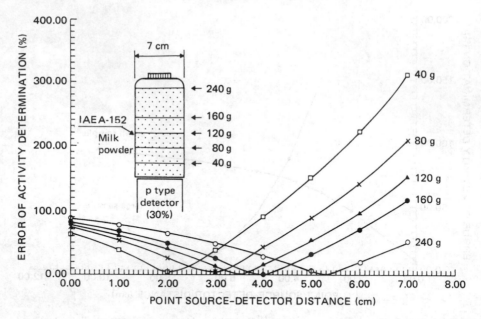

FIG. 3. Errors of ^{137}Cs activity determination in IAEA-152 milk powder (standard reference material) samples of mass in the range 40–240 g, obtained using point source efficiency calibration curves for activity calculation.

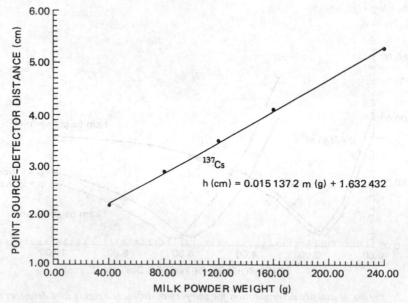

FIG. 4. The optimum point source–detector distance for detector efficiency calibration that can be used for volume milk powder sample measurements in the mass range 40–240 g with heights of up to 12 cm in a cylindrical container of 7 cm diameter.

Conclusion

If a detector is calibrated for a particular container for one specified sample height, this calibration can be used for any other sample height in this container and an error no greater than 300% will be obtained with a very rapid method of activity determination. Table I can be used for comparison.

3. DETECTOR EFFICIENCY CALIBRATION WITH A POINT SOURCE

In the work described here, each detector was calibrated for a point source geometry located at a distance of 0–10 cm from the detector surface. This type of calibration can be performed very quickly at any laboratory using germanium detectors as point sources are normally available. The point source calibration reference curves were used for quick activity estimation of selected radionuclides in volume sources (each of 31 calibrated geometries). The goal of this work was the calculation of errors of activity estimation using point source calibration for samples of different sizes (2–2000 g) for each container (Figs 1–4).

Results

Quick efficiency calibration of a Ge detector using a point source can be performed with rapid methods for radioactivity estimation of small (~2 g) and large (~1000 g) environmental and food samples. To achieve an accuracy of ~50% detectors should be calibrated using at least two distances from the detector surface: 2–3 cm and 5–6 cm. Calibration at 2–3 cm should be performed for small samples (<200 g). Calibration at 5–6 cm should be performed for samples of >200 g and high density samples.

ACKNOWLEDGEMENT

This work was performed with the support of the International Atomic Energy Agency under Research Contract No. 5453/RB.

IAEA-SM-306/14P

METHOD FOR RAPID CALCULATION OF THE EFFICIENCY OF SEMICONDUCTOR DETECTORS IN MEASUREMENTS OF VOLUMINOUS SAMPLES

M. KORUN, R. MARTINČIČ
Jožef Stefan Institute,
Ljubljana, Yugoslavia

A method for the determination of the efficiency of solid state gamma ray detectors is presented. With appropriate software, developed at the Jožef Stefan Institute, the method has proven to be rapid and convenient for emergency situations as well as for routine laboratory work. The parameters of the sample, such as diameter, thickness, density and attenuation, are the input data for the calculation.

To determine efficiency accurately in the case of voluminous samples one usually needs standard solutions and/or solid standards of mixed radionuclides in several geometric configurations identical to the sample containers. These standards of certified mixed radionuclides are available from several reputable suppliers but they are costly and not always at hand. Therefore, a method for efficiency determination based on standard point source efficiency calibration has been developed.

Full energy peak efficiency is defined as the ratio between the number of photons recorded in the total absorption peak and the number of photons emitted by the source. It may be expressed in the following way:

$$\eta_V(E) = \frac{1}{V} \int_V \eta_{PS}(\vec{r}, E) \, \exp(-\mu(E) \, \bar{s}(\vec{r})) \, d^3\vec{r}$$

where

$\eta_V(E)$ is the efficiency of a voluminous sample at energy E,
V is the volume of the sample,
$\eta_{PS}(\vec{r}, E)$ is the efficiency of a point source at position $\vec{r}$,
$\mu(E)$ is the attenuation coefficient at energy E,
$\bar{s}(\vec{r})$ is the mean distance travelled by a gamma ray emitted at position $\vec{r}$ that produces a pulse in the photoabsorption peak.

The first term in the integrand represents the point source efficiency at position $\vec{r}$, and the second term the self-absorption of gamma rays in the sample; the integration is performed over the sample volume. Thus the calibration of the detector is performed by measuring the spatial dependence of the detector efficiency for standard point sources and the corrections due to self-absorption are subsequently taken into account.

The spatial dependence of the efficiency of a certain standard point source is determined experimentally by measurements of the efficiency in a suitably chosen grid of points in the space where the samples are to be placed. The measurements are repeated with several other calibration sources.

Experimentally determined axial and radial dependences of detector efficiency for a point source are then approximated by simple analytical functions. Because the parameters of these functions are smooth functions of energy the spatial dependence of the efficiency at any gamma ray energy can be reconstructed simply by interpolation.

In an absorbing medium the spatial dependence of detector efficiency is altered and the parameters of the analytical functions have to be properly adjusted. The functional dependences of the parameters on the absorption are based on their geometric and physical meanings and the parameter modifications are made using simple geometric considerations. The parameters are linear functions of the absorption coefficient and represent estimations which have proven to be rapid in calculation and at the same time sufficient for all practical cases.

Once the spatial and energy dependences are known they are used as a database for the software which performs the energy interpolation, adjustment for the absorption and integration over the sample thickness. Since the approximation of the radial dependence is chosen such that integration is done analytically, only the integration over the axial co-ordinate is performed numerically. Thus the calculation time required for the efficiency calculation becomes negligible.

The practical data needed to construct the efficiency curve for a specific type of sample are sample geometry and density. The only assumption about the sample which is necessary is on the chemical composition of the sample, which determines the absorption coefficients.

This method of detector efficiency determination could be used on a computerized multichannel analyser or any other small computer, including programmable calculators.

IAEA-SM-306/148P

USING STANDARD SPECTRA TO DEVELOP AND TEST GAMMA RAY ANALYSIS SOFTWARE

R.M. KEYSER
EG&G Ortec,
Oak Ridge, Tennessee,
United States of America

Presented by W. Hanstein

1.   INTRODUCTION

Beginning in the early 1970s, several groups, notably the International Atomic Energy Agency, determined a need for a way to test the ability of gamma ray spectrum reduction software to perform the functions for which it was designed. Over the past several years, the International Electrotechnical Commission (IEC) has been working on the updating of a standard that will define the parameters to be tested and provide a set of spectra to be used for the test. This proposed IEC standard [1] is designed to test the ability of the software to locate and determine the area of singlets and to locate and resolve doublets.

2.   SPECTRA

The spectra included in the standard consist of a calibration spectrum, a singlet spectrum and two spectra of doublets. The spectra were all collected with an n type HPGe detector with a resolution of 1.95 keV at 1332.5 keV. The individual nuclide spectra were collected separately so as to minimize the contamination of any line. It is believed that these spectra have the energy variation of peak shape that represents the actual function in currently available detectors. The background under the peak areas also represents expected background for such systems.

The calibration spectrum consists of eight isolated, large, well formed peaks covering the energy range of interest. The singlet spectrum contains 21 peaks with varying amplitude over the energy range of 40–1500 keV. The variation in amplitude was accomplished by varying the collection time for the spectrum sections.

The two doublet spectra each contain seven doublets. The separation ranges from greater than the FWHM to the FWHM to less than the FWHM, and the amplitude ratio ranges from 1:1 to 1:1000. The small peak can be on either side of the large peak. It is assumed that these cases will test the limits of current software.

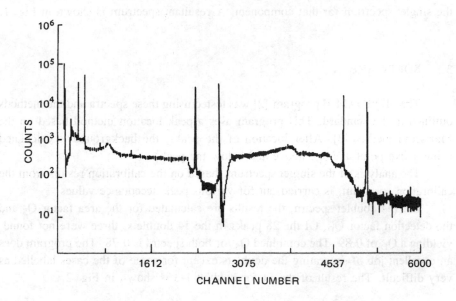

FIG. 1. First doublet spectrum.

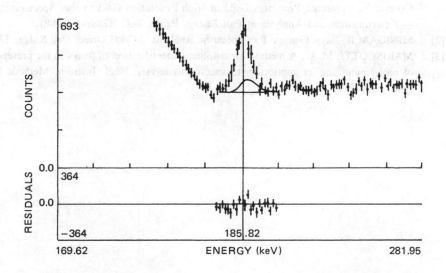

FIG. 2. Results of deconvolution for doublet 1, peak 3.

The position and amplitude of each individual component were derived from the singlet spectrum for that component. A resultant spectrum is shown in Fig. 1.

3.   SOFTWARE

The MINIGAM-II program [2] was tested using these spectra and the methods outlined in the standard. This program uses a peak location method based on the Mariscotti method [3]. After location of the peaks, the background is calculated using a five point average above and below the peak.

The analysis of the singlet spectrum, based on the calibration results from the calibration spectrum, is carried out for various peak acceptance values.

For the doublet spectra, the results are calculated for the area factor Q_S and the detection factor Q_D. Of the 28 peaks in the 14 doublets, three were not found, yielding a Q_D of 0.89. The combined Q_S for both spectra is 0.78. The program does an excellent job of separating the doublets except for some of the cases labelled as very difficult. The result of the fit for doublet 1-3 is shown in Fig. 2.

REFERENCES

[1]   INTERNATIONAL ELECTROTECHNICAL COMMISSION, Definition of Quality Criteria for Computer Programs Used in High Resolution Gamma-Ray Spectrometry — Determination and Analysis of Full Energy Peaks, IEC, Geneva (1988).

[2]   MINIGAM II, Basic Gamma-Ray Spectrum Analysis, EG&G Ortec, Oak Ridge, TN.

[3]   MARISCOTTI, M.A., A method for automatic identification of peaks in the presence of background and its application to spectrum analysis, Nucl. Instrum. Methods **50** (1967) 309–320.

CHAIRMEN OF SESSIONS

Session 1	Chairman	V.N. PETROV	Union of Soviet Socialist Republics
	Co-Chairman	F. LUYKX	CEC
Session 2	Chairman	A. LAMBRECHTS	France
	Co-Chairman	I.P. LOS'	Union of Soviet Socialist Republics
Session 3	Chairman	M.C. BELL	United States of America
	Co-Chairman	J. ESPINOSA GONZALEZ	Panama
Session 4	Chairman	I. OTHMAN	Syrian Arab Republic
	Co-Chairman	J. GELEIJNS	Netherlands
Session 5	Chairman	R. MARTINČIČ	Yugoslavia
	Co-Chairman	R.J.C. KIRCHMANN	Belgium
Session 6	Chairman	A.W. RANDELL	FAO
	Co-Chairman	Z. PIETRZAK-FLIS	Poland
Session 7	Chairman	O. PAAKKOLA	Finland
	Co-Chairman	U.H. BÄVERSTAM	Sweden
Session 8	Chairman	P.J. WAIGHT	WHO
	Co-Chairman	A.S. MOLLAH	Bangladesh
Session 9	Chairman	R. SCHELENZ	IAEA
	Co-Chairman	P. STRAND	Norway
Session 10	Chairman	J.C. TJELL	FAO/IAEA
	Co-Chairman	A.D. HORRILL	United Kingdom
Session 11	Chairman	F.P.W. WINTERINGHAM	United Kingdom
	Co-Chairman	B.G. BENNETT	UNSCEAR
Special Session on Hot Particles	Chairman	B. SALBU	Norway

SECRETARIAT OF THE SYMPOSIUM

J.C. TJELL	Scientific Secretary (FAO/IAEA)
R. SCHELENZ	Scientific Co-Secretary (IAEA)
T. WATABE	Scientific Co-Secretary (IAEA)
H. SCHMID	Symposium Organizer (IAEA)
S.P. FLITTON	Proceedings Editor (IAEA)
E. KATZ	Proceedings Editor (IAEA)
J.-N. AQUISTAPACE	French Editor (IAEA)
O.I. MELNIK	Russian Editor (IAEA)
L. HERRERO	Spanish Editor (IAEA)

HOW TO ORDER IAEA PUBLICATIONS

 An exclusive sales agent for IAEA publications, to whom all orders and inquiries should be addressed, has been appointed in the following country:

UNITED STATES OF AMERICA   UNIPUB, 4611-F Assembly Drive, Lanham, MD 20706-4391

 In the following countries IAEA publications may be purchased from the sales agents or booksellers listed or through major local booksellers. Payment can be made in local currency or with UNESCO coupons.

ARGENTINA	Comisión Nacional de Energía Atómica, Avenida del Libertador 8250, RA-1429 Buenos Aires
AUSTRALIA	Hunter Publications, 58 A Gipps Street, Collingwood, Victoria 3066
BELGIUM	Service Courrier UNESCO, 202, Avenue du Roi, B-1060 Brussels
CHILE	Comisión Chilena de Energía Nuclear, Venta de Publicaciones, Amunategui 95, Casilla 188-D, Santiago
CHINA	IAEA Publications in Chinese: China Nuclear Energy Industry Corporation, Translation Section, P.O. Box 2103, Beijing IAEA Publications other than in Chinese: China National Publications Import & Export Corporation, Deutsche Abteilung, P.O. Box 88, Beijing
CZECHOSLOVAKIA	S.N.T.L., Mikulandska 4, CS-116 86 Prague 1 Alfa, Publishers, Hurbanovo námestie 3, CS-815 89 Bratislava
FRANCE	Office International de Documentation et Librairie, 48, rue Gay-Lussac, F-75240 Paris Cedex 05
HUNGARY	Kultura, Hungarian Foreign Trading Company, P.O. Box 149, H-1389 Budapest 62
INDIA	Oxford Book and Stationery Co., 17, Park Street, Calcutta-700 016 Oxford Book and Stationery Co., Scindia House, New Delhi-110 001
ISRAEL	Heiliger & Co. Ltd. 23 Keren Hayesod Street, Jerusalem 94188
ITALY	Libreria Scientifica, Dott. Lucio de Biasio "aeiou", Via Meravigli 16, I-20123 Milan
JAPAN	Maruzen Company, Ltd, P.O. Box 5050, 100-31 Tokyo International
PAKISTAN	Mirza Book Agency, 65, Shahrah Quaid-e-Azam, P.O. Box 729, Lahore 3
POLAND	Ars Polona-Ruch, Centrala Handlu Zagranicznego, Krakowskie Przedmiescie 7, PL-00-068 Warsaw
ROMANIA	Ilexim, P.O. Box 136-137, Bucharest
SOUTH AFRICA	Van Schaik Bookstore (Pty) Ltd, P.O. Box 724, Pretoria 0001
SPAIN	Díaz de Santos, Lagasca 95, E-28006 Madrid Díaz de Santos, Balmes 417, E-08022 Barcelona
SWEDEN	AB Fritzes Kungl. Hovbokhandel, Fredsgatan 2, P.O. Box 16356, S-103 27 Stockholm
UNITED KINGDOM	Her Majesty's Stationery Office, Publications Centre, Agency Section, 51 Nine Elms Lane, London SW8 5DR
USSR	Mezhdunarodnaya Kniga, Smolenskaya-Sennaya 32-34, Moscow G-200
YUGOSLAVIA	Jugoslovenska Knjiga, Terazije 27, P.O. Box 36, YU-11001 Belgrade

---

 Orders from countries where sales agents have not yet been appointed and requests for information should be addressed directly to:

 Division of Publications
International Atomic Energy Agency
Wagramerstrasse 5, P.O. Box 100, A-1400 Vienna, Austria